W0080622

BOLYAI SOCIETY
MATHEMATICAL STUDIES 24

BOLYAI SOCIETY MATHEMATICAL STUDIES

Imre Bárány
Károly J. Böröczky
Gábor Fejes Tóth
János Pach
(Eds.)

Geometry –
Intuitive, Discrete, and Convex

A Tribute to László Fejes Tóth

 Springer

 JÁNOS BOLYAI MATHEMATICAL SOCIETY

Imre Bárány
Károly J. Böröczky
Gábor Fejes Tóth
János Pach

Alfréd Rényi Institute of Mathematics
Hungarian Academy of Sciences
Reáltanoda u. 13-15
Budapest 1053
Hungary
E-mail: barany.imre@renyi.mta.hu
        boroczky.karoly.j@renyi.mta.hu
        fejes.toth.gabor@renyi.mta.hu
        pach.janos@renyi.mta.hu

Mathematics Subject Classification (2010): 05-06, 52-06

Library of Congress Control Number: 2013950614

ISSN 1217-4696
ISBN 978-3-642-41497-8 Springer Berlin Heidelberg New York
ISBN 978-963-9453-17-3 János Bolyai Mathematical Society, Budapest

Springer is a part of Springer Science+Business Media
springer.com

© 2013 János Bolyai Mathematical Society and Springer-Verlag
Printed in Hungary

The use of general descriptive names, registered names, trademarks, etc. in this publication does not imply, even in the absence of a specific statement, that such names are exempt from the relevant protective laws and regulations and therefore free for general use.

The photo on the back cover is courtesy of Włodzimierz Kuperberg

Cover design: WMXDesign GmbH, Heidelberg

Printed on acid-free paper   44/3142/db – 5 4 3 2 1 0

CONTENTS

PREFACE

In the summer of 2008, the editors of the present volume organized a well attended conference at the Rényi Institute of the Hungarian Academy of Sciences to commemorate the highly influential work and character of László Fejes Tóth (1915–2005), one of the best known Hungarian geometers of all time, who served as a director of the Institute for thirteen years. The idea of publishing a collection of essays dedicated to him was conceived shortly after the conference, and it was embraced by a number of outstanding colleagues.

Hungary produced many famous mathematicians in the twentieth century, but only few of them sparked the interest of a large number of talented students in his subject and created an active school. László Fejes Tóth was one of the exceptions.

Like Paul Erdős, John von Neumann, and Paul Turán, he defended his thesis under the supervision of Leopold Fejér at Pázmány University, Budapest. In his thesis, he solved a problem in Fourier analysis. He found the problem and its solution entirely by himself, when reading a classical monograph of Francesco Tricomi.

After his army service, he started teaching in Kolozsvár (Cluj). He met Dezső Lázár, a mathematics teacher at the local Jewish High School, who told him about an exciting open problem that had a huge impact on Fejes Tóth's later work: How should one arrange n points in the unit square so as to maximize the minimum distance between them? In other words, what is the maximum density of a packing of n congruent disks in a square? Unaware of earlier work by Axel Thue, László Fejes Tóth found an asymptotically tight answer to this question. He generalized the problem in many different directions. A few years later he showed that the maximum density of a packing of congruent copies of any centrally symmetric convex body in the plane is attained for a lattice packing. He also solved another important open problem. Steiner conjectured that among all convex polytopes of unit surface area that are combinatorially equivalent to a given Platonic body, the regular polytope has the largest possible volume. For tetrahedra and octahedra, the conjecture can be easily verified by symmetrization. Fejes Tóth developed a new technique using sums of spherical moments to prove Steiner's conjecture for the cube and for the dodecahedron. (The problem is still open for the icosahedron.) The method has turned out to have many other interesting consequences.

Several years of systematic research in this area resulted in Fejes Tóth's monograph, *"Lagerungen in der Ebene, auf der Kugel und im Raum,"* which appeared in the prestigious series of Springer-Verlag, *Grundlehren der mathematischen Wissenschaften* in 1953. The book became an instant classic. As Ambrose Rogers wrote one decade later, "Until recently, the theory of packing and covering was not sufficiently well developed to justify the publication of a book devoted exclusively to it. After the publication of L. Fejes Tóth's excellent book in 1953, there would be no need for a second work on the subject, but for the fact that he confines his attention mainly to two and three dimensions." The higher dimensional problems and results inspired by Fejes Tóth's work have led to important discoveries in coding theory, in combinatorics, and in many other areas. Tom Hales' solution of "Kepler's conjecture," the sphere packing problem in three dimensions, was motivated by Fejes Tóth's program outlined in his book. Sixty years after the publication of its first edition, the *Lagerungen* is still considered a basic work in geometry. Its annotated English translation will appear soon.

One of László Fejes Tóth's most impressive abilities was to ask beautiful and deep mathematical questions. Many of these questions can be found in the short communications listed at the end of Fejes Tóth's complete bibliography, included in this volume. His work and his modest, unassuming personality had a lasting impact on the professional life and development of the editors as well as many of the contributors of this volume.

Imre Bárány
Károly Böröczky, Jr.
Gábor Fejes Tóth
János Pach

LIST OF PUBLICATIONS OF LÁSZLÓ FEJES TÓTH

Until 1946, László Fejes Tóth published under the name of László Fejes.

1935

Des séries exponentielles de Cauchy. *C. R. Acad. Sci., Paris* **200** (1935), 1712–1714.
JFM 62.1191.03

1937

Sokszögekre vonatkozó szélsőérték feladatokról. (Hungarian) [On extremum problems concerning polygons.] *Középiskolai Matematikai és Fizikai Lapok* **13** (1937), 1–4.

1938

A Cauchy-féle exponenciális sor. (Hungarian) [The exponential series of Cauchy.] *Mat. Fiz. Lapok* **45** (1938), 115–132. JFM 64.0284.04

Poliéderekre vonatkozó szélsőértékfeladatok. (Hungarian) [Extremum problems concerning polyhedra.] *Mat. Fiz. Lapok* **45** (1938), 191–199. JFM 64.0732.02

1939

Über die Approximation konvexer Kurven durch Polygonfolgen. *Compositio Math., Groningen* **6** (1939), 456–467. JFM 65.0822.03

Two inequalities concerning trigonometric polynomials. *J. London Math. Soc.* **14** (1939), 44–46. JFM 65.0254.01

Über zwei Maximumaufgaben bei Polyedern. *Tôhoku Math. J.* **46** (1939), 79–83. MR0002194

A simuló n-lapról. (Hungarian) [On the approximating n-hedron.] *Mat. Fiz. Lapok* **46** (1939), 141–145. JFM 65.0827.01

1940

Über einen geometrischen Satz. *Math. Z.* **46** (1940), 83–85. MR0001587

Eine Bemerkung zur Approximation durch *n*-Eckringe. *Compositio Math.* **7** (1940), 474–476. MR0001588

Sur un théorème concernant l'approximation des courbes par des suites de polygones. *Ann. Scuola Norm. Super. Pisa (2)* **9** (1940), 143–145. MR0004993

Egy extremális soklapról. (Hungarian) [On an extremal polyhedron.] *Math. Naturwiss. Anz. Ungar. Akad. Wiss.* **59** (1940), 476–479. MR0015831

1942

Az egyenlőoldalú háromszögrács, mint szélsőértékfeladatok megoldása. (Hungarian) [The regular triangular lattice as the solution of extremum problems.] *Mat. Fiz. Lapok* **49** (1942), 238–248. MR0017924

A szabályos testek, mint szélsőértékfeladatok megoldásai. (Hungarian) [The regular polyhedra as the solution of extremum problems.] *Mat. Termeszett. Értes.* **61** (1942), 471–477. Zbl 0028.07604

A lehűlés Fourier-soráról. (Hungarian) [On the Fourier series of the cooling.] *Math.-naturw. Anz. Ungar. Akad. Wiss.* **61** (1942), 478–495. JFM 68.0144.03

1943

Einige Extremaleigenschaften des Kreisbogens bezüglich der Annäherung durch Polygone. *Acta Sci. Math., Szeged* **10** (1943), 164–173. Zbl 0028.09201

Über eine Extremaleigenschaft der Kegelschnittbogen. *Monatsh. Math. Phys.* **50** (1943), 317–326. MR0010420

Über die dichteste Kugellagerung. *Math. Z.* **48** (1943), 676–684. MR0009129

Über eine Abschätzung des kürzesten Abstandes zweier Punkte eines auf einer Kugelfläche liegenden Punktsystems. *Jber. Deutsch. Math. Verein.* **53** (1943), 66–68. MR0017539

Egy gömbfelület befedése egybevágó gömbsüvegekkel. (Hungarian) [Covering the sphere with congruent caps.] *Mat. Fiz. Lapok* **50** (1943), 40–46. Zbl 0060.34808

Az ellipszis izoperimetrikus és az ellipszoid izepifán tulajdonságáról. (Hungarian) [On the isoperimertric property of the ellipse and the isepiphane property of the ellipsoid.] *Math.-Naturw. Anz. Ungar. Akad. Wiss.* **62** (1943), 88–94. Zbl 0061.38207

A gömbfelületet egyenlő felszínű konvex részekre osztó legrövidebb görbehálózat. (Hungarian) [The shortest net dividing the sphere into convex parts of equal area.] *Math. Naturwiss. Anz. Ungar. Akad. Wiss.* **62** (1943), 349–354. MR0024155

1944

Über einige Extremaleigenschaften der regulären Polyeder und des gleichseiti-
gen Dreiecksgitters. *Ann. Scuola Norm. Super. Pisa (2)* **13** (1944), 51–58.
MR0024156

A térnek egy szélsőértékfeltételt kielégítő felbontása soklapokra. (Hungarian) [Par-
tition of the spacc into polyhedra satisfying an extremum property.] *Mat. Fiz.
Lapok* **51** (1944), 19 pp. MR0025171

Extremális pontrendszerek a síkban, a gömbfclületen és a térben. (Hungarian)
[Extremal distributions of points in the plane, on the surface of the sphere
and in space.] Univ. Francisco-Josephina. Kolozsvár. Acta Sci. Math. Nat.,
(1944). no. 23, iv+54 pp. MR0019927

1945

Einige Bemerkungen über die dichteste Lagerung inkongruenter Kreise. *Comment.
Math. Helv.* **17** (1945), 256–261. MR0014229

1946

Über die Fouriersche Reihe der Abkühlung. *Acta Univ. Szeged. Sect. Sci. Math.* **11**
(1946), 28–36. MR0018250

Eine Bemerkung über die Bedeckung der Ebene durch Eibereiche mit Mittelpunkt.
Acta Univ. Szeged. Sect. Sci. Math. **11** (1946), 93–95. MR0017559

1947

———— and Hadwiger, H., Mittlere Trefferzahlen und geometrische Wahrschein-
lichkeiten. *Experientia* **3** (1947), 366–369. MR0022691

New proof of a minimum property of the regular n-gon. *Amer. Math. Monthly* **54**
(1947), 589. MR0024154

1948

———— and Hadwiger, H., Über Mittelwerte in einem Bereichsystem. *Bull. école
Polytech. Jassy [Bul. Politehn. Gh. Asachi]* **3** (1948), 29–35. MR0026351

An inequality concerning polyhedra. *Bull. Amer. Math. Soc.* **54** (1948), 139–146.
MR0024639

Approximation by polygons and polyhedra. *Bull. Amer. Math. Soc.* **54** (1948),
431–438. MR0024640

On ellipsoids circumscribed and inscribed to polyhedra. *Acta Univ. Szeged. Sect.
Sci. Math.* **11** (1948), 225–228. MR0026815

The isepiphan problem for n-hedra. *Amer. J. Math.* **70** (1948), 174–180. Zbl
0038.35603 MR0024157

Inequalities concerning polygons and polyhedra. *Duke Math. J.* **15** (1948), 817–822. MR0027545

Über die mittlere Schnittpunktszahl konvexer Kurven und Isoperimetrie. *Elemente der Math.* **3** (1948), 113–114. MR0027546

On the densest packing of convex domains. *Nederl. Akad. Wetensch., Proc.* **51** (1948), 544–547. (Indagationes Math. **10** (1948), 188–192.) MR0025753

On the total length of the edges of a polyhedron. *Norske Vid. Selsk. Forh., Trondhjem* **21** (1948), 32–34. MR0033073

1949

On the densest packing of circles in a convex domain. *Norske Vid. Selsk. Forh., Trondhjem* **21** (1949), 68–71. MR0033551

Über dichteste Kreislagerung und dünnste Kreisüberdeckung. *Comment. Math. Helv.* **23** (1949), 342–349. MR0033550

On the densest packing of spherical caps. *Amer. Math. Monthly* **56** (1949), 330–331. MR0030217

Ausfüllung eines konvexen Bereiches durch Kreise. *Publ. Math. Debrecen* **1** (1949), 92–94. MR0036526

1950

Extremum properties of the regular polyhedra. *Canadian J. Math.* **2** (1950), 22–31. MR0033072

Some packing and covering theorems. *Acta Sci. Math. Szeged* **12** (1950), Leopoldo Fejér et Frederico Riesz LXX annos natis dedicatus, Pars A, 62–67. MR0038086

Covering by dismembered convex discs. *Proc. Amer. Math. Soc.* **1** (1950), 806–812. MR0042148

A Brunn-Minkowski-féle tételről. (Hungarian) [On the Brunn-Minkowski theorem.] *Mat. Lapok* **1** (1950), 211–217. MR0037002

Az izoperimetrikus probléma I. (Hungarian) [The isoperimetric problem I.] *Mat. Lapok* **1** (1950), 363–383. MR0042146

Elementarer Beweis einer isoperimetrischen Ungleichung. *Acta Math. Acad. Sci. Hungar.* **1** (1950), 273–276. MR0048061

1951

Az izoperimetrikus probléma II. (Hungarian) [The isoperimetric problem. II.] *Mat. Lapok* **2** (1951), 34–45. MR0042147

Über gesättigte Kreissysteme. *Math. Nachr.* **5** (1951), 253–258. MR0043490

Über den Affinumfang. *Math. Nachr.* **6** (1951), 51–64. MR0045404

1952

A legsűrűbb gömbelhelyezésről. (Hungarian) [On the densest packing of balls.] In: *Comptes Rendus du Premier Congrès des Mathématiciens Hongrois, 27 Août–2 Septembre 1950*, 619–642. Akadémiai Kiadó, Budapest, (1952). MR0055710

Ein Beweisansatz für die isoperimetrische Eigenschaft des Ikosaeders. *Acta Math. Acad. Sci. Hungar.* **3** (1952), 155–163. MR0054269

1953

Lagerungen in der Ebene, auf der Kugel und im Raum. Die Grundlehren der Mathematischen Wissenschaften in Einzeldarstellungen mit besonderer Berücksichtigung der Anwendungsgebiete, Band **65**. Springer-Verlag, Berlin-Göttingen-Heidelberg, (1953). x+197 pp. MR0057566

Kreisausfüllungen der hyperbolischen Ebene. *Acta Math. Acad. Sci. Hungar.* **4** (1953), 103–110. MR0058230

Kreisüberdeckungen der hyperbolischen Ebene. *Acta Math. Acad. Sci. Hungar.* **4** (1953), 111–114. MR0058231

On close-packings of spheres in spaces of constant curvature. *Publ. Math. Debrecen* **3** (1953), 158–167. MR0061401

1954

Über die dichteste Horozyklenlagerung. *Acta Math. Acad. Sci. Hungar.* **5** (1954), 41–44. MR0063062

Annäherung von Kurven durch Kurvenbogenzüge. *Publ. Math. Debrecen* **3** (1954), 273–280. MR0072501

1955

Extremum properties of the regular polytopes. *Acta Math. Acad. Sci. Hungar.* **6** (1955), 143–146. MR0071029

Megjegyzések Dowker sokszögtételeihez. (Hungarian) [Remarks on polygon theorems of Dowker.] *Mat. Lapok* **6** (1955), 176–179. MR0074017

1956

Characterisation of the nine regular polyhedra by extremum properties. *Acta Math. Acad. Sci. Hungar.* **7** (1956), 31–48. MR0079289

On the volume of a polyhedron in non-Euclidean spaces. *Publ. Math. Debrecen* **4** (1956), 256–261. MR0079284

Über die dünnste Horozyklenüberdeckung. *Acta Math. Acad. Sci. Hungar.* **7** (1956), 95–98. MR0079287

Triangles inscrits et circonscrits à une courbe convexe sphérique. *Acta Math. Acad. Sci. Hungar.* **7** (1956), 163–167. MR0080322

On the sum of distances determined by a pointset. *Acta Math. Acad. Sci. Hungar.* **7** (1956), 397–401. MR0107212

Erdős, P. and ———, Pontok elhelyezése egy tartományban. (Hungarian) [The distribution of points in a region.] *Magyar Tud. Akad. Mat. Fiz. Oszt. Közl.* **6** (1956), 185–190. MR0095451

1957

Szabályos alakzatok. (Hungarian) [Regular configurations.] *Magyar Tud. Akad. Mat. Fiz. Oszt. Közl.* **7** (1957), 39–47. MR0097030

Filling of a domain by isoperimetric discs. *Publ. Math. Debrecen* **5** (1957), 119–127. MR0090070

1958

Über eine Extremaleigenschaft des fünf- und sechseckigen Sternes. *Elem. Math.* **13** (1958), 32–34. MR0092175

——— and Molnár, J., Unterdeckung und Überdeckung der Ebene durch Kreise. *Math. Nachr.* **18** (1958), 235–243. MR0096174

Annäherung von Eibereichen durch Polygone. *Math.-Phys. Semesterber.* **6** (1958/1959), 253–261. MR0124815

1959

Über Extremaleigenschaften der regulären polyeder. *Bulgar. Akad. Nauk Izv. Mat. Inst.* **4** (1959), 121–130. MR0110051

Über eine Punktverteilung auf der Kugel. *Acta Math. Acad. Sci. Hungar.* **10** (1959), 13–19. MR0105654

An extremal distribution of great circles on a sphere. *Publ. Math. Debrecen* **6** (1959), 79–82. MR0105065

Eräitä "kauniita" extremaalikuvioita. (Finnish) [On some "nice" extremal figures.] *Arkhimedes* **2** (1959), 1–10. MR0116265

Kugelunterdeckungen und Kugelüberdeckungen in Räumen konstanter Krümmung. *Arch. Math.* **10** (1959), 307–313. MR0106437

An arrangement of two-dimensional cells. *Ann. Univ. Sci. Budapest. Eötvös. Sect. Math.* **2** (1959), 61–64. MR0116264

Körbe és kör köré írt sokszögekről. (Hungarian) [On polygons inscribed and circumscribed a circle.] *Mat. Lapok* **10** (1959), 23–25. MR0110979

Sur la représentation d'une population infinie par un nombre fini d'éléments. *Acta. Math. Acad. Sci. Hungar.* **10** (1959), 299–304. MR0113178

Verdeckung einer Kugel durch Kugeln. *Publ. Math. Debrecen* **6** (1959), 234–240. MR0113179

1960

_____ and Heppes, A., Filling of a domain by equiareal discs. *Publ. Math. Debrecen* **7** (1960), 198–203. MR0123239

Neuere Ergebnisse in der diskreten Geometrie. *Elem. Math.* **15** (1960), 25–36. MR0117655

On shortest nets with meshes of equal area. *Acta Math. Acad. Sci. Hungar.* **11** (1960), 363–370. MR0126214

Über eine Volumenabschätzung für Polyeder. *Monatsh. Math.* **64** (1960), 374–377. MR0117656

On the stability of a circle packing. *Ann. Univ. Sci. Budapest. Eötvös Sect. Math.* **3-4** (1960/1961), 63–66. MR0140005

1962

Dichteste Kreispackungen auf einem Zylinder. *Elem. Math.* **17** (1962), 30–33. MR0133738

On primitive polyhedra. *Acta Math. Acad. Sci. Hungar.* **13** (1962), 379–382. MR0145411

1963

_____ and Heppes, A., Über stabile Körpersysteme. *Compositio Math.* **15** (1963), 119–126. MR0161227

Isoperimetric problems concerning tessellations. *Acta Math. Acad. Sci. Hungar.* **14** (1963), 343–351. MR0157294

Coxeter, H. S. M. and _____, The total length of the edges of a non-Euclidean polyhedron with triangular faces. *Quart. J. Math. Oxford Ser. (2)* **14** (1963), 273–284. MR0158307

On the isoperimetric property of the regular hyperbolic tetrahedra. *Magyar Tud. Akad. Mat. Kutató Int. Közl.* **8** (1963), 53–57. MR0166681

Mi a diszkrét geometria? (Hungarian) [What is discrete geometry?] *Magyar Tud. Akad. Mat. Fiz. Oszt. Közl.* **13** (1963), 229–238. MR0161213

Újabb eredmények a diszkrét geometriában. (Hungarian) [Recent results in discrete geometry.] *Magyar Tud. Akad. Mat. Fiz. Oszt. Közl.* **13** (1963), 341–354. MR0234354

1964

Regular figures. A Pergamon Press Book The Macmillan Co., New York (1964) xi+339 pp. MR0165423

Über eine Extremaleigenschaft der affin-regulären Vielecke. *Magyar Tud. Akad. Mat. Kutató Int. Közl.* **8** (1964), 299–302. MR0167898

What the bees know and what they do not know. *Bull. Amer. Math. Soc.* **70** (1964), 468–481. MR0163221

Bleicher, M. N. and ———, Circle-packings and circle-coverings on a cylinder. *Michigan Math. J.* **11** (1964), 337–341. MR0169140

1965

Reguläre Figuren. Adadémiai Kiadó, Verlag der Ungarischen Akademie der Wissenschaften, Budapest, (1965), 316 pp. MR0173195

On the total area of the faces of a four-dimensional polytope. *Canad. J. Math.* **17** (1965), 93–99. MR0173201

Bleicher, M. N. and ———, Two-dimensional honeycombs. *Amer. Math. Monthly* **72** (1965), 969–973. MR0185512

Minkowskian distribution of discs. *Proc. Amer. Math. Soc.* **16** (1965), 999–1004. MR0180921

Distribution of points in the elliptic plane. *Acta Math. Acad. Sci. Hungar.* **16** (1965), 437–440. MR0184139

1966

On the permeability of a circle-layer. *Studia Sci. Math. Hungar.* **1** (1966), 5–10. MR0206824

——— and Heppes, A., Regions enclosed by convex domains. *Studia Sci. Math. Hungar.* **1** (1966), 413–417. MR0203582

Mehrfache Kreisunterdeckungen und Kreisüberdeckungen auf der Kugel. *Elem. Math.* **21** (1966), 34–35. MR0192414

1967

Packings and coverings in the plane. In: *Proc. Colloquium on Convexity (Copenhagen, 1965)* 78–87 Kobenhavns Univ. Mat. Inst., Copenhagen, (1967). MR0221386

Minkowskian circle-aggregates. *Math. Ann.* **171** (1967), 97–103. MR0217702

Eine Kennzeichnung des Kreises. *Elem. Math.* **22** (1967), 25–27. MR0212653

On the arrangement of houses in a housing estate. *Studia Sci. Math. Hungar.* **2** (1967), 37–42. MR0215188

Close packing of segments. *Ann. Univ. Sci. Budapest. Eötvös Sect. Math.* **10** (1967), 57–60. MR0240721

On the number of equal discs that can touch another of the same kind. *Studia Sci. Math. Hungar.* **2** (1967), 363–367. MR0221388

———— and Heppes, A., A variant of the problem of the thirteen spheres. *Canad. J. Math.* **19** (1967), 1092–1100. MR0216371

1968

Über das Didosche Problem. *Elem. Math.* **23** (1968), 97–101. MR0236820

On the permeability of a layer of parallelograms. *Studia Sci. Math. Hungar.* **3** (1968), 195–200. MR0232285

Solid circle-packings and circle-coverings. *Studia Sci. Math. Hungar.* **3** (1968), 401–409. MR0238192

Három lemez Dido-helyzetéről. (Hungarian) [On the Dido-position of three disks.] *Mat. Lapok* **19** (1968), 9–12. MR0256268

Egy sokszög oldalainak hatványösszegéről. (Hungarian) [On the power sum of the sides of a polygon.] *Mat. Lapok* **19** (1968), 55–58. MR0238191

1969

Über die Nachbarschaft eines Kreises in einer Kreispackung. *Studia Sci. Math. Hungar.* **4** (1969), 93–97. MR0253163

Remarks on a theorem of R. M. Robinson. *Studia Sci. Math. Hungar.* **4** (1969), 441–445. MR0254744

Scheibenpackungen konstanter Nachbarnzahl. *Acta Math. Acad. Sci. Hungar.* **20** (1969), 375–381. MR0257887

Sokszögekre vonatkozó iterációs eljárások. (Hungarian) [Iteration methods for convex polygons.] *Mat. Lapok* **20** (1969), 15–23. MR0254733

1970

Über eine affininvariante Maßzahl bei Eipolyedern. *Studia Sci. Math. Hungar.* **5** (1970), 173–180. MR0268776

1971

Perfect distribution of points on a sphere. *Period. Math. Hungar.* **1** (1971), 25–33. MR0290252

Über Scheiben mit richtungsinvarianter Packungsdichte. *Elem. Math.* **26** (1971), 58–60. MR0291957

Punktverteilungen in einem Quadrat. *Studia Sci. Math. Hungar.* **6** (1971), 439–442. MR0296815

Lencsék legsűrűbb elhelyezése a síkon. (Hungarian) [The densest packing of lenses in the plane.] *Mat. Lapok* **22** (1971), 209–213. MR0323725

1972

Lagerungen in der Ebene auf der Kugel und im Raum. Zweite verbesserte und erweiterte Auflage. Die Grundlehren der mathematischen Wissenschaften, Band **65**. Springer-Verlag, Berlin-New York, (1972). xi+238 pp. MR0353117

Egy záródási tétel. (Hungarian) [A closing theorem.] *Mat. Lapok* **23** (1972), 9–12. MR0350617

1973

Distribution of points on convex polyhedra. *Period. Math. Hungar.* **4** (1973), 29–38. MR0331226

Fejes Tóth, G. and _____, Remark on a paper of C. H. Dowker. *Period. Math. Hungar.* **3** (1973), 271–274. MR0333990

Fejes Tóth, G. and _____, On totally separable domains. *Acta Math. Acad. Sci. Hungar.* **24** (1973), 229–232. MR0322690

On the density of a connected lattice of convex bodies. *Acta Math. Acad. Sci. Hungar.* **24** (1973), 373–376. MR0331228

Five-neighbour packing of convex discs. *Period. Math. Hungar.* **4** (1973), 221–229. MR0345006

Diszkrét geometriai kutatások Magyarországon. (Hungarian) [Research in discrete geometry in Hungary.] *Mat. Lapok* **24** (1973), 201-210. Zbl 0362.52001

1974

_____ and Makai, E., Jr., On the thinnest non-separable lattice of convex plates. *Stud. Sci. Math. Hungar.* **9** (1974), 191–193. MR0370369

Megjegyzések Tarski sávproblémájának egy duálisához. (Hungarian) [Remarks on the dual to Tarski's plank problem.] *Mat. Lapok* **25** (1974), 13–20 (1977). MR0442827

1975

Some remarks on saturated sets of points. *Studia Sci. Math. Hungar.* **10** (1975), 75–79. MR0458313

On Hadwiger numbers and Newton numbers of a convex body. *Studia Sci. Math. Hungar.* **10** (1975), 111–115. MR0440469

1976

Close packing and loose covering with balls. *Publ. Math. Debrecen* **23** (1976), 323–326. MR0428199

1977

———— and Sauer, N., Thinnest packing of cubes with a given number of neighbours. *Canad. Math. Bull.* **20** (1977), 501–507. MR0478017

Illumination of convex discs. *Acta Math. Acad. Sci. Hungar.* **29** (1977), 355–360. MR0464065

———— and Heppes, A., A remark on the Hadwiger numbers of a convex disc. *Studia Sci. Math. Hungar.* **12** (1977), 409–412. MR0607095

Dichteste Kugelpackung. Eine Idee von Gauss. In: *Festschrift zur 200. Wiederkehr des Geburtstages von Carl Friedrich Gauss. Abh. Braunschweig. Wiss. Gesellsch.* **27** (1977), 311–321. MR0497687

Megjegyzések Rényi és Sulanke egy tételéhez. (Hungarian) [Remarks on a theorem of Rényi and Sulanke.] *Mat. Lapok* **29** (1977/81), 33–38. MR0664003

1978

Remarks on the closest packing of convex discs. *Comment. Math. Helv.* **53** (1978), 536–541. MR0514025

Egy kört eltakaró körökről. (Hungarian) [On circles covering a given circle.] *Mat. Lapok* **30** (1978/82), 317–320. MR0740498

Bezdek, A. and ————, Adott sugáreloszlású körök sűrű elhelyezése. (Hungarian) [Dense packing of circles with given radius-distribution.] *Mat. Lapok* **30** (1978/82), 321–327. MR0740499

1980

Unterdeckungen und Überdeckungen der Ebene durch konvexe Scheiben. In: *Twelfth Styrian Mathematical Symposium (Graz, 1980)*, Ber. No. 142, 23 pp., Bericht, 140–150, Forschungszentrum Graz, (1980). MR0625868

Some density-bounds for packing and covering with incongruent circles. *Studia Sci. Math. Hungar.* **15** (1980), 63–70. MR0681425

Packing and covering with convex discs. *Studia Sci. Math. Hungar.* **15** (1980), 93–100. MR0681430

Approximation of convex domains by polygons. *Studia Sci. Math. Hungar.* **15** (1980), 133–138. MR0681434

Solid packing of circles in the hyperbolic plane. *Studia Sci. Math. Hungar.* **15** (1980), 299–302. MR0681451

_____ and Heppes, A., Multisaturated packings of circles. *Studia Sci. Math. Hungar.* **15** (1980), 303–307. MR0681452

Fejes Tóth, G. and _____, Dictators on a planet. *Studia Sci. Math. Hungar.* **15** (1980), 313–316. MR0681454

1981

Some researches inspired by H. S. M. Coxeter. In: *The geometric vein* 271–277, Springer, New York-Berlin, (1981). MR0661784

1982

Packing of r-convex discs. *Studia Sci. Math. Hungar.* **17** (1982), 449–452. MR0761560

Packing and covering with r-convex discs. *Studia Sci. Math. Hungar.* **18** (1982), 69–73. MR0759317

_____ and Florian, A., Packing and covering with convex discs. *Mathematika* **29** (1982), 181–193. MR0696874

1983

Fejes Tóth, G. and _____, Packing the plane with Minkowskian sums of convex sets. *Studia Sci. Math. Hungar.* **18** (1983), 461–464. MR0787950

On the densest packing of convex discs. *Mathematika* **30** (1983), 1–3. MR0720944

1984

Compact packing of circles. *Studia Sci. Math. Hungar.* **19** (1984), 103–107. MR0787791

_____ and Wills, J. M., Enclosing a convex body by homothetic bodies. *Geom. Dedicata* **15** (1984), 279–287. MR0739931

Density bounds for packing and covering with convex discs. *Exposition. Math.* **2** (1984), 131–153. MR0783129

1985

Packing of homothetic discs of n different sizes. *Studia Sci. Math. Hungar.* **20** (1985), 217–221. MR0886023

Densest packing of translates of a domain. *Acta Math. Hungar.* **45** (1985), 437–440. MR0791463

Isoperimetric problems for tilings. *Mathematika* **32** (1985), 10–15. MR0817101, Corrigendum: *Mathematika* **33** (1986), 189–191. MR0882491

Isoperimetric problems for tilings. In: Diskrete Geometrie, 3. Kolloq., Salzburg (1985), 101–110. Zbl 0569.52013

Stable packing of circles on the sphere. Dual French-English text. *Structural Topology* **11** (1985), 9–14. MR0804975

1986

Symmetry induced by economy. *Symmetry: unifying human understanding, I. Comput. Math. Appl. Part B* **12** (1986), 83–91. MR0838138

Packing of ellipses with continuously distributed area. *Discrete Math.* **60** (1986), 263–267. MR0852113

Densest packing of translates of the union of two circles. *Discrete Comput. Geom.* **1** (1986), 307–314. MR0866366

1987

On spherical tilings generated by great circles. *Geom. Dedicata* **23** (1987), 67–71. MR0886775

1989

Fejes Tóth, G. and _____, A geometrical analogue of the phase transformation of crystals. *Symmetry 2: unifying human understanding, Part 1. Comput. Math. Appl.* **17** (1989), 251–254. MR0994202

1991

Fejes Tóth, G. and _____, Remarks on 5-neighbor packings and coverings with circles. In: *Applied geometry and discrete mathematics 275–288, DIMACS Ser. Discrete Math. Theoret. Comput. Sci.* **4**, Amer. Math. Soc., Providence, RI, (1991). MR1116354

1993

Flight in a packing of disks. *Discrete Comput. Geom.* **9** (1993), 1–9. MR1184690

1999

Minkowski circle packings on the sphere. *Discrete Comput. Geom.* **22** (1999), 161–166. MR1698538

Research problems

A problem of illumination. *Am. Math. Mon.* **77** (1970), 869–870.

Parasites on the stem of a planet. *Am. Math. Mon.* **78** (1971), 528–529.

A problem concerning sphere-packing and sphere-covering. *Am. Math. Mon.* **79** (1972), 62–63.

Exploring a planet. *Am. Math. Mon.* **80** (1973), 1043–1044. Zbl 0274.52014

Research problem No. 4. *Period. Math. Hung.* **4** (1973), 81–82.

Research problem No. 6. Period. Math. Hung. **4** (1973), 231–232.

A covering problem. *Am. Math. Mon.* **81** (1974), 632. Zbl 0287.52010

Research problem No. 13. *Period. Math. Hung.* **6** (1975), 197–199.

Tessellation of the plane with convex polygons having a constant number of neighbors. *Am. Math. Mon.* **82** (1975), 273–276. MR0375084

Research problem No. 14. *Period. Math. Hung.* **6** (1975), 277–178.

A combinatorial problem concerning oriented lines in the plane. *Am. Math. Mon.* **82** (1975), 387–389. Zbl 0304.52002

Research problem No. 17. *Period. Math. Hung.* **7** (1976), 87–90.

Research problems No. 21. *Period. Math. Hung.* **8** (1977), 103–104. Zbl 0415.52009

Research problem No. 24. *Period. Math. Hung.* **9** (1978), 173–174. Zbl 0386.52008

Sequences of polyhedra. Am. Math. Mon. **88** (1981), 145–146. Zbl 0458.52001

Research problem No. 34. *Period. Math. Hung.* **14** (1984), 309–314.

Research problem No. 37. *Period. Math. Hung.* **15** (1984), 249–250. Zbl 0543.52013

Research problem No. 38. *Period. Math. Hung.* **16** (1985), 61–63. Zbl 0573.52014

Research problem No. 41. *Period. Math. Hung.* **18** (1987), 251–254.

Research problem No. 43. *Period. Math. Hung.* **19** (1988), 91–92. Zbl 0642.52014

Research problems. No. 44. *Period. Math. Hung.* **20** (1989), 89–91. Zbl 0671.52013

Research problems. No. 45. *Period. Math. Hung.* **20** (1989), 169–171. Zbl 0683.52013

Research problem. *Period. Math. Hung.* **31** (1995), 163–164. Zbl 0867.52006

Research problem. *Period. Math. Hung.* **31** (1995), 165–166. Zbl 0867.52005

BOLYAI SOCIETY
MATHEMATICAL STUDIES, 24

Geometry –
Intuitive, Discrete, and Convex
pp. 23–43.

Atoms for Parallelohedra

JIN AKIYAMA, MIDORI KOBAYASHI, HIROSHI NAKAGAWA,
GISAKU NAKAMURA and IKURO SATO

A parallelohedron is a convex polyhedron which tiles 3-dimensional space by translations only. A polyhedron σ is said to be an atom for the set Π of parallelohedra if for each parallelohedron P in Π, there exists an affine-stretching transformation $A : \mathbb{R}^3 \longrightarrow \mathbb{R}^3$ such that $A(P)$ is the union of a finite number of copies of σ. In this paper, we will present two different atoms for the parallelohedra, and determine the number of these atoms used to make up each parallelohedron. We will also show an arrangement of the parallelohedra in lattice-like order and introduce the notion of indecomposability.

1. INTRODUCTION

A **parallelohedron** is a convex polyhedron which tiles 3-dimensional space (*i.e.*, **space-filling**) by translations only. In 1885, a Russian crystallographer, Evgraf Fedorov [5], established that there are exactly five families of parallelohedra, namely, parallelepiped, rhombic dodecahedron, hexagonal prism, elongated rhombic dodecahedron, and truncated octahedron (Figure 1). Note that each family contains infinitely many different shapes of polyhedra since applying affine transformation to any of them will not affect the space-filling property.

This paper deals with convex polyhedra only. We say that two polyhedra P and Q are **congruent** if either P and Q are identical, or one is a mirror image of the other. In what follows, we do not distinguish between two **congruent** polyhedra. Minkowski [7] obtained the following results for a general d-dimensional parallelohedron.

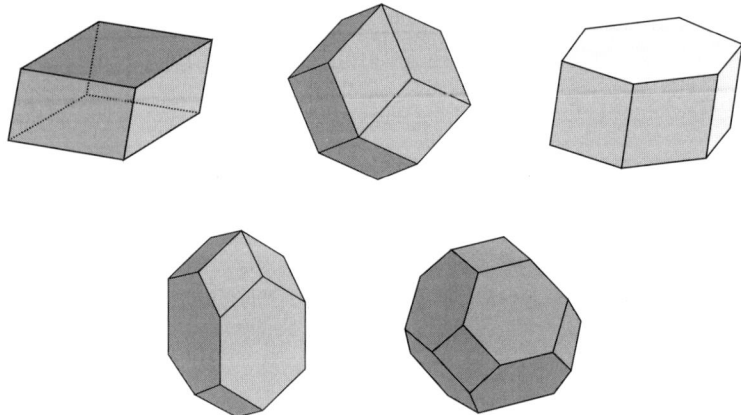

Fig. 1. Fedorov's Parallelohedra

Theorem A (Minkowski). *If P is a d-dimensional parallelohedron, then*

(1) *P is centrally symmetric,*

(2) *All faces of P are centrally symmetric, and*

(3) *The projection of P along any of its (d − 2)-faces onto the complementary 2-plane is either a parallelogram or a centrally symmetric hexagon.*

Theorem B (Minkowski). *The number $f_{d-1}(P)$ of faces in a d-parallelohedron P does not exceed $2(2^d − 1)$ and there is a parallelohedron P with $f_{d-1} = 2(2^d − 1)$.*

Dolbilin [4] extended Minkowski's theorems for non-face-to-face tilings of space. There are also numerous studies on parallelohedra discussed by Alexandrov [3] and Gruber [6].

Let Π be the set of all **parallelohedra**. A polyhedron σ is said to be an **atom** for Π if for each parallelohedron P in Π, there exists an affine-stretching transformation $A : \mathbb{R}^3 \longrightarrow \mathbb{R}^3$ such that $A(P)$ is the union of a finite number of copies of σ. *i.e.,*

$$\forall P \in \Pi, \qquad A(P) = \bigcup \sigma.$$

Note that every parallelohedron has the **Dehn invariant** zero, and this property is preserved under affine-stretching transformations [1, 2, 8].

In this paper, we will determine two different **atoms** for the set Π above and show an inclusion property among the parallelohedra. Let $F_1, F_2, \ldots, F_5$

denote the five families of parallelohedra, namely, parallelepiped, rhombic dodecahedron, hexagonal prism, elongated rhombic dodecahedron, and truncated octahedron, respectively. For each of these five families F_i $(i = 1, \ldots, 5)$, choose a polyhedron $P_i \in F_i$ to be a **representative** in any manner. The resulting set $\Sigma = \{P_1, \ldots, P_5\}$ consists of five parallelohedra, and we call it a **representative set of parallelohedra.**

Let $\Sigma = \{P_1, \ldots, P_5\}$ be a representative set. Given any parallelohedron $P \in \Pi$, there exists an affine-stretching transformation $A : \mathbb{R}^3 \longrightarrow \mathbb{R}^3$ such that $A(P) \in \Sigma$. Hence, in order to determine an atom for the set Π, it suffices to find a polyhedron σ and a representative set $\Sigma = \{P_1, \ldots, P_5\}$ such that each parallelohedron $P_i \in \Sigma$ is the union of a finite number of copies of σ. Therefore, in this paper we consider the **representative set** that consists of the cube, the rhombic dodecahedron, the skewed hexagonal prism, the elongated rhombic dodecahedron, and the truncated octahedron.

2. An Atom for the Set of Parallelohedra

Throughout the paper, we denote the set of all parallelohedra by Π, and we use Σ to denote the **representative set** of parallelohedra that consists of the cube, the rhombic dodecahedron, the skewed hexagonal prism, the elongated rhombic dodecahedron, and the truncated octahedron. In the following theorem, we will show that a special kind of pentahedron, which we call a **pentadron**, is an **atom** for the set Π. A **pentadron** is either one of the pentahedra shown in Figure 2(a), with developments shown in Figure 2(b).

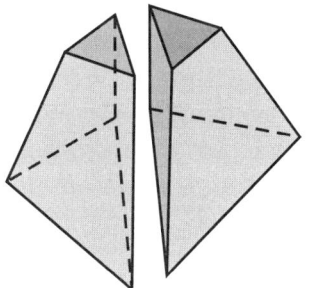

(a) Two Congruent Pentadra

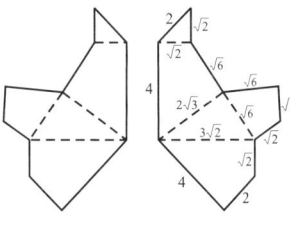

(b) Developments of Pentadra

Fig. 2. Pentahedra as an Atom

Theorem 1. *A pentadron is an atom for* Π, *and the decomposition of each parallelohedron in* Σ *into pentadra is summarized in Table 1.*

Table 1: The Number of Pentadra in the Parallelohedra

Parallelohedron	Number of Pentadra
Cube	96
Rhombic Dodecahedron	192
Skewed Hexagonal Prism	144
Elongated Rhombic Dodecahedron	384
Truncated Octahedron	48

Proof. We consider four cases.

(*1*) **Cube and Rhombic Dodecahedron.** Consider the cube in Figure 3(a). The cube can be decomposed into 6 congruent pyramids by using 4 planes, each of which passes through the center of the cube and contains two opposite edges (Figure 3(b)).

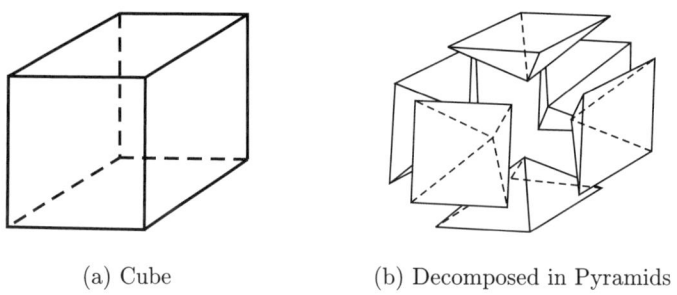

(a) Cube   (b) Decomposed in Pyramids

Fig. 3. Cube Decomposed in Pyramids

One such pyramid consists of a square base and four isosceles triangles joined at the peak (Figure 4). We now divide this pyramid further into 4 congruent parts by two planes passing through the peak and through each of the two diagonals of the base. Each part is a tetrahedron with three right angles (Figure 4). We will call this tetrahedron a **right tetra.** Thus, we have shown that a cube can be decomposed into 24 congruent **right tetras.**

On the other hand, if a square-based pyramid, such as the one shown in Figure 4 is attached to each face of a cube, a rhombic dodecahedron is formed (Figure 5). This implies that a rhombic dodecahedron can be decomposed into 48 congruent **right tetras.**

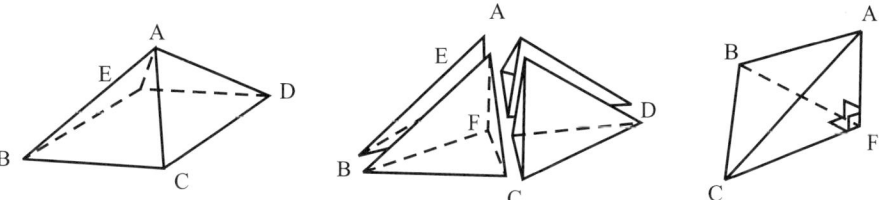

Fig. 4. Pyramid Decomposed in Right Tetra

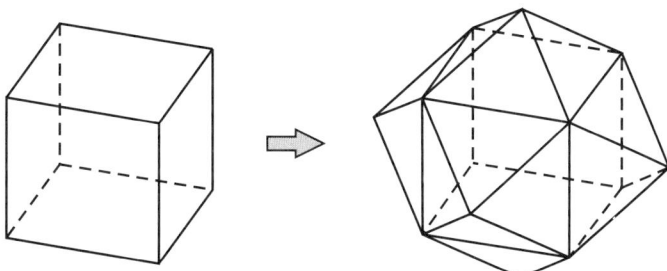

Fig. 5. Rhombic Dodecahedron Composed with a Cube and 6 Pyramids

(*2*) **Skewed Hexagonal Prism.** The base of the right tetra is a right isosceles triangle. Two right tetras glued at the base will form yet another tetrahedron (Figure 6(a)). We call this tetrahedron a **tetrapak.** If 3 such **tetrapaks** are glued appropriately, a triangular prism with an equilateral triangle at its cross section is formed. (Figure 6(b)). Finally, a skewed hexagonal prism is formed using three pairs of triangular prisms, where each pair consists of a triangular prism and its mirror image (Figure 6(c)). Therefore a skewed hexagonal prism can be formed using 18 **tetrapaks.**

(*3*) **Elongated Rhombic Dodecahedron.** An elongated dodecahedron is formed using two solids: a rhombic dodecahedron and a concave solid which can be obtained from the rhombic dodecahedron. In order to obtain the second, we consider any vertex of degree 4 of the rhombic dodecahedron. We cut the dodecahedron along the edges containing this vertex and open the solid at that vertex. The resulting solid resembles a helmet (Figure 7). Figure 8 illustrate how the elongated dodecahedron is obtained from a rhombic dodecahedron and this helmet-shaped solid.

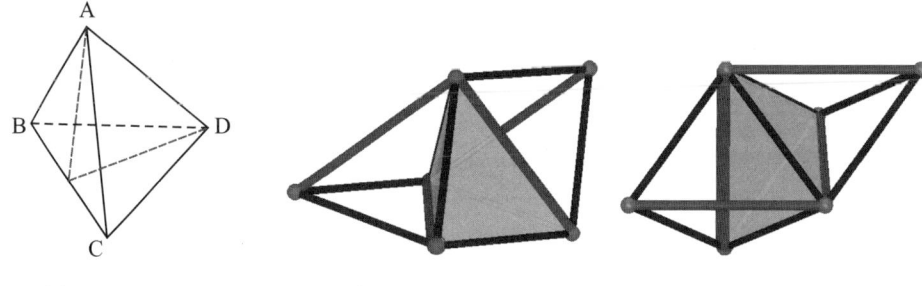

(a) A Tetrapak   (b) A Triangular Prism Viewd from Different Angles

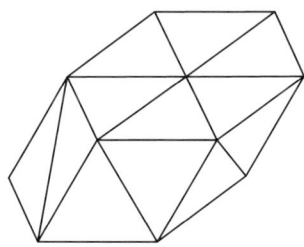

(c) A Skewed Hexagonal Prism

Fig. 6. Skewed Hexagonal Prism Composed with Tetrapak

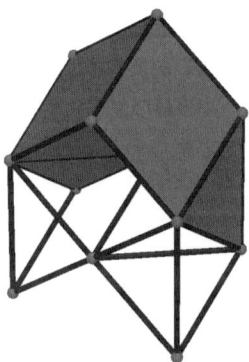

Fig. 7. The Helmet-shaped Solid

As we have mentioned, a rhombic dodecahedron consists of 48 right tetras. Therefore, the elongated rhombic dodecahedron can be formed using 96 right tetras.

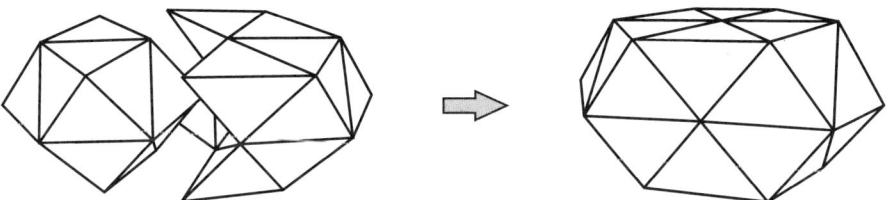

Fig. 8. Obtaining the Elongated Rhombic Dodecahedron

(*4*) **Truncated Octahedron.** A tetrapak has another important property. It can be divided into 4 congurent hexahedra using 6 planes
(Figure 9). Each plane should pass through the midpoint of an edge
and through the center of the tetrapak and must be perpendicular to
the edge. We will call each of the resulting hexahedron a **c-squadron.**

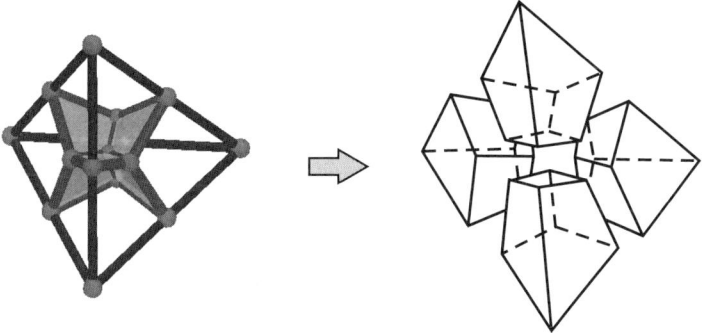

Fig. 9. Tetrapak Decomposed in c-squadrons

If we rearrange these 4 **c-squadra** in an appropriate way, we can
form the a diamond-shaped solid (Figure 10(a)). Then, using 6 of
these solids, we can construct a truncated octahedron (Figure 10(b)).
Thus a truncated octahedron can be constructed using 24 **c-squadra**.

A summary of the results we have obtained so far is given in Table 2.
The remaining problem is as follows: Is there a single polyhedron upon
which a **right tetra**, a **tetrapak** and **c-squadron** may be built? To settle
this problem, we consider a c-squadron (Figure 11(a)).

If a c-squadron is divided as shown in Figure 11(b), two congruent
pentahedra are formed, which were the **pentadra** we introduced in the
beginning of this section. Four **pentadra** can be used to build a right
tetra (Figure 11(c)) and hence, 8 **pentadra** make up a tetrapak. These

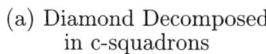

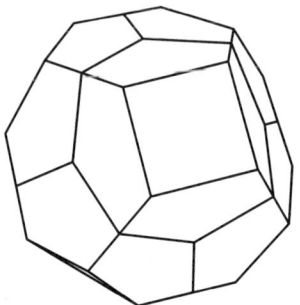

(a) Diamond Decomposed
in c-squadrons

(b) Truncated Octahedron

Fig. 10. Truncated Octahedron Composed with Diamonds

Table 2: Decomposition of the Parallelohedra

Parallelohedra	Decomposition
Cube	24 right tetras
Rhombic Dodecahedron	48 right tetras
Skewed Hexagonal Prism	18 tetrapaks
Elongated Rhombic Dodecahedron	96 right tetras
Truncated Octahedron	24 c-squadra

facts, combined with the information in Table 2 will enable us to obtain Table 1. We have therefore shown that each of the parallelohedra in our representative set can be constructed by a finite number of **pentadra.** This completes the proof. ∎

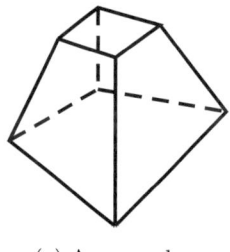

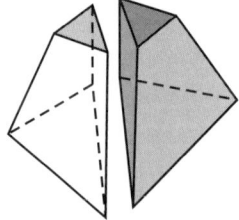

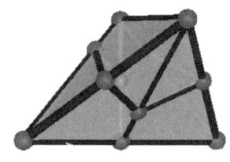

(a) A c-squadron

(b) Two Pentadra

(c) A Right Tetra

Fig. 11. Right Tetra Decomposed in c-squadrons

Remark 1. There is a way to build a cube by using only 12 pentadra (Figure 12).

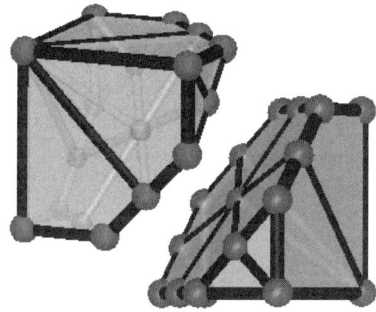

Fig. 12. Cube Decomposed into 12 Pentadra

3. EMBEDDINGS IN THE ELONGATED RHOMBIC DODECAHEDRON

In the previous section, we found that the number of pentadra needed to construct each type of parallelohedron is a multiple of 48. We now look into an inclusive property among the different types of parallelohedra. Let Σ be the **representative set** as in Section 2, and denote each parallelohedron in Σ as in Table 3.

Table 3: Denoting the Parallelohedra

Parallelohedra	Symbols
Cube	P_1
Rhombic Dodecahedron	P_2
Skewed Hexagonal Prism	P_3
Elongated Rhombic Dodecahedron	P_4
Truncated Octahedron	P_5

Theorem 2. *Let* $\Sigma = \{P_1, P_2, \ldots, P_5\}$ *be the representative set as above. Then we have the following inclusive property among the elements of* Σ:

$$P_5 \subset P_1 \subset P_2 \subset P_4 \qquad \text{and} \qquad P_3 \subset P_4,$$

where the notation "$Q_1 \subset Q_2$" has the meaning "Q_1 is embedded in Q_2" in such a way that when viewing Q_2 as the union of pentadra as in Section 2, Q_1 is the union of a subset of the pentadra used for Q_2.

Proof. There are four cases to consider since the symbol "$\subset$" appears four times in the statement of Theorem 2.

(1) $\mathbf{P_2} \subset \mathbf{P_4}$. From **Case** (*3*) of the proof of Theorem 1 and from Figure 8, it is clear that a rhombic dodecahedron is **embedded in** an elongated rhombic dodecahedron.

(2) $\mathbf{P_1} \subset \mathbf{P_2}$. Also from **Case** (*1*) of the proof of Theorem 1, we have shown that a cube is **embedded in** a rhombic dodecahedron (Figure 5).

(3) $\mathbf{P_5} \subset \mathbf{P_1}$. We now show that there is a special embedding of a truncated octahedron inside a cube. Glue a pair of congruent pentadra in such a way that the resulting hexahedron will have a right isosceles triangular face (Figure 13). Let us call this hexahedron a **tripenquadron**. This polyhedron has two other right triangular faces and has a unique vertex of degree 4.

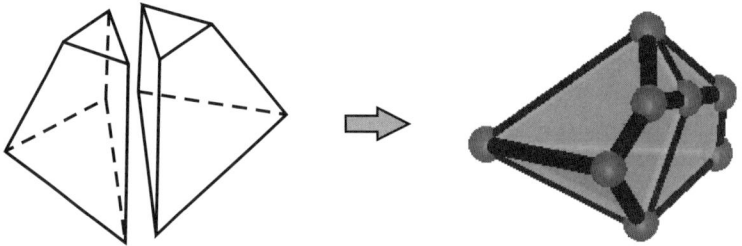

Fig. 13. Tripenquadron Composed with Two Pentadra

Suppose we now take three **tripenquadra** and glue them along their right triangular faces so that they all meet at the vertex whose degree is 4 (Figure 14). This meeting point is the peak of the resulting solid whose base is a regular hexagon.

By taking 8 such solids and gluing them along the right isosceles triangular faces, we obtain a punctured cube having 6 square holes at the center of each face (Figure 15).

To examine the interior of this punctured cube, let us open it by a plane perpendicular to two opposite faces through their diagonals. The cross section reveals that the faces in the interior of this punctured cube consist of regular hexagons and that the hole is obtained by carving out half of a truncated octahedron. The truncated octahedron and the result of fitting it into the punctured cube is shown in (Figure 16). It follows that a truncated octahedron is embedded in a special way inside a cube.

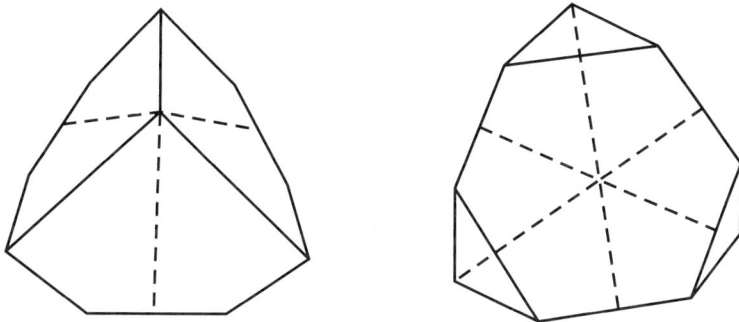

Fig. 14. Tripenquadra Composition

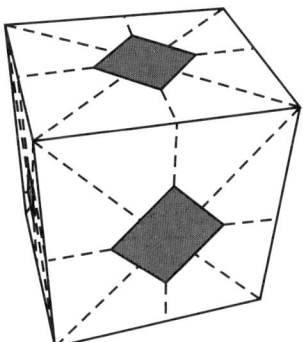

Fig. 15. Punctured Cube

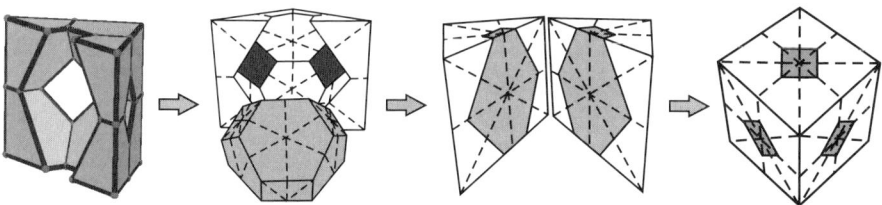

Fig. 16. Octahedron inside a Punctured Cube

(*4*) **P₃ ⊂ P₄.** Finally, we show that a skewed hexagonal prism is embedded inside an elongated rhombic dodecahedron. Consider again Figure 8 where a rhombic dodecahedron and a helmet-shaped solid are combined to form an elongated rhombic dodecahedron.

First, we divide the rhombic dodecahedron into two congruent parts by a plane which passes through six vertices, and consider only one part (Figure 17).

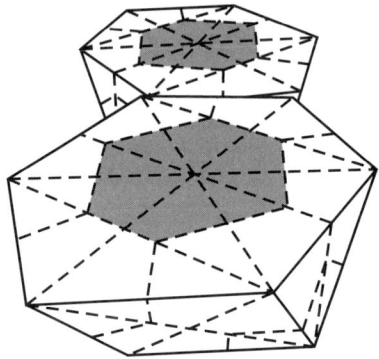

Fig. 17. Rhombic Dodecahedron Cut into Two Congruent Parts

Next, we take the helmet-shaped solid and cut off a part as shown in Figure 18. The right-hand solid in Figure 18 is V-shaped and consists of 2 skewed triangular prisms which are mirror images of each other. Each triangular prism is formed by gluing 3 tetrapaks as in Figure 6.

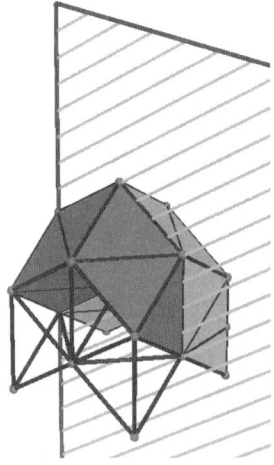

Fig. 18. Cutting the Helmet-shaped Solid

Now if this part is attached to the lower half of the rhombic dodeca-hedron, we obtain the skewed hexagonal prism as shown in Figure 19.

This shows the special embedding of the skewed hexagonal prism inside an elongated rhombic dodecahedron.

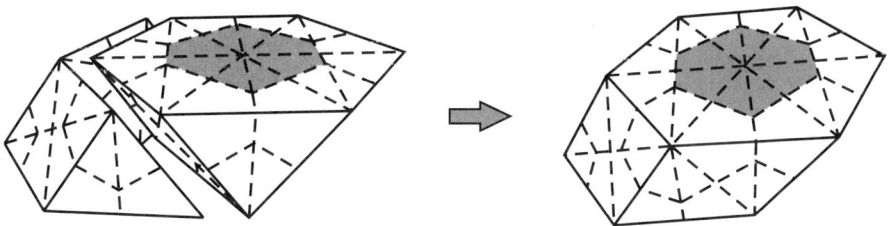

Fig. 19. Embedding the Skewed Hexagonal Prism

This completes the proof. ∎

Consider the tessellation by translates of the 3-dimensional space $\mathbb{R}^3$ by the elongated rhombic dodecahedron P_4. Since P_4 is the union of copies of **pentadra**, we obtain immediately the tessellation of the space $\mathbb{R}^3$ by **pentadra**. Now, Theorem 2 tells us that each of the other four parallelohedra $P_i \in \Sigma$ ($i = 1, 2, 3, 5$) is embeded in P_4 in such a way that P_i is the union of a subset of pentadra used for P_4. Therefore, in the tessellation by translates of the space $\mathbb{R}^3$ by P_4 (consequently by **pentadra**), each of the other four parallelohedra $P_i \in \Sigma$ will appear periodically. This is summarized in the following Corollary.

Corollary 1. *In space, there is an arrangement of pentadra such that each of the elements of Σ appears in lattice-like order.*

4. A Second Atom for the Set of Parallelohedra

In this section, we present a second atom for the set of all parallelohedra Π. We call this second atom **tetradron**, and it is either one of the tetrahedra shown in Figure 20 along with its development.

Before getting into the theorem, we first illustrate how to obtain a **tetradron** from a cube. Cut a cube of unit length by passing a plane through two opposite edges to obtain two congruent triangular prisms (Figure 21).

Take one of the prisms and label its vertices with the letters A, B, C, D, E and F. By dividing this prism into three parts using the planes BCD

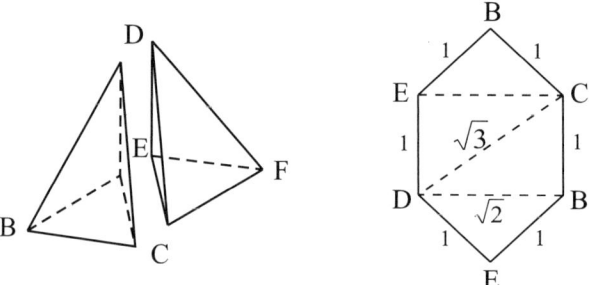

Fig. 20. Tetradron as an Atom

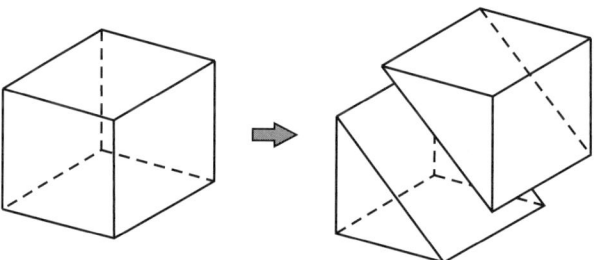

Fig. 21. Cube Composed with Two Prisms

and CDE, we obtain three congruent tetrahedra all of whose faces are right triangles (Figure 22).

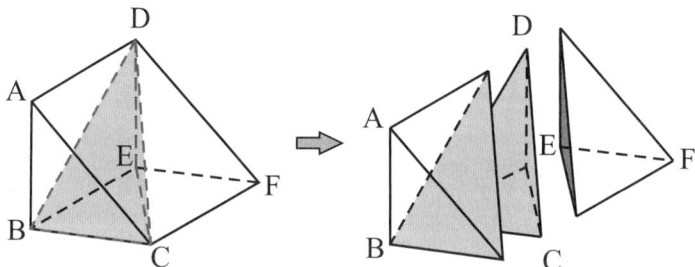

Fig. 22. Prism Composed with Three Tetrahedra

Tetrahedra $ABCD$ and $CDEF$ are identical while tetrahedron $BCDE$ is a mirror image of each of the previous two, and thus, these tetrahedra are congruent. Any tetrahedron such as $ABCD$ or its mirror image is a **tetradron.** We are ready to state the main theorem of Section 4.

Theorem 3. *A tetradron is an atom for* Π, *and the decomposition of each parallelohedron in* Σ *into tetradra is summarized in Table 4.*

Table 4: The Number of Tetradra in the Parallelohedra

Parallelohedron	Number of Tetradra
Cube	6
Rhombic Dodccahedron	96
Skewed Hexagonal Prism	18
Elongated Rhombic Dodecahedron	144
Skewed Truncated Octahedron	384

Proof. Since the case for the **cube** is obvious from the above illustrations, we will only consider the other four cases, each corresponding to a remaining parallelohedron in Σ.

(*1*) **Skewed Hexagonal Prism.** The construction of a skewed hexagonal prism using congruent copies of a tetradron is shown in Figure 23. First, we take three identical tetradra and glue some faces to form a skewed triangular prism (Figure 23(a)). Figure 23(b) shows a mirror image of the triangular prism and can be constructed by a similar gluing of the tetradra which are mirror images of the previous. A gluing of three pairs of each type of triangular prism gives rise to a skewed hexagonal prism (Figure 23(c)).

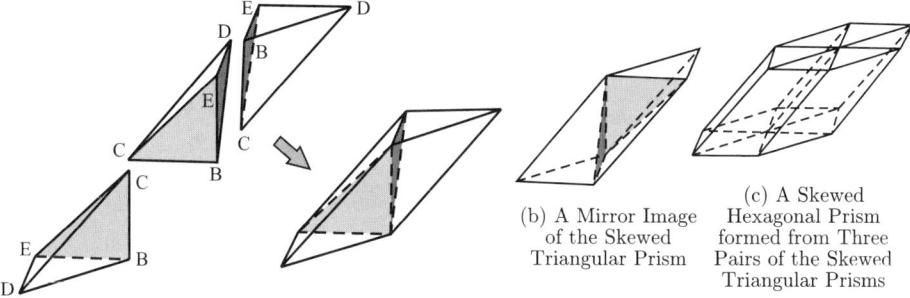

(b) A Mirror Image of the Skewed Triangular Prism

(c) A Skewed Hexagonal Prism formed from Three Pairs of the Skewed Triangular Prisms

(a) Obtaining a Skewed Triangular Prism from Tetradra

Fig. 23. Skewed Hexagonal Prism Composed with Tetradra

(*2*) **Rhombic Dodecahedron.** Next, we will illustrate how to construct a rhombic dodecahedron using tetradra. We join a pair of tetradra, one a mirror image of the other, to form a right tetra of unit height whose base is a right isosceles triangle with hypotenuse of length 2 units. The square-based pyramid is obtained by gluing four right tetras (Figure 24). This pyramid is centrally symmetric and also has a height of 1 unit. A rhombic dodecahedron is formed when each of the faces of a $2 \times 2 \times 2$ cube is capped by this square-based pyramid.

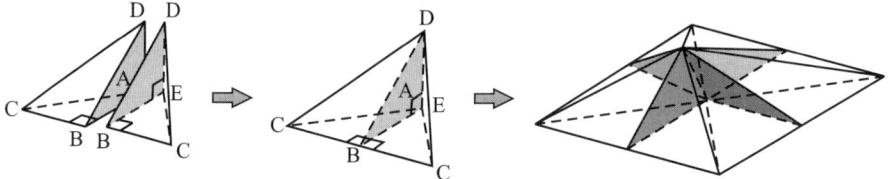

Fig. 24. Squared-base Pyramid Composed with Tetradra

(*3*) **Elongated Rhombic Dodecahedron.** An elongated rhombic do-decahedron can be constructed by first dividing the rhombic dodecahe-dron into two congruent nanohedra. We insert a $2 \times 2 \times 1$ rectangular cuboid between the two nanohedra to obtain an elongated rhombic dodecahedron (Figure 25). Note that we can insert a $2 \times 2 \times k$ rect-angular cuboid for any positive integer k; that is, we can elongate the dodecahedron to any length.

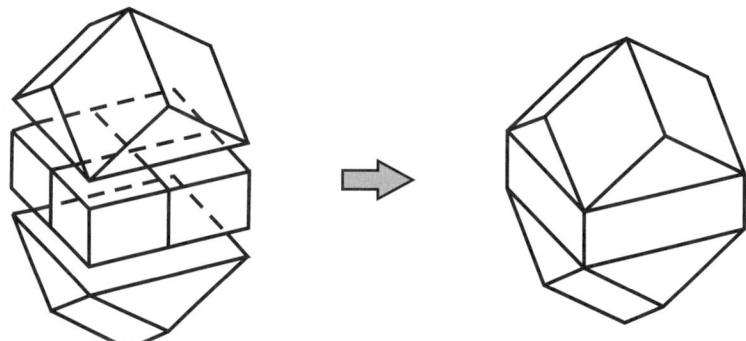

Fig. 25. Elongated Rhombic Dodecahedron Obtained from a Rhombic Dodecahedron

(*4*) **Truncated Octahedron.** Note that a tetradron is a tetrahedral reptile, that is, it replicates itself. For instance, by using 8 tetradra consisting of two identical ones and three pairs of mirror images, we can construct a similar tetradron whose size is double the original one. Similarly, a tetradron which is triple in size can be constructed from 9 tetradra and 9 pairs of mirror images (Figure 26).

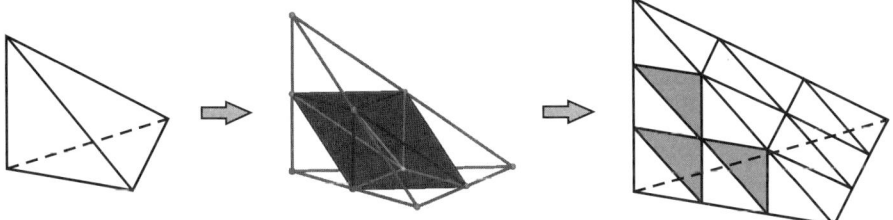

Fig. 26. Tetradron is a Tetrahedral Reptile

Using this self-replicating property, it is possible to construct a pyramid which is thrice the size of the one shown in Figure 24. In Figures 27, we obtain an octahedron by gluing two such pyramids.

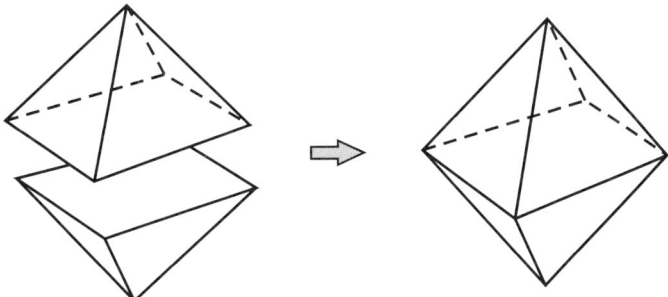

Fig. 27. Octahedron Composed of Squared Pyramids

Figure 28 shows the same octahedron being made up of tetradra. Remove a portion of the upper pyramid, one-third unit away from a vertex. If such a cut is done to every vertcx, we get the truncated octahedron (Figure 28).

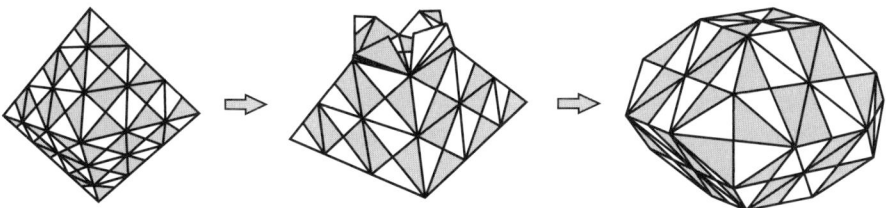

Fig. 28. Truncated Octahedron Composed of Tetradra

This completes the proof. ∎

5. Indecomposability of the Atoms

A polyhedron is said to be **indecomposable** if it cannot be decomposed into two or more congruent polyhedral pieces. In the previous sections, we have presented two different atoms for the set of all parallelohedra Π: a **pentadron** and a **tetradron**. However, if two identical **pentadra** are glued along their kite faces, we obtain a **tetradron** (Figure 29). In other words, a **tetradron** can be decomposed into two congruent parts, and therefore is not **indecomposable**. In this sense, we would like to find an **indecomposable atom** for Π; that is, one which cannot be decomposed into two or more congruent polyhedra.

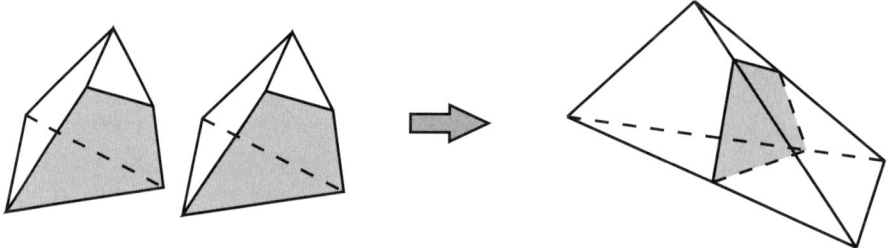

Fig. 29. Obtaining a Tetradron by Gluing Two Pentadra

The following proposition shows that it is not possible to decompose a **pentadron** into two congruent parts.

Proposition 1. *A pentadron σ cannot be decomposed into two congruent polyhedra.*

Proof. Let α and β be two polyhedra obtained by dissecting σ by a plane π (Figure 30(a)), and let k be the number of sides of its polygonal cross-section. Since the total number of faces of α and β is $7 + k$, k must be either 3 or 5 to make α and β congruent. We denote the vertices of σ by A, B, C, P, Q and R as illustrated in Figure 30(b).

(1) **Suppose k = 3.** In this case either α or β is a tetrahedron and the other is a hexahedron. Thus, we can disregard this case.

(2) **Suppose k = 5.** In this case both α and β are hexahedra and the dissecting plane π has to cut all the faces of σ. This implies that no vertices of σ are on π. Let α be the polyhedron containing P. Note that σ has only one right isosceles triangle ABC, where $\angle B = 90°$.

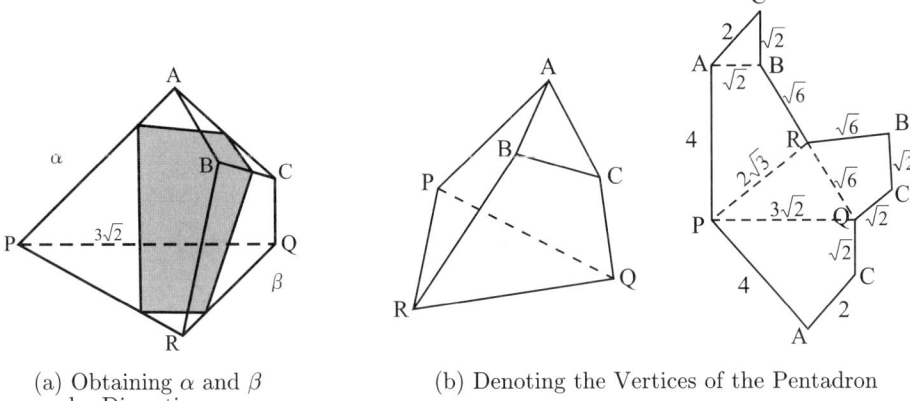

(a) Obtaining α and β
by Dissecting σ

(b) Denoting the Vertices of the Pentadron

Fig. 30. Pentadron as an Atom

The dissecting plane π cuts exactly two sides of $\triangle ABC$. We divide our proof into three subcases depending on which sides of $\triangle ABC$ π intersects.

(i) **Suppose π intersects the sides BC and AC.** Note that the side PQ has the largest length $3\sqrt{2}$ among all sides of σ. There are two distinct cases according to whether α contains PQ as its side or α not (Figure 31).

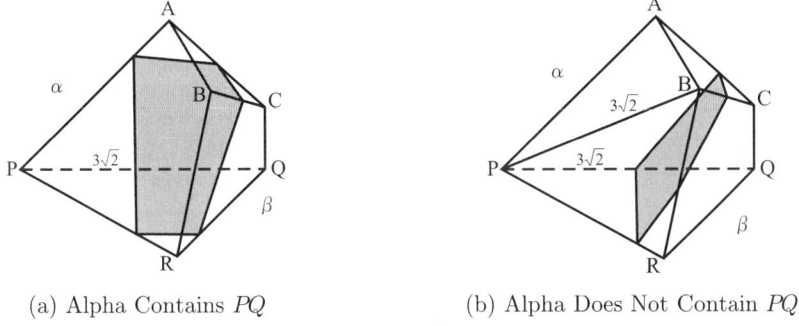

(a) Alpha Contains PQ

(b) Alpha Does Not Contain PQ

Fig. 31. π Intersects the Sides BC and AC

The first case implies that α is not congruent to β since β cannot contain any side of length $3\sqrt{2}$. In the second case, both α and β contain exactly one pentagonal face distinct from the cutting surface. One of diagonals of the pentagonal face on α

has length $3\sqrt{2}$, but the pentagonal face of β does not contain such a diagonal.

(ii) **Suppose π intersects the sides AB and AC.** The hexahedron α contains the side AP of length 4 as a side of a pentagonal face, but β does not have a pentagonal face with a side of the same length (Figure 32(a)). Thus α is not congruent to β.

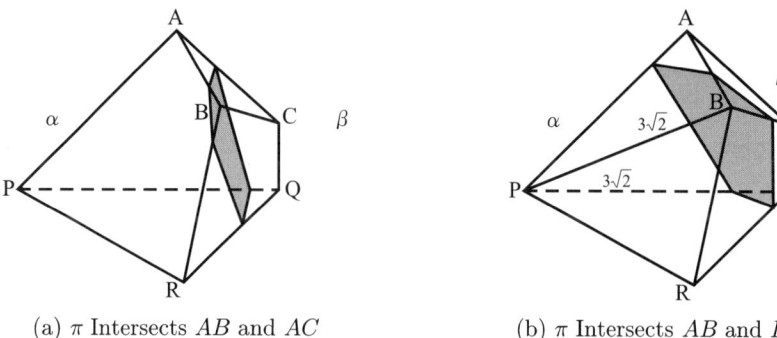

(a) π Intersects AB and AC (b) π Intersects AB and BC

Fig. 32. π Intersects the Side AB

(iii) **Suppose π intersects the sides AB and BC.** Note that $\triangle BPR$ and $\triangle PQR$ have the maximum area $3\sqrt{2}$ among all right triangles which can be taken on the surface of σ and whose vertices are those of σ (Figure 32(b)). This case implies that α contains $\triangle BPR$ on its surface, but since $\triangle PQR$ is cut by π, β cannot have a right triangle with the same area.

This completes the proof. ∎

Proposition 1 suggests the following conjecture.

Conjecture 1. *A pentadron is an indecomposable atom.*

Acknowledgements. The authors would like to thank Xin Chen, Agnes Garciano, and Toshinori Sakai for their assistance in the preparation of this manuscript. We would also like to thank the referee for his helpful comments and for providing us many figures used in this paper.

REFERENCES

[1] J. Akiyama, I. Sato and H. Seong, *On Reversibility among Parallelohedra*, Computational Geometry, LNCS 7579 (2012), 14–28.

[2] J. Akiyama and H. Seong, *Transformations which Preserve Reversibility*, to appear in the Proc. of TJJCCGG 2012, LNCS.

[3] A. D. Alexandrov, *Convex Polyhedra*, Springer Monographs in Mathematics, Springer, Berlin, 2005.

[4] N. P. Dolbilin, *Minkowski's Theorems on Parallelohedra and Their Generalizations*, Communications of the Moscow Mathematical Society (2007), 793–795.

[5] E. S. Fedorov, *An Introduction to the Theory of Figures*, Notices of the Imperial Mineralogical Society (St. Petersburg) Ser. 2, Vol. 21. 1–279 (1885). Republished with comments by Akad. Nanak. SSSR, Moscow, 1953, in Russian.

[6] P. M. Gruber, *Convex and Discrete Geometry*, A Series of Comprehensive Studies in Mathematics (Berlin), vol. 336, Springer-Verlag, 2007.

[7] H. Minkowski, *Allgemeine Lehrsätze über Konvexe Polyeder*, Nachr. Ges. Wiss. Göttingen (1897), 198–219.

[8] M. I. Stogrin, *Regular Dirichlet–Voronoi Partitions for the Second Triclinic Group*, Proceedings of the Steklov Institute of Mathematics, no. 123, Amer. Math. Soc., 1973.

Jin Akiyama, Gisaku Nakamura

Research Center for Math Education,
Tokyo University of Science,
Tokyo 162-8601,
Japan.

e-mail: ja@jin-akiyama.com

Midori Kobayashi

School of Administration and
Informatics,
University of Shizuoka,
Shizuoka 422-8526,
Japan

e-mail:
midori@u-shizuoka-ken.ac.jp

Hiroshi Nakagawa

e-mail: hiro-4@do6.enjoy.ne.jp

Ikuro Sato

Department of Pathology,
Research Institute,
Miyagi Cancer Center,
Miyagi 981-1293,
Japan

e-mail: sato-ik510@miyagi-pho.jp

BOLYAI SOCIETY
MATHEMATICAL STUDIES, 24

Geometry –
Intuitive, Discrete, and Convex
pp. 45–64.

Tarski's Plank Problem Revisited

KÁROLY BEZDEK*

In the 1930's, Tarski introduced his plank problem at a time when the field Discrete Geometry was about to born. It is quite remarkable that Tarski's question and its variants continue to generate interest in the geometric and analytic aspects of coverings by planks in the present time as well. The paper is of a survey type with some new results and with a list of open research problems on the discrete geometric side of the plank problem.

1. Introduction

Tarski's plank problem has generated a great interest in understanding the geometry of coverings by planks. There have been a good number of results published in connection with the plank problem of Tarski that are surveyed in this paper. The goal of this paper is to survey the state of the art of Tarski's plank problem from the point of view of discrete geometry and to prove some new results and to list some relevant research problems as well. The topics discussed include not only coverings by planks but also coverings by cylinders and the sets to be covered include balls as well as lattice points. For some natural reason, a good subcollection of the research problems listed raises challanging questions on balls, that are the most symmetric bodies still central for research in (discrete) geometry. Last but not least we mention that the partial covering problem by planks introduced in this paper connects Tarski's plank problem to the Kakeya–Pál as well as to the

---

*Partially supported by a Natural Sciences and Engineering Research Council of Canada Discovery Grant and by the Hung. Acad. Sci. Found. (OTKA), grant no. K72537. (This survey is partially based on the author's talk delivered at the meeting "Intuitive Geometry, in Memoriam László Fejes Tóth", June 30–July 4, 2008, Budapest, Hungary.)

Blaschke–Lebesgue problems. In this way, ball-polyhedra are investigated as well. The rest of the paper studies the topics outlined in six consecutive sections.

2. Plank Theorems

A *convex body* of the Euclidean space $\mathbb{E}^d$ is a compact convex set with non-empty interior. Let $\mathbf{C} \subset \mathbb{E}^d$ be a convex body, and let $H \subset \mathbb{E}^d$ be a hyperplane. Then the distance $\mathrm{w}(\mathbf{C}, H)$ between the two supporting hyperplanes of $\mathbf{C}$ parallel to H is called the *width of* $\mathbf{C}$ *parallel to* H. Moreover, the smallest width of $\mathbf{C}$ is called the *minimal width* of $\mathbf{C}$ and is denoted by $\mathrm{w}(\mathbf{C})$.

Recall that in the 1930's, Tarski posed what came to be known as the plank problem. A *plank* $\mathbf{P}$ in $\mathbb{E}^d$ is the (closed) set of points between two distinct parallel hyperplanes. The *width* $\mathrm{w}(\mathbf{P})$ of $\mathbf{P}$ is simply the distance between the two boundary hyperplanes of $\mathbf{P}$. Tarski conjectured that if a convex body of minimal width w is covered by a collection of planks in $\mathbb{E}^d$, then the sum of the widths of these planks is at least w. This conjecture was proved by Bang in his memorable paper [8]. (In fact, the proof presented in that paper is a simplification and generalization of the proof published by Bang somewhat earlier in [7].) Thus, the following statement we call the plank theorem of Bang.

Theorem 2.1. *If the convex body $\mathbf{C}$ is covered by the planks $\mathbf{P}_1, \mathbf{P}_2, \ldots,$ $\mathbf{P}_n$ in $\mathbb{E}^d$ (i.e. $\mathbf{C} \subset \mathbf{P}_1 \cup \mathbf{P}_2 \cup \cdots \cup \mathbf{P}_n \subset \mathbb{E}^d$), then*

$$\sum_{i=1}^{n} \mathrm{w}(\mathbf{P}_i) \geq \mathrm{w}(\mathbf{C}).$$

In [8], Bang raised the following stronger version of Tarski's plank problem called the affine plank problem. We phrase it via the following definition. Let $\mathbf{C}$ be a convex body and let $\mathbf{P}$ be a plank with boundary hyperplanes parallel to the hyperplane H in $\mathbb{E}^d$. We define the $\mathbf{C}$-*width* of the plank $\mathbf{P}$ as $\frac{\mathrm{w}(\mathbf{P})}{\mathrm{w}(\mathbf{C},H)}$ and label it by $\mathrm{w}_{\mathbf{C}}(\mathbf{P})$. (This notion was introduced by Bang [8] under the name "relative width".)

Conjecture 2.2. If the convex body $\mathbf{C}$ is covered by the planks $\mathbf{P}_1, \mathbf{P}_2, \ldots,$ $\mathbf{P}_n$ in $\mathbb{E}^d$, then

$$\sum_{i=1}^{n} \mathrm{w_C}(\mathbf{P}_i) \geq 1.$$

The special case of Conjecture 2.2, when the convex body to be covered is centrally symmetric, has been proved by Ball in his celebrated paper [2]. Thus, the following is the plank theorem of Ball.

Theorem 2.3. *If the centrally symmetric convex body $\mathbf{C}$ is covered by the planks $\mathbf{P}_1, \mathbf{P}_2, \ldots, \mathbf{P}_n$ in $\mathbb{E}^d$, then*

$$\sum_{i=1}^{n} \mathrm{w_C}(\mathbf{P}_i) \geq 1.$$

From the point of view of discrete geometry it seems natural to mention that after proving Theorem 2.3 Ball [3] used Bang's proof of Theorem 2.1 to derive a new argument for an improvement of the Davenport–Rogers lower bound on the density of economical sphere lattice packings.

It was Alexander [1] who noticed that Conjecture 2.2 is equivalent to the following generalization of a problem of Davenport.

Conjecture 2.4. If a convex body $\mathbf{C}$ in $\mathbb{E}^d$ is sliced by $n - 1$ hyperplane cuts, then there exists a piece that covers a translate of $\frac{1}{n}\mathbf{C}$.

We note that the paper [10] of A. Bezdek and the author proves Conjecture 2.4 for successive hyperplane cuts (i.e. for hyperplane cuts when each cut divides one piece). Also, the same paper ([10]) introduced two additional equivalent versions of Conjecture 2.2. As they seem to be of independent interest we recall them following the terminology used in [10].

Let $\mathbf{C}$ and $\mathbf{K}$ be convex bodies in $\mathbb{E}^d$ and let H be a hyperplane of $\mathbb{E}^d$. The $\mathbf{C}$-*width of* $\mathbf{K}$ *parallel to* H is denoted by $\mathrm{w_C}(\mathbf{K}, H)$ and is defined as $\frac{\mathrm{w}(\mathbf{K},H)}{\mathrm{w}(\mathbf{C},H)}$. The *minimal* $\mathbf{C}$-*width of* $\mathbf{K}$ is denoted by $\mathrm{w_C}(\mathbf{K})$ and is defined as the minimum of $\mathrm{w_C}(\mathbf{K}, H)$, where the minimum is taken over all possible hyperplanes H of $\mathbb{E}^d$. Recall that the inradius of $\mathbf{K}$ is the radius of the largest ball contained in $\mathbf{K}$. It is quite natural then to introduce the $\mathbf{C}$-*inradius of* $\mathbf{K}$ as the factor of the largest (positively) homothetic copy of $\mathbf{C}$, a translate of which is contained in $\mathbf{K}$. We need to do one more step to introduce the so-called successive $\mathbf{C}$-inradii of $\mathbf{K}$ as follows. Let r be the $\mathbf{C}$-inradius of $\mathbf{K}$. For any $0 < \rho \leq r$ let the $\rho\mathbf{C}$-*rounded body of* $\mathbf{K}$ be denoted by $\mathbf{K}^{\rho\mathbf{C}}$ and be defined as the union of all translates of $\rho\mathbf{C}$ that are covered by $\mathbf{K}$. Now, take a fixed integer $n \geq 1$. On the one hand, if $\rho > 0$ is sufficiently small, then $\mathrm{w}_{\mathbf{C}}(\mathbf{K}^{\rho\mathbf{C}}) > n\rho$. On the other hand, $\mathrm{w}_{\mathbf{C}}(\mathbf{K}^{r\mathbf{C}}) = r \leq nr$. As $\mathrm{w}_{\mathbf{C}}(\mathbf{K}^{\rho\mathbf{C}})$ is a decreasing continuous function of $\rho > 0$ and $n\rho$ is a strictly increasing continuous function of ρ there exists a uniquely determined $\rho > 0$ such that

$$\mathrm{w}_{\mathbf{C}}(\mathbf{K}^{\rho\mathbf{C}}) = n\rho.$$

This uniquely determined ρ is called the *n-th successive* $\mathbf{C}$*-inradius of* $\mathbf{K}$ and is denoted by $r_{\mathbf{C}}(\mathbf{K}, n)$. Notice that $r_{\mathbf{C}}(\mathbf{K}, 1) = r$. Now, the two equivalent versions of Conjecture 2.2 and Conjecture 2.4 introduced in [10] can be phrased as follows.

Conjecture 2.5. If a convex body $\mathbf{K}$ in $\mathbb{E}^d$ is covered by the planks $\mathbf{P}_1$, $\mathbf{P}_2, \ldots, \mathbf{P}_n$, then $\sum_{i=1}^{n} \mathrm{w}_{\mathbf{C}}(\mathbf{P}_i) \geq \mathrm{w}_{\mathbf{C}}(\mathbf{K})$ for any convex body $\mathbf{C}$ in $\mathbb{E}^d$.

Conjecture 2.6. Let $\mathbf{K}$ and $\mathbf{C}$ be convex bodies in $\mathbb{E}^d$. If $\mathbf{K}$ is sliced by $n-1$ hyperplanes, then the minimum of the greatest $\mathbf{C}$-inradius of the pieces is equal to the n-th successive $\mathbf{C}$-inradius of $\mathbf{K}$, i.e. it is $r_{\mathbf{C}}(\mathbf{K}, n)$.

A. Bezdek and the author [10] proved the following theorem that (under the condition that $\mathbf{C}$ is a ball) answers a question raised by Conway ([9]) as well as proves Conjecture 2.6 for successive hyperplane cuts.

Theorem 2.7. Let $\mathbf{K}$ and $\mathbf{C}$ be convex bodies in $\mathbb{E}^d, d \geq 2$. If $\mathbf{K}$ is sliced into n pieces by $n-1$ successive hyperplane cuts (i.e. when each cut divides one piece), then the minimum of the greatest $\mathbf{C}$-inradius of the pieces is the n-th successive $\mathbf{C}$-inradius of $\mathbf{K}$, i.e. $r_{\mathbf{C}}(\mathbf{K}, n)$. An optimal partition is achieved by $n-1$ parallel hyperplane cuts equally spaced along the $\mathbf{C}$-width of the $r_{\mathbf{C}}(\mathbf{K}, n)\mathbf{C}$-rounded body of $\mathbf{K}$.

3. COVERING CONVEX BODIES BY CYLINDERS

In his paper [8], Bang by describing a concrete example and writing that it may be extremal proposes to investigate a quite challanging question that can be phrased as follows.

Problem 3.1. Prove or disprove that the sum of the base areas of finitely many cylinders covering a 3-dimensional convex body is at least half of the minimum area 2-dimensional projection of the body.

If true, then the estimate of Problem 3.1 is a sharp one due to a covering of a regular tetrahedron by two cylinders described in [8]. A very recent paper of Litvak and the author ([16]) investigates Problem 3.1 as well as its higher dimensional analogue. Their main result can be summarized as follows.

Given $0 < k < d$ define a k-codimensional cylinder $\mathbf{C}$ in $\mathbb{E}^d$ as a set which can be presented in the form $\mathbf{C} = H + B$, where H is a k-dimensional linear subspace of $\mathbb{E}^d$ and B is a measurable set (called the base) in the orthogonal complement $H^\perp$ of H. For a given convex body $\mathbf{K}$ and a k-codimensional cylinder $\mathbf{C} = H + B$ we define the cross-sectional volume $\mathrm{crv}_{\mathbf{K}}(\mathbf{C})$ of $\mathbf{C}$ with respect to $\mathbf{K}$ as follows

$$\mathrm{crv}_{\mathbf{K}}(\mathbf{C}) := \frac{\mathrm{vol}_{d-k}(\mathbf{C} \cap H^\perp)}{\mathrm{vol}_{d-k}(P_{H^\perp}\mathbf{K})} = \frac{\mathrm{vol}_{d-k}(P_{H^\perp}\mathbf{C})}{\mathrm{vol}_{d-k}(P_{H^\perp}\mathbf{K})} = \frac{\mathrm{vol}_{d-k}(B)}{\mathrm{vol}_{d-k}(P_{H^\perp}\mathbf{K})},$$

where $P_{H^\perp} : \mathbb{E}^d \to H^\perp$ denotes the orthogonal projection of $\mathbb{E}^d$ onto $H^\perp$. Notice that for every invertible affine map $T : \mathbb{E}^d \to \mathbb{E}^d$ one has $\mathrm{crv}_{\mathbf{K}}(\mathbf{C}) = \mathrm{crv}_{T\mathbf{K}}(T\mathbf{C})$. The following theorem is proved in [16].

Theorem 3.2. *Let $\mathbf{K}$ be a convex body in $\mathbb{E}^d$. Let $\mathbf{C}_1, \ldots, \mathbf{C}_N$ be k-codimensional cylinders in $\mathbb{E}^d$, $0 < k < d$ such that $\mathbf{K} \subset \bigcup_{i=1}^N \mathbf{C}_i$. Then*

$$\sum_{i=1}^N \mathrm{crv}_{\mathbf{K}}(\mathbf{C}_i) \geq \frac{1}{\binom{d}{k}}.$$

Moreover, if $\mathbf{K}$ is an ellipsoid and $\mathbf{C}_1, \ldots, \mathbf{C}_N$ are 1-codimensional cylinders in $\mathbb{E}^d$ such that $\mathbf{K} \subset \bigcup_{i=1}^N \mathbf{C}_i$, then

$$\sum_{i=1}^N \mathrm{crv}_{\mathbf{K}}(\mathbf{C}_i) \geq 1.$$

The case $k = d - 1$ of Theorem 3.2 corresponds to Conjecture 2.2 i.e. to the affine plank problem. Theorem 3.2 for $k = d-1$ implies the lower bound $1/d$ that can be somewhat further improved (for more details see [16]).

As an immediate corollary of Theorem 3.2 we get the following estimate for Problem 3.1.

Corollary 3.3. *The sum of the base areas of finitely many (1-codimensional) cylinders covering a 3-dimensional convex body is always at least one third of the minimum area 2-dimensional projection of the body.*

Also, note that the inequality of Theorem 3.2 on covering ellipsoids by 1-codimensional cylinders is best possible. By looking at this result from the point of view of k-codimensional cylinders we are led to ask the following quite natural question. Unfortunately, despite its elementary character it is still open.

Problem 3.4. Let $0 < c(d, k) \leq 1$ denote the largest real number with the property that if $\mathbf{K}$ is an ellipsoid and $\mathbf{C}_1, \ldots, \mathbf{C}_N$ are k-codimensional cylinders in $\mathbb{E}^d$, $1 \leq k \leq d-1$ such that $\mathbf{K} \subset \bigcup_{i=1}^{N} \mathbf{C}_i$, then $\sum_{i=1}^{N} \mathrm{crv}_{\mathbf{K}}(\mathbf{C}_i) \geq c(d, k)$. Determine $c(d, k)$ for given d and k.

On the one hand, Theorem 2.1 and Theorem 3.2 imply that $c(d, d-1) = 1$ and $c(d, 1) = 1$ moreover, $c(d, k) \geq \frac{1}{\binom{d}{k}}$.

4. COVERING LATTICE POINTS BY HYPERPLANES

In their paper [13], Hausel and the author have established the following discrete version of Tarski's plank problem.

Recall that the lattice width of a convex body $\mathbf{K}$ in $\mathbb{E}^d$ is defined as

$$w(\mathbf{K}, \mathbb{Z}^d) = \min \left\{ \max_{x \in \mathbf{K}} \langle x, y \rangle - \min_{x \in \mathbf{K}} \langle x, y \rangle \mid y \in \mathbb{Z}^d, \ y \neq 0 \right\},$$

where $\mathbb{Z}^d$ denotes the integer lattice of $\mathbb{E}^d$. It is well-known that if $y \in \mathbb{Z}^d$, $y \neq 0$ is chosen such that $\lambda y \notin \mathbb{Z}^d$ for any $0 < \lambda < 1$ (i.e. y is a primitive lattice point), then

$$\max_{x \in \mathbf{K}} \langle x, y \rangle - \min_{x \in \mathbf{K}} \langle x, y \rangle$$

is equal to the Euclidean width of $\mathbf{K}$ in the direction y divided by the Euclidean distance between two consecutive lattice hyperplanes of $\mathbb{Z}^d$ that are orthogonal to y. Thus, if $\mathbf{K}$ is the convex hull of finitely many points of $\mathbb{Z}^d$, then

$$\max_{x \in \mathbf{K}} \langle x, y \rangle - \min_{x \in \mathbf{K}} \langle x, y \rangle$$

is an integer namely, it is less by one than the number of lattice hyperplanes of $\mathbb{Z}^d$ that intersect $\mathbf{K}$ and are orthogonal to y. Now, we are ready to state the following conjecture of Hausel and the author ([13]).

Conjecture 4.1. Let $\mathbf{K}$ be a convex body in $\mathbb{E}^d$. Let $H_1, \ldots, H_N$ be hyperplanes in $\mathbb{E}^d$ such that

$$\mathbf{K} \cap \mathbb{Z}^d \subset \bigcup_{i=1}^{N} H_i.$$

Then

$$N \geq w(\mathbf{K}, \mathbb{Z}^d) - d.$$

Properly translated copies of cross-polytopes, described in [13], show that if true, then the above inequality is best possible.

The special case, when $N = 0$, is of independent interest. (In particular, this case seems to be "responsible" for the term d in the inequality of Conjecture 4.1.) Namely, it seems reasonable to conjecture (see also [6]) that if $\mathbf{K}$ is an integer point free convex body in $\mathbb{E}^d$, then $w(\mathbf{K}, \mathbb{Z}^d) \leq d$. On the one hand, this has been proved by Banaszczyk [5] for ellipsoids. On the other hand, for general convex bodies containing no integer points, Banaszczyk, Litvak, Pajor and Szarek [6] have proved the inequality $w(\mathbf{K}, \mathbb{Z}^d) \leq C \cdot d^{\frac{3}{2}}$, where C is an absolute positive constant. This improves an earlier result of Kannan and Lovász [29].

Although Conjecture 4.1 is still open we have the following partial results published recently. Improving the estimates of [13], Talata [34] has succeeded in deriving a proof of the following inequality.

Theorem 4.2. *Let $\mathbf{K}$ be a convex body in $\mathbb{E}^d$. Let $H_1, \ldots, H_N$ be hyperplanes in $\mathbb{E}^d$ such that*

$$\mathbf{K} \cap \mathbb{Z}^d \subset \bigcup_{i=1}^{N} H_i.$$

Then

$$N \geq c \cdot \frac{w(\mathbf{K}, \mathbb{Z}^d)}{d} - d,$$

where c is an absolute positive constant.

In the paper [16], Litvak and the author have shown that the plank theorem of Ball [2] implies a slight improvement on the above inequality for centrally symmetric convex bodies whose lattice width is at most quadratic in dimension. (Actually, this approach is different from Talata's technique and can lead to a somewhat even stronger inequality in terms of the relevant basic measure of the given convex body. For more details on this we refer the interested reader to [16].)

Theorem 4.3. *Let* **K** *be a centrally symmetric convex body in* $\mathbb{E}^d$. *Let* H_1, ..., H_N *be hyperplanes in* $\mathbb{E}^d$ *such that*

$$\mathbf{K} \cap \mathbb{Z}^d \subset \bigcup_{i=1}^{N} H_i.$$

Then

$$N \geq c \cdot \frac{w(\mathbf{K}, \mathbb{Z}^d)}{d \ln(d+1)},$$

where c *is an absolute positive constant.*

Motivated by Conjecture 4.1 and by a conjecture of Corzatt [23] (according to which if in the plane the integer points of a convex domain can be covered by N lines, then those integer points can also be covered by N lines having at most four different slopes) Brass [20] has raised the following related question.

Problem 4.4. For every positive integer d find the smallest constant $c(d)$ such that if the integer points of a convex body in $\mathbb{E}^d$ can be covered by N hyperplanes, then those integer points can also be covered by $c(d) \cdot N$ parallel hyperplanes.

Theorem 4.2 implies that $c(d) \leq c \cdot d^2$ for convex bodies in general and for centrally symmetric convex bodies Theorem 4.3 yields the somewhat better upper bound $c \cdot d \ln(d+1)$. As a last note we mention that the problem of finding good estimates for the constants of Theorems 4.2 and 4.3 is an interesting open question as well.

5. Partial Coverings by Planks

It seems that the following variant of Tarski's plank problem hasn't yet been considered: Let $\mathbf{C}$ be a convex body of minimal width $w > 0$ in $\mathbb{E}^d$. Moreover, let $w_1 > 0, w_2 > 0, \ldots, w_n > 0$ be given with $w_1 + w_2 + \cdots + w_n < w$. Then find the arrangement of n planks say, of $\mathbf{P}_1, \mathbf{P}_2, \ldots, \mathbf{P}_n$, of width $w_1, w_2, \ldots, w_n$ in $\mathbb{E}^d$ such that their union covers the largest volume subset of $\mathbf{C}$, that is, for which $\mathrm{vol}_d((\mathbf{P}_1 \cup \mathbf{P}_2 \cup \cdots \cup \mathbf{P}_n) \cap \mathbf{C})$ is as large as possible. As the following special case is the most striking form of the above problem, we are putting it forward as the main question of this section.

Problem 5.1. Let $\mathbf{B}^d$ denote the unit ball centered at the origin $\mathbf{o}$ in $\mathbb{E}^d$. Moreover, let $w_1, w_2, \ldots, w_n$ be positive real numbers satisfying the inequality $w_1 + w_2 + \cdots + w_n < 2$. Then prove or disprove that the union of the planks $\mathbf{P}_1, \mathbf{P}_2, \ldots, \mathbf{P}_n$ of width $w_1, w_2, \ldots, w_n$ in $\mathbb{E}^d$ covers the largest volume subset of $\mathbf{B}^d$ if and only if $\mathbf{P}_1 \cup \mathbf{P}_2 \cup \cdots \cup \mathbf{P}_n$ is a plank of width $w_1 + w_2 + \cdots + w_n$ with $\mathbf{o}$ as a center of symmetry.

As an immediate remark we note that it would not come as a surprise to us if it turned out that the answer to Problem 5.1 is positive in proper low dimensions and negative in (sufficiently) high dimensions. In what follows we discuss some partial results.

Clearly, there is an affirmative answer to Problem 5.1 for $n = 1$. Also, we have the following positive results. For the sake of completeness we include their short proofs.

Theorem 5.2. *If $\mathbf{P}_1$ and $\mathbf{P}_2$ are planks in $\mathbb{E}^d$, $d \geq 2$ of width w_1 and w_2 having $0 < w_1 + w_2 < 2$, then $\mathbf{P}_1 \cup \mathbf{P}_2$ covers the largest volume subset of $\mathbf{B}^d$ if and only if $\mathbf{P}_1 \cup \mathbf{P}_2$ is a plank of width $w_1 + w_2$ possessing $\mathbf{o}$ as a center of symmetry.*

Proof. The following is an outline of a quite elementary proof. First, let us consider the case when $\mathbf{P}_1$ and $\mathbf{P}_2$ are planks in $\mathbb{E}^2$ of width w_1 and w_2 having $0 < w_1 + w_2 < 2$. We say, that $(\mathbf{P}_1 \cup \mathbf{P}_2) \cap \mathbf{B}^2$ is a crossing subset of $\mathbf{B}^2$, if $\mathbf{B}^2 \setminus (\mathbf{P}_1 \cup \mathbf{P}_2)$ consists of 4 connected components. Now, it is not hard to see that among the crossing subsets (resp., non-crossing subsets) the only extremal configuration with respect to maximizing the area is the one with $\mathbf{P}_1$ and $\mathbf{P}_2$ being perpendicular to each other and having $\mathbf{o}$ as a center of symmetry (resp., the one with $\mathbf{P}_1 \cup \mathbf{P}_2$ being a plank of width $w_1 + w_2$ and having $\mathbf{o}$ as a center of symmetry). Second, it is easy to check that between the two critical configurations the non-crossing one possesses a larger area, finishing the proof Theorem 5.2 for $d = 2$. Finally, if $\mathbf{P}_1$ and $\mathbf{P}_2$ are planks in $\mathbb{E}^d$, $d \geq 3$ of width w_1 and w_2 having $0 < w_1 + w_2 < 2$, then an application of the 2-dimensional case of Theorem 5.2, just proved, to the 2-dimensional flats of $\mathbb{E}^d$ that are parallel to the normal vectors of $\mathbf{P}_1$ and $\mathbf{P}_2$ followed by an integration of the areas of the corresponding sets sitting on the 2-flats in question, yield the desired claim. ∎

Theorem 5.3. *Let $w_1, w_2, \ldots, w_n$ be positive real numbers satisfying the inequality $w_1 + w_2 + \cdots + w_n < 2$. Then the union of the planks $\mathbf{P}_1, \mathbf{P}_2, \ldots, \mathbf{P}_n$ of width $w_1, w_2, \ldots, w_n$ in $\mathbb{E}^3$ covers the largest volume subset of $\mathbf{B}^3$ if and only if $\mathbf{P}_1 \cup \mathbf{P}_2 \cup \cdots \cup \mathbf{P}_n$ is a plank of width $w_1 + w_2 + \cdots + w_n$ with $\mathbf{o}$ as a center of symmetry.*

Proof. Let $\mathbf{P}_1, \mathbf{P}_2, \ldots, \mathbf{P}_n$ be an arbitrary family of planks of width $w_1, w_2, \ldots, w_n$ in $\mathbb{E}^3$ and let $\mathbf{P}$ be a plank of width $w_1 + w_2 + \cdots + w_n$ with $\mathbf{o}$ as a center of symmetry. Moreover, let $S(x)$ denote the sphere of radius x centered at $\mathbf{o}$. Now, recall the well-known fact that if $\mathbf{P}(y)$ is a plank of width y whose both boundary planes intersect $S(x)$, then $\mathrm{sarea}(S(x) \cap \mathbf{P}(y)) = 2\pi x y$, where $\mathrm{sarea}(\,.\,)$ refers to the surface area measure on $S(x)$. This implies in a straightforward way that

$$\mathrm{sarea}[(\mathbf{P}_1 \cup \mathbf{P}_2 \cup \cdots \cup \mathbf{P}_n) \cap S(x)] \leq \mathrm{sarea}(\mathbf{P} \cap S(x)),$$

and so,

$$\mathrm{vol}_3((\mathbf{P}_1 \cup \mathbf{P}_2 \cup \cdots \cup \mathbf{P}_n) \cap \mathbf{B}^3)$$

$$= \int_0^1 \mathrm{sarea}[(\mathbf{P}_1 \cup \mathbf{P}_2 \cup \cdots \cup \mathbf{P}_n) \cap S(x)] \, dx$$

$$\leq \int_0^1 \mathrm{sarea}(\mathbf{P} \cap S(x)) \, dx = \mathrm{vol}_3(\mathbf{P} \cap \mathbf{B}^3),$$

finishing the proof of the "if" part of Theorem 5.3. Actually, a closer look of the above argument gives a proof of the "only if" part as well. ∎

As an immediate corollary we get the following statement.

Corollary 5.4. *If $\mathbf{P}_1$, $\mathbf{P}_2$ and $\mathbf{P}_3$ are planks in $\mathbb{E}^d$, $d \geq 3$ of widths w_1, w_2 and w_3 satisfying $0 < w_1 + w_2 + w_3 < 2$, then $\mathbf{P}_1 \cup \mathbf{P}_2 \cup \mathbf{P}_3$ covers the largest volume subset of $\mathbf{B}^d$ if and only if $\mathbf{P}_1 \cup \mathbf{P}_2 \cup \mathbf{P}_3$ is a plank of width $w_1 + w_2 + w_3$ having $\mathbf{o}$ as a center of symmetry.*

Proof. Indeed an application of Theorem 5.3 to the 3-dimensional flats of $\mathbb{E}^d$ that are parallel to the normal vectors of $\mathbf{P}_1$, $\mathbf{P}_2$ and $\mathbf{P}_3$ followed by an integration of the volumes of the corresponding sets lying in the 3-flats in question, yield the desired claim. ∎

In general, we have the following estimate that one can derive from Bang's paper [8] as follows. In order to state it properly we introduce two definitions.

Definition 5.5. Let $\mathbf{C}$ be a convex body in $\mathbb{E}^d$ and let m be a positive integer. Then let $\mathcal{T}^m_{\mathbf{C},d}$ denote the family of all sets in $\mathbb{E}^d$ that can be obtained as the intersection of at most m translates of $\mathbf{C}$ in $\mathbb{E}^d$.

Definition 5.6. Let $\mathbf{C}$ be a convex body of minimal width $w > 0$ in $\mathbb{E}^d$ and let $0 < x \leq w$ be given. Then for any non-negative integer n let

$$v_d(\mathbf{C}, x, n) := \min\{\mathrm{vol}_d(\mathbf{Q}) \mid \mathbf{Q} \in \mathcal{T}^{2^n}_{\mathbf{C},d} \text{ and } w(\mathbf{Q}) \geq x\}.$$

Now, we are ready to state the theorem which although was not published by Bang in [8], it follows from his proof of Tarski's plank conjecture.

Theorem 5.7. *Let $\mathbf{C}$ be a convex body of minimal width $w > 0$ in $\mathbb{E}^d$. Moreover, let $\mathbf{P}_1, \mathbf{P}_2, \ldots, \mathbf{P}_n$ be planks of width $w_1, w_2, \ldots, w_n$ in $\mathbb{E}^d$ with $w_0 = w_1 + w_2 + \cdots + w_n < w$. Then*

$$\mathrm{vol}_d(\mathbf{C} \setminus (\mathbf{P}_1 \cup \mathbf{P}_2 \cup \cdots \cup \mathbf{P}_n)) \geq v_d(\mathbf{C}, w - w_0, n),$$

that is

$$\mathrm{vol}_d((\mathbf{P}_1 \cup \mathbf{P}_2 \cup \cdots \cup \mathbf{P}_n) \cap \mathbf{C}) \leq \mathrm{vol}_d(\mathbf{C}) - v_d(\mathbf{C}, w - w_0, n).$$

Clearly, the first inequality above implies (via an indirect argument) that if the planks $\mathbf{P}_1, \mathbf{P}_2, \ldots, \mathbf{P}_n$ of width $w_1, w_2, \ldots, w_n$ cover the convex body $\mathbf{C}$ in $\mathbb{E}^d$, then $w_1 + w_2 + \cdots + w_n \geq w$. Also, as an additional observation we mention the following statement, that on the one hand, can be derived from Theorem 5.7 in a straightforward way, on the other hand, represents the only case when the estimate in Theorem 5.7 is sharp.

Corollary 5.8. *Let $\mathbf{T}$ be an arbitrary triangle of minimal width (i.e. of minimal height) $w > 0$ in $\mathbb{E}^2$. Moreover, let $w_1, w_2, \ldots, w_n$ be positive real numbers satisfying the inequality $w_1 + w_2 + \cdots + w_n < w$. Then the union of the planks $\mathbf{P}_1, \mathbf{P}_2, \ldots, \mathbf{P}_n$ of width $w_1, w_2, \ldots, w_n$ in $\mathbb{E}^2$ covers the largest area subset of $\mathbf{T}$ if $\mathbf{P}_1 \cup \mathbf{P}_2 \cup \cdots \cup \mathbf{P}_n$ is a plank of width $w_1 + w_2 + \cdots + w_n$ sitting on the side of $\mathbf{T}$ with height w.*

6. Linking the Kakeya–Pál and the Blaschke–Lebesgue Problems to the Partial Covering Problem

Recall that the *Kakeya–Pál problem* is about minimizing the volume of convex bodies of given minimal width $w > 0$ in $\mathbb{E}^d$. For short reference let $\mathbf{K}_{KP}^{w,d}$ denote any of the minimal volume convex bodies in the Kakeya–Pál problem. (Actually, Kakeya phrased his question in 1917 as follows: what is the smallest area of a convex set within which one can rotate a needle by 180°.) Pál [31] has solved this problem for $d = 2$ by showing that the smallest area convex domain of minimal width $w > 0$ is a regular triangle of height w. As it is well-known, the Kakeya–Pál problem is unsolved in higher dimensions (for more details on this see for example [21]). Thus, the following is an immediate corollary of Theorem 5.7.

Corollary 6.1. *Let* $\mathbf{C}$ *be a convex body of minimal width* $w > 0$ *in* $\mathbb{E}^d$. *Moreover, let* $\mathbf{P}_1, \mathbf{P}_2, \ldots, \mathbf{P}_n$ *be planks of width* $w_1, w_2, \ldots, w_n$ *in* $\mathbb{E}^d$ *with* $w_0 = w_1 + w_2 + \cdots + w_n < w$. *Then*

$$\mathrm{vol}_d((\mathbf{P}_1 \cup \mathbf{P}_2 \cup \cdots \cup \mathbf{P}_n) \cap \mathbf{C}) \leq \mathrm{vol}_d(\mathbf{C}) - \mathrm{vol}_d(\mathbf{K}_{KP}^{w-w_0,d}).$$

It seems that the best lower bound for the Kakeya–Pál problem is due to Firey [24] claiming that $\mathrm{vol}_d(\mathbf{K}_{KP}^{w,d}) \geq f(d)w^d$ with $f(d) = \frac{2}{\sqrt{3}\cdot d!}$. Corollary 6.1 suggests to further investigate and improve Firey's inequality for $d \geq 3$. (For $d = 2$ the inequality in question is identical to Pál's result [31] and so, it is optimal.) Here, we claim the following improvement.

Theorem 6.2. *Let* $\mathbf{C}$ *be a convex body of minimal width* $w > 0$ *in* $\mathbb{E}^d$. *Moreover, for each odd integer* $d \geq 3$ *let* $g(d) = \sqrt{\frac{3\cdot\pi^{d-3}\cdot(d+1)!!}{2^{d-2}\cdot(d!!)^5}}$ *and for each even integer* $d \geq 4$ *let* $g(d) = \sqrt{\frac{3\cdot\pi^{d-3}\cdot(d+2)!!}{(d+1)^2\cdot(d!!)^2\cdot((d-1)!!)^3}}$.

Then

$$\mathrm{vol}_d(\mathbf{C}) \geq g(d)w^d > f(d)w^d$$

for all $d \geq 3$.

Proof. We outline the proof by describing its main idea and by leaving out the more or less straightforward but somewhat lengthy computations. First, we need the following result of Steinhagen [33]. Let $\mathbf{C}$ be a convex body of minimal width $w > 0$ in $\mathbb{E}^d$. Moreover, for each odd integer $d \geq 3$ let $r(d) = \frac{1}{2\sqrt{d}}$ and for each even integer $d \geq 2$ let $r(d) = \frac{\sqrt{d+2}}{2(d+1)}$. Then the inradius r of $\mathbf{C}$ (which is the radius of the largest ball lying in $\mathbf{C}$) is always at least as large as $r(d)w$. Second, recall Kubota's formula [19] according to which

$$\mathrm{svol}_{d-1}(\mathrm{bd}(\mathbf{C})) = \frac{1}{\mathrm{vol}_{d-1}(\mathbf{B}^{d-1})} \int_{\mathbb{S}^{d-1}} \mathrm{vol}_{d-1}(\mathbf{C}|\mathbf{x}) \, dx,$$

where $\mathrm{bd}(\, . \,)$ (resp., $\mathrm{svol}_{d-1}(\, . \,)$) stands for the boundary (resp., $(d-1)$-dimensional surface volume) of the corresponding set and $\mathbb{S}^{d-1} = \mathrm{bd}(\mathbf{B}^d)$ moreover, $\mathbf{C} \mid \mathbf{x}$ denotes the orthogonal projection of $\mathbf{C}$ onto the hyperplane passing through $\mathbf{o}$ with normal vector $\mathbf{x}$ and the integration on $\mathbb{S}^{d-1}$ is with respect to the surface area measure. Thus, Steinhagen's theorem and Kubota's formula imply in a straightforward way

$$\mathrm{vol}_d(\mathbf{C}) \geq \frac{r(d)w}{d} \, \mathrm{svol}_{d-1}(\mathrm{bd}(\mathbf{C})) \geq \frac{r(d)w \, \mathrm{vol}_d(\mathbf{B}^d)}{\mathrm{vol}_{d-1}(\mathbf{B}^{d-1})} \min_{\mathbf{x} \in \mathbb{S}^{d-1}} \{\mathrm{vol}_{d-1}(\mathbf{C} \mid \mathbf{x})\}.$$

Finally, as $\mathbf{C}|\mathbf{x}$ is a $(d-1)$-dimensional convex body of minimal width at least w for all $\mathbf{x} \in \mathbb{S}^{d-1}$, therefore the above inequality, repeated in a recursive way for lower dimensions, leads to the desired inequality claimed in Theorem 6.2. ∎

Remark 6.3. For comparison we mention that $g(3) = \frac{2}{9} = 0.2222\cdots >$ $f(3) = \frac{1}{3\sqrt{3}} = 0.1924\ldots$ (resp., $g(4) = \sqrt{\frac{2\pi}{75}} = 0.2894\cdots > f(4) = \frac{1}{12\sqrt{3}} = 0.0481\ldots$). Also, recall that Heil [26] has constructed a convex body in $\mathbb{E}^3$ of minimal width 1 and of volume $0.298\ldots$.

Corollary 6.1 can be further improved when $\mathbf{C}$ is a unit ball and the sum of the widths of the planks is at most one. The details are as follows.

First, recall that the *Blaschke–Lebesgue problem* is about finding the minimum volume convex body of constant width $w > 0$ in $\mathbb{E}^d$. In particular, the Blaschke Lebesgue theorem states that among all convex domains of constant width w, the Reuleaux triangle of width w has the smallest area, namely $\frac{1}{2}(\pi - \sqrt{3}) \, w^2$. W. Blaschke [18] and H. Lebesgue [30] were the first to show this and the succeding decades have seen other works published on different proofs of that theorem. For a most recent new proof, and for a survey on the state of the art of different proofs of the Blaschke–Lebesgue theorem, see the elegant paper of E. M. Harrell [25]. Here we note that the Blaschke–Lebesgue problem is unsolved in three and more dimensions. Even finding the 3-dimensional set of least volume presents formidable difficulties. On the one hand, Chakerian [22] proved that any convex body of constant width 1 in $\mathbb{E}^3$ has volume at least $\frac{\pi(3\sqrt{6}-7)}{3} = 0.365\ldots$. On the other hand, it has been conjectured by Bonnesen and Fenchel [19] that Meissner's 3-dimensional generalizations of the Reuleaux triangle of volume $\pi\left(\frac{2}{3} - \frac{1}{4}\sqrt{3}\arccos\left(\frac{1}{3}\right)\right) = 0.420\ldots$ are the only extramal sets in $\mathbb{E}^3$.

For our purposes it will be useful to introduce the notation $\mathbf{K}_{BL}^{w,d}$ (resp., $\overline{\mathbf{K}}_{BL}^{w,d}$) for a convex body of constant width w in $\mathbb{E}^d$ having minimum volume (resp., surface volume). One may call $\mathbf{K}_{BL}^{w,d}$ (resp., $\overline{\mathbf{K}}_{BL}^{w,d}$) a Blaschke–Lebesgue-type convex body with respect to volume (resp., surface volume). Note that for $d = 2,3$ one may choose $\mathbf{K}_{BL}^{w,d} = \overline{\mathbf{K}}_{BL}^{w,d}$ however, this is likely not to happen for $d \geq 4$. (For more details on this see [22].) As an important note we mention that Schramm [32] has proved the inequality

$$\mathrm{vol}_d(\mathbf{K}_{BL}^{w,d}) \geq \left(\sqrt{3 + \frac{2}{d+1}} - 1\right)^d \left(\frac{w}{2}\right)^d \mathrm{vol}_d(\mathbf{B}^d),$$

which gives the best lower bound for all $d > 4$. By observing that the orthogonal projection of a convex body of constant width w in $\mathbb{E}^d$ onto any hyperplane of $\mathbb{E}^d$ is a $(d-1)$-dimensional convex body of constant width w one obtains from the previous inequality of Schramm the following one:

$$\mathrm{svol}_{d-1}(\mathrm{bd}(\overline{\mathbf{K}}_{BL}^{w,d})) \geq d\left(\sqrt{3 + \frac{2}{d}} - 1\right)^{d-1} \left(\frac{w}{2}\right)^{d-1} \mathrm{vol}_d(\mathbf{B}^d).$$

Second, let us recall that if X is a finite (point) set lying in the interior of a unit ball in $\mathbb{E}^d$, then the intersection of the (closed) unit balls of $\mathbb{E}^d$ centered at the points of X is called a ball-polyhedron and it is denoted by $\mathbf{B}[X]$. (For an extensive list of properties of ball-polyhedra see the recent paper [15].) Of course, it also makes sense to introduce $\mathbf{B}[X]$ for sets X that are not finite but in those cases we get sets that are typically not ball-polyhedra.

Now, we are ready to state our theorem.

Theorem 6.4. *Let* $\mathbf{B}[X] \subset \mathbb{E}^d$ *be a ball-polyhedron of minimal width* x *with* $1 \leq x < 2$. *Then*

$$\mathrm{vol}_d(\mathbf{B}[X]) \geq \mathrm{vol}_d(\mathbf{K}_{BL}^{2-x,d}) + \mathrm{svol}_{d-1}(\mathrm{bd}(\overline{\mathbf{K}}_{BL}^{2-x,d})) \cdot (x-1) + \mathrm{vol}_d(\mathbf{B}^d) \cdot (x-1)^d.$$

Proof. Recall that if X is finite set lying in the interior of a unit ball in $\mathbb{E}^d$, then we can talk about its spindle convex hull $\text{conv}_s(X)$, which is simply the intersection of all (closed) unit balls of $\mathbb{E}^d$ that contain X (for more details see [15]). The following statement can be obtained by combining Corollary 3.4 of [15] and Proposition 1 of [14].

Lemma 6.5. *Let X be a finite set lying in the interior of a unit ball in $\mathbb{E}^d$. Then*

(i) $\text{conv}_s(X) = \mathbf{B}\big[\mathbf{B}[X]\big]$ *and therefore* $\mathbf{B}[X] = \mathbf{B}\big[\text{conv}_s(X)\big]$;

(ii) *the Minkowski sum* $\mathbf{B}[X] + \text{conv}_s(X)$ *is a convex body of constant width 2 in $\mathbb{E}^d$ and so, $w(\mathbf{B}[X]) + \text{diam}\big(\text{conv}_s(X)\big) = 2$, where $\text{diam}(\,.\,)$ stands for the diameter of the corresponding set in $\mathbb{E}^d$.*

By part (ii) of Lemma 6.5, $\text{diam}\big(\text{conv}_s(X)\big) \leq 2 - x$. This implies, via a classical theorem of convexity (see for example [19]), the existence of a convex body $\mathbf{L}$ of constant width $(2 - x)$ in $\mathbb{E}^d$ with $\text{conv}_s(X) \subset \mathbf{L}$. Hence, using part (i) of Lemma 6.5, we get that $\mathbf{B}[\mathbf{L}] \subset \mathbf{B}[X] = \mathbf{B}\big[\text{conv}_s(X)\big]$. Finally, notice that as $\mathbf{L}$ is a convex body of constant width $(2 - x)$ therefore $\mathbf{B}[\mathbf{L}]$ is in fact, the outer-parallel domain of $\mathbf{L}$ having radius $(x - 1)$ (that is $\mathbf{B}[\mathbf{L}]$ is the union of all d-dimensional (closed) balls of radii $(x - 1)$ in $\mathbb{E}^d$ that are

$$\text{vol}_d(\mathbf{B}[X]) \geq \text{vol}_d\big(\mathbf{B}[\mathbf{L}]\big)$$

$$= \text{vol}_d(\mathbf{L}) + \text{svol}_{d-1}(\text{bd}(\mathbf{L})) \cdot (x - 1) + \text{vol}_d(\mathbf{B}^d) \cdot (x - 1)^d.$$

The inequality above together with the following obvious ones

$$\text{vol}_d(\mathbf{L}) \geq \text{vol}_d(\mathbf{K}_{BL_d}^{2-x}) \quad \text{and} \quad \text{svol}_{d-1}(\text{bd}(\mathbf{L})) \geq \text{svol}_{d-1}(\text{bd}(\overline{\mathbf{K}}_{BL_d}^{2-x}))$$

imply Theorem 6.4 in a straightforward way. ∎

Thus, Theorem 5.7 and Theorem 6.4 imply the following immediate estimate.

Corollary 6.6. *Let $\mathbf{B}^d$ denote the unit ball centered at the origin $\mathbf{o}$ in $\mathbb{E}^d$, $d \geq 2$. Moreover, let $\mathbf{P}_1, \mathbf{P}_2, \ldots, \mathbf{P}_n$ be planks of width $w_1, w_2, \ldots, w_n$ in $\mathbb{E}^d$ with $w_0 = w_1 + w_2 + \cdots + w_n \leq 1$. Then*

$$\text{vol}_d((\mathbf{P}_1 \cup \mathbf{P}_2 \cup \cdots \cup \mathbf{P}_n) \cap \mathbf{B}^d) \leq \text{vol}_d(\mathbf{B}^d) - v_d(\mathbf{B}^d, 2 - w_0, n)$$

$$\leq \big(1 - (1 - w_0)^d\big)\,\text{vol}_d(\mathbf{B}^d) - \text{vol}_d(\mathbf{K}_{BL}^{w_0,d}) - \text{svol}_{d-1}(\text{bd}(\overline{\mathbf{K}}_{BL}^{w_0,d})) \cdot \big(1 - w_0\big)$$

Corollary 6.6 leaves open the following question (even in dimension two).

Problem 6.7. Improve the upper bound of Corollary 6.1 for the unit ball when $1 < w_0 < 2$.

7. STRENGTHENING THE PLANK THEOREMS OF BALL AND BANG

Recall that Ball ([2]) generalized the plank theorem of Bang ([7], [8]) for coverings of balls by planks in Banach spaces (where planks are defined with the help of linear functionals instead of inner product). This theorem was further strengthened by Kadets [28] for real Hilbert spaces as follows. Let $\mathbf{C}$ be a closed convex subset with non-empty interior in the real Hilbert space $\mathbb{H}$ (finite or infinite dimensional). We call $\mathbf{C}$ a *convex body* of $\mathbb{H}$. Then let $r(\mathbf{C})$ denote the supremum of the radii of the balls contained in $\mathbf{C}$. (One may call $r(\mathbf{C})$ the *inradius* of $\mathbf{C}$.) Planks and their widths in $\mathbb{H}$ are defined with the help of the inner product of $\mathbb{H}$ in the usual way. Thus, if $\mathbf{C}$ is a convex body in $\mathbb{H}$ and $\mathbf{P}$ is a plank of $\mathbb{H}$, then the width $w(\mathbf{P})$ of $\mathbf{P}$ is always at least as large as $2r(\mathbf{C} \cap \mathbf{P})$. Now, the main result of [28] is the following.

Theorem 7.1. *Let the ball $\mathbf{B}$ of the real Hilbert space $\mathbb{H}$ be covered by the convex bodies $\mathbf{C}_1, \mathbf{C}_2, \ldots, \mathbf{C}_n$ in $\mathbb{H}$. Then*

$$\sum_{i=1}^{n} r(\mathbf{C}_i \cap \mathbf{B}) \geq r(\mathbf{B}).$$

We note that an independent proof of the 2-dimensional Euclidean case of Theorem 7.1 can be found in [12]. Kadets ([28]) proposes to investigate the analogue of Theorem 7.1 in Banach spaces. Thus, an affirmative answer to the following problem would improve the plank theorem of Ball.

Problem 7.2. Let the ball $\mathbf{B}$ be covered by the convex bodies $\mathbf{C}_1, \mathbf{C}_2, \ldots, \mathbf{C}_n$ in an arbitrary Banach space. Prove or disprove that

$$\sum_{i=1}^{n} r(\mathbf{C}_i \cap \mathbf{B}) \geq r(\mathbf{B}).$$

It is well-known that Bang's plank theorem holds in complex Hilbert spaces as well. However, for those spaces Ball [4] was able to prove the following much stronger theorem. (In fact, [4] was published a number of years before [28].)

Theorem 7.3. *If the planks of widths $w_1, w_2, \ldots, w_n$ cover a ball of diameter w in a complex Hilbert space, then*

$$\sum_{i=1}^{n} w_i^2 \geq w^2.$$

In order to complete the picture on plank-type results (from the point of view of discrete geometry) in spaces other than Euclidean we mention the statement below, proved by Schneider and the author [17]. It is an extension of Theorem 7.1 for coverings of large balls in spherical spaces. Needless to say that it would be desirable to extend some other plank-type results as well to spherical spaces.

Theorem 7.4. *If the spherically convex bodies $\mathbf{K}_1, \mathbf{K}_2, \ldots, \mathbf{K}_n$ with inradii* $r(\mathbf{K}_1), r(\mathbf{K}_2), \ldots, r(\mathbf{K}_n)$ *cover the spherical ball of radius $r(\mathbf{B}) \geq \frac{\pi}{2}$ in a spherical space, then*

$$\sum_{i=1}^{n} r(\mathbf{K}_i) \geq r(\mathbf{B}).$$

We close our survey with another strengthening of the plank theorem of Bang in $\mathbb{E}^2$. Namely, in [11], by proving some partial results, A. Bezdek asked which convex domains in $\mathbb{E}^2$ have the property that whenever an annulus consisting of the domain less a sufficiently small scaled copy of itself, is covered by planks the sum of the widths of the planks must still be at least the minimal width of the domain. In [35], White and Wisewell characterized the polygons for which this is so. However, the following perhaps most striking case of A. Bezdek's conjecture remains open.

Conjecture 7.5. Let $\mathbf{B}$ be a unit disk in $\mathbb{E}^2$. Then there exists an $\varepsilon > 0$ such that if $\varepsilon \mathbf{B}$ lies in the interior of $\mathbf{B}$ and the annulus $\mathbf{B} \setminus \varepsilon \mathbf{B}$ is covered by finitely many planks, then the sum of the widths of the planks is at least two.

Acknowledgements. The author wishes to thank the referee for the detailed comments on an earlier version of this paper.

REFERENCES

[1] R. Alexander, *A problem about lines and ovals*, Amer. Math. Monthly **75** (1968), 482–487.

[2] K. Ball, *The plank problem for symmetric bodies*, Invent. Math. **104** (1991), 535–543.

[3] K. Ball, *A lower bound for the optimal density of lattice packings*, Internat. Math. Res. Notices **10** (1992), 217–221.

[4] K. Ball, *The complex plank problem*, Bull. London Math. Soc. **33/4** (2001), 433–442.

[5] W. Banaszczyk, *Inequalities for convex bodies and polar reciprocal lattices in R^n II: Application of K-convexity*, Discrete Comput. Geom. **16** (1996), 305–311.

[6] W. Banaszczyk, A. E. Litvak, A. Pajor and S. J. Szarek, *The flatness theorem for nonsymmetric convex bodies via the local theory of Banach spaces*, Math. Oper. Res. **24/3** (1999), 728–750.

[7] T. Bang, *On covering by parallel-strips*, Mat. Tidsskr. B. **1950** (1950), 49–53.

[8] T. Bang, *A solution of the "Plank problem"*, Proc. Am. Math. Soc. **2** (1951), 990–993.

[9] A. Bezdek and K. Bezdek, *A solution of Conway's fried potato problem*, Bull. London Math. Soc. **27/5** (1995), 492–496.

[10] A. Bezdek and K. Bezdek, *Conway's fried potato problem revisited*, Arch. Math. **66/6** (1996), 522–528.

[11] A. Bezdek, *Covering an annulus by strips*, Discrete Comput. Geom. **30** (2003), 177–180.

[12] A. Bezdek, *On a generalization of Tarski's plank problem*, Discrete Comput. Geom. **38** (2007), 189–200.

[13] K. Bezdek and T. Hausel, *On the number of lattice hyperplanes which are needed to cover the lattice points of a convex body*, Colloq. Math. Soc. János Bolyai **63** (1994), 27–31.

[14] K. Bezdek, R. Connelly and B. Csikós, *On the perimeter of the intersection of congruent disks*, Beiträge Algebra Geom. **47/1** (2006), 53–62.

[15] K. Bezdek, Zs. Lángi, M. Naszódi, and P. Papez, *Ball-polyhedra*, Discrete Comput. Geom. **38/2** (2007), 201–230.

[16] K. Bezdek and A. E. Litvak, *Covering convex bodies by cylinders and lattice points by flats*, J. Geom. Anal. **19/2** (2009), 233–243.

[17] K. Bezdek and R. Schneider, *Covering large balls with convex sets in spherical space,* Beiträge zur Alg. und Geom. 51/1 (2010), 229–235.

[18] W. Blaschke, *Konvexe Bereiche gegebener konstanter Breite und kleinsten Inhalts,* Math. Ann. **76** (1915), 504–513.

[19] T. Bonnesen and W. Fenchel, *Theory of Convex Bodies,* (English translation), BCS Associates (Moscow, Idaho, USA), 1987.

[20] P. Brass, W. Moser and J. Pach, *Research problems in discrete geometry,* Springer, New York, 2005.

[21] S. Campi, A. Colesanti and P. Gronchi, *Minimum problems for volumes of convex bodies,* Lecture Notes in Pure and Appl. Math., 177, Dekker, New York (1996), 43–55.

[22] G. D. Chakerian, *Sets of constant width,* Pacific J. Math. **19** (1966), 13–21.

[23] C. E. Corzatt, *Covering convex sets of lattice points with straight lines,* Proceedings of the Sundance conference on combinatorics and related topics (Sundance, Utah, 1985), Congr. Numer. **50** (1985), 129–135.

[24] W. J. Firey, *Lower bounds for volumes of convex bodies,* Arch. Math. **16** (1965), 69–74.

[25] E. M. Harrell, *A direct proof of a theorem of Blaschke and Lebesgue,* J. Geom. Anal. **12/1** (2002), 81–88.

[26] E. Heil, *Kleinste konvexe Körper gegebener Dicke,* Technische Hochschule Darmstadt, Preprint **453** (1978), 1–2.

[27] V. Kadets, *Weak cluster points of a sequence and coverings by cylinders,* Mat. Fiz. Anal. Geom. **11/2** (2004), 161–168.

[28] V. Kadets, *Coverings by convex bodies and inscribed balls,* Proc. Amer. Math. Soc. **133/5** (2005), 1491–1495.

[29] R. Kannan and L. Lovász, *Covering minima and lattice-point-free convex bodies,* Ann. Math. **128** (1988), 577–602.

[30] H. Lebesgue, *Sur le problemedes isoperimetres at sur les domaines de larguer constante,* Bull. Soc. Math. France C.R. **7** (1914), 72–76.

[31] J. Pál, *Ein minimumproblem für Ovale,* Math. Ann. **83** (1921), 311–319.

[32] O. Schramm, *On the volume of sets having constant width,* Israel J. Math. **63/2** (1988), 178–182.

[33] P. Steinhagen, *Über die grösste Kugel in einer konvexen Punktmenge,* Abh. Math. Sem. Hamburg **1** (1922), 15–26.

[34] I. Talata, *Covering the lattice points of a convex body with affine subspaces*, Bolyai Soc. Math. Stud. **6** (1997), 429–440.

[35] S. White and L. Wisewell, *Covering polygonal annuli by strips*, Discrete Comput. Geom. **37/4** (2007), 577–585.

Károly Bezdek

Department of Mathematics and Statistics,
University of Calgary,
Canada,
Department of Mathematics,
University of Pannonia,
Veszprém, Hungary,
and
Institute of Mathematics,
Eötvös University,
Budapest, Hungary

e-mail: bezdek@math.ucalgary.ca

BOLYAI SOCIETY
MATHEMATICAL STUDIES, 24

Geometry –
Intuitive, Discrete, and Convex
pp. 65–90.

DENSE PACKING OF SPACE WITH VARIOUS CONVEX SOLIDS

ANDRÁS BEZDEK and WŁODZIMIERZ KUPERBERG*

Dedicated to Professor László Fejes Tóth

One of the basic problems in discrete geometry is to determine the most efficient packing of congruent replicas of a given convex set K in the plane or in space. The most commonly used measure of efficiency is density. Several types of the problem arise depending on the type of isometries allowed for the packing: packing by translates, lattice packing, translates and point reflections, or all isometries. Due to its connections with number theory, crystallography, etc., lattice packing has been studied most extensively. In two dimensions the theory is fairly well developed, and there are several significant results on lattice packing in three dimensions as well. This article surveys the known results, focusing on the most recent progress. Also, many new problems are stated, indicating directions in which future development of the general packing theory in three dimensions seems feasible.

1. DEFINITIONS AND PRELIMINARIES

A d-dimensional *convex body* is a compact convex subset of $\mathbb{R}^n$, contained in a d-dimensional flat and with non-void interior relative to the flat. A 2-dimensional convex body is called a *convex disk*. The (d-dimensional) volume of a d-dimensional convex body K will be denoted by $\mathrm{Vol}(K)$, but

*Both authors gratefully acknowledge research support: A. Bezdek was supported by the Hungarian Research Foundation OTKA, grant #068398; W. Kuperberg was supported in part by the DiscConvGeo (Discrete and Convex Geometry) project, in the framework of the European Community's "Structuring the European Research Area" programme.

for $d = 2$ we will sometimes alternately use the term "area" and the notation Area(K).

The *Minkowski sum* of sets A and B in $\mathbb{R}^d$ is defined as the set

$$A + B = \{x + y : x \in A, \ y \in B\}.$$

If A consists of a single point a, we write simply $a + B$ instead of $\{a\} + B$.

For every convex body K in $\mathbb{R}^d$ and every real number λ, the set λK is defined as $\{\lambda x : x \in K\}$. We usually write $-K$ instead of $(-1)K$, and $K - L$ instead of $K + (-1)L$. A convex body K in $\mathbb{R}^d$ is *centrally symmetric* if there is a point $c \in \mathbb{R}^d$ (the *center* of K) such that $K = 2c - K$. For each convex body K, the centrally symmetric convex body $\mathbf{D}K = \frac{1}{2}(K - K)$ is called *the difference body* of K.

A *packing* of $\mathbb{R}^d$ is a family of d-dimensional convex bodies K_i whose interiors are mutually disjoint. A packing is a *tiling* if the union of its members is the whole space $\mathbb{R}^d$.

In what follows, we consider mostly packings with congruent replicas of a convex body K. If the family $\mathcal{P} = \{K_i\}$ $(i = 1, 2, \ldots)$ of congruent replicas K_i of a d-dimensional convex body K is a packing, then *density* of $\mathcal{P}$ is defined as

$$d(\mathcal{P}) = \limsup_{r \to \infty} \frac{1}{\text{Vol}(\mathbf{B}(r))} \sum_{i=1}^{\infty} \text{Vol}(K_i \cap \mathbf{B}(r)),$$

where $\mathbf{B}(r)$ is the ball of radius r, centered at the origin. The supremum of $d(\mathcal{P})$ taken over all packings $\mathcal{P}$ with congruent replicas of K is called *the packing density of K* and is denoted by $\delta(K)$. The supremum is actually the maximum, as a densest packing with replicas of K exists (see Groemer [23]). In case the allowed replicas of K are restricted to translates of K or to translates of K by a lattice, the corresponding packing densities of K are denoted by $\delta_T(K)$ and by $\delta_L(K)$, respectively. We also consider packings in which translates of K and translates of $-K$ are used; the corresponding packing density is denoted by $\delta_{T^*}(K)$. The lattice-like version requires that each packing consists of translates of a non-overlapping pair $K \cup (v - K)$ by the vectors of a lattice; the corresponding density is denoted by $\delta_{L^*}(K)$ (here both the lattice L and the vector v are chosen so that the resulting packing is of maximum density). Naturally, the more restrictions are imposed on the type of the allowed packing arrangements, the smaller is the corresponding packing density, therefore

$$0 < \delta_L(K) \le \delta_T(K) \le \delta_{T^*}((K) \le \delta(K) \le 1$$

and

$$0 < \delta_{L^*}(K) \le \delta_{T^*}((K) \le \delta(K) \le 1.$$

Obviously, if space $\mathbb{R}^d$ can be tiled by congruent replicas of K, then $\delta(K) = 1$. The converse is less obvious, but not very difficult to prove: If $\delta(K) = 1$, then $\mathbb{R}^d$ can be tiled by congruent replicas of K. Similarly, if $\delta_T(K) = 1$, then K can tile space by its translated replicas; and if $\delta_{T^*}(K) = 1$, then space can be tiled by translates of K combined with translates of $-K$.

It is well-known that a family $\mathcal{P} = \{K + v_i\}$ of translates of a convex body K is a packing if and only if the family $\mathcal{P}' = \{\mathbf{D}K + v_i\}$ is a packing (see [43], also [16], [41] and [24]). This implies immediately that

$$(1.1) \qquad \delta_T(K) = \frac{\mathrm{Vol}(K)}{\mathrm{Vol}(\mathbf{D}K)} \delta_T(\mathbf{D}K) \le \frac{\mathrm{Vol}(K)}{\mathrm{Vol}(\mathbf{D}K)},$$

which gives a meaningful (*i.e.,* smaller than 1) upper bound in case K is not centrally symmetric. The analogous statement and bound hold for the lattice packing density δ_L.

For more details, definitions, and basic properties on these notions, see [19]. For an overview of lattices and lattice packings, see [16], [41] and [24].

2. INTRODUCTION

In contrast to the well developed theory of packing in two dimensions, there are not many results about packing densities of convex bodies in $\mathbb{R}^3$. With few exceptions, most of such results simply provide the value of the packing density $\delta_L(K)$ for a specific convex body K, usually obtained by means of a classical method described by Minkowski [43]. In the next section we review those results, occasionally citing and describing some relevant results about packing the plane $\mathbb{R}^2$ with congruent replicas of a *convex disk* (a convex body of dimension 2).

In Sections 6, 7, and 8 we consider two simple types of convex bodies in $\mathbb{R}^3$, namely cones and cylinders. Given a convex disk K in $\mathbb{R}^3$ and a point v not in the plane of K, the *cone with base K and apex v*, denoted by $\mathbf{C}_v(K)$, is the union of all line segments with one end at v and the other one in K. Given a convex disk K in $\mathbb{R}^3$ and a line segment s not parallel to the plane of K, the *cylinder with base K and generating segment s*, denoted by $\mathbf{\Pi}_s(K)$, is the Minkowski sum $s + K$. (Observe that, with the exception of tetrahedra, the base and apex of a cone are uniquely determined by the cone itself; likewise, with the exception of the parallelepipeds, a cylinder has two bases exactly - one is a translate of the other, and its generating segment is determined uniquely up to translation.) These two simple types of convex bodies we suggest to investigate first as a first step towards building a systematic theory of packing in dimension three. The plan is particularly suitable for the study of densities δ_T, δ_L, $\delta_{T^*}(K)$, and $\delta_{L^*}(K)$ because of the affine invariance of the corresponding problems. Both for the cone and for the cylinder, each of the packing densities mentioned above depends only on the affine class of the base. In Section 5 we describe in detail the nature of the affine invariance, we draw some immediate conclusions concerning those suitable densities, and we state a few fundamental open problems.

3. LATTICE PACKING IN SPACE

We begin with the following table listing a few convex bodies in $\mathbb{R}^3$ whose lattice packing densities δ_L have been explicitly computed.

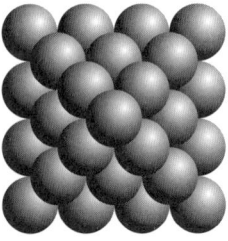 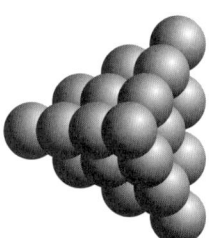

Fig. 1. Two clusters in the densest lattice packing of balls: a "square pyramid" and a "regular tetrahedron"

Comments to Table 1.

1. The densest lattice arrangements of spheres (balls) in $\mathbb{R}^3$ (see Fig. 1) was described already by Kepler [39], but unsupported by proof, Kepler's assertion can only be considered as a conjecture. The first one to prove that $\delta_L(\mathbf{B}^3) = \frac{\pi}{\sqrt{18}}$ was Gauss [21]. Actually, Kepler asserted that the lattice arrangement shown in Fig. 1 is of maximum density

#	Body	Packing Density δ_L	Author & Reference
1	Ball $\{x : |x| \leq 1\}$	$\frac{\pi}{\sqrt{18}} = 0.74048\ldots$	Gauss [21]
2	Regular octahedron	$\frac{18}{19} = 0.9473\ldots$	Minkowski [43]
3	Cylinder $C = \mathbf{\Pi}_s(K)$	$\delta_L(C) = \delta_L(K)$	Chalk & Rogers [10], also Yeh [57]
4	Slab of a cube (see definition below)	(see formula below)	Whitworth [54]
5	Slab of a ball (see definition below)	(see formula below)	Chalk [9]
6	Double cone (see definition below)	$\pi\sqrt{6}/9 = 0.85503\ldots$	Whitworth [55]
7	Tetrahedron	$\frac{18}{49} = 0.3673\ldots$	Hoylman [37]

Table 1.

among *all* sphere packings. This stronger conjecture, however, turned out to be extremely difficult to prove (see Section 8, subsection 8.1).

2. The regular octahedron is also called the regular 3-dimensional *cross-polytope* and is denoted by $\mathbf{X}^3$. Using his method for computing lattice packing density of a centrally symmetric convex body, Minkowski [43] proved that $\delta_L(\mathbf{X}^3) = \frac{18}{19}$. He applied the same method to the tetrahedron, but without success, for in the process he made a mistake in assuming that the difference body of the regular tetrahedron is the regular octahedron (see Comment 7 below).

3. The seemingly obvious equality $\delta_L(C) = \delta_L(K)$ is not trivial at all. The trivial part is the inequality $\delta_L(C) \geq \delta_L(K)$, obtained by stacking layers of cylinders erected over the densest lattice packing of the plane with translates of the base, but the opposite inequality is quite nontrivial, since a cross-section of a lattice packing of the cylinders by a plane parallel to the cylinders' bases need not be a lattice packing of the bases in the plane, and, *a priori,* the density of such a packing could be greater than in any lattice packing. A result of L. Fejes Tóth [17], independently discovered also by Rogers [46], says that this in

fact cannot happen, *i.e.*, the density of a packing with translates of a convex disk cannot exceed the maximum density attained in a lattice arrangement.

4. The λ-slab of a cube $(0 < \lambda \leq 3)$ is defined as

$$K_\lambda = \{x \in \mathbb{R}^3 : |x_i| \leq 1, i = 1, 2, 3; |x_1 + x_2 + x_3| \leq \lambda\},$$

and its lattice packing density is given by the formula

$$\delta_L(K_\lambda) = \begin{cases} \dfrac{1}{9}(9 - \lambda^2) & \text{if } 0 < \lambda \leq \dfrac{1}{2}, \\ \dfrac{1}{4}\lambda(9 - \lambda^2)/(-\lambda^3 - 3\lambda^2 + 24\lambda - 1) & \text{if } \dfrac{1}{2} \leq \lambda \leq 1, \\ \dfrac{9}{8}(\lambda^3 - 9\lambda^2 + 27\lambda - 3)/\lambda(\lambda^2 - 9\lambda + 27) & \text{if } 1 \leq \lambda \leq 1. \end{cases}$$

Whitworth uses Minkowski's method, and his result generalizes the case of the regular octahedron $(\lambda = 1)$, item 2 in the Table.

5. The λ-slab of a ball $(0 < \lambda \leq 1)$ is defined as

$$B_\lambda = \{x \in \mathbb{R}^3 : |x| \leq 1, |x_3| \leq \lambda\},$$

and its lattice packing density is given by the formula

$$\delta_L(B_\lambda) = \frac{\pi}{6}\sqrt{3 - \lambda^2}.$$

Chalk uses Minkowski's method, and his result generalizes the case of the ball $(\lambda = 1)$, item 1 in the Table.

6. The double cone (see Fig. 2) is the set

$$K = \{x \in \mathbb{R}^3 : \sqrt{x_1^2 + x_2^2} + |x_3| \leq 1\}.$$

As in item 4, Whitworth uses Minkowski's method to establish lattice packing density of K.

7. Minkowski's error in computing the lattice packing density of the tetrahedron was noticed by Groemer [22], who proved that $\frac{18}{49}$ is a lower bound for the density. Then Douglas and Hoylman proved that Groemer's bound is in fact the tetrahedron's lattice packing density.

Fig. 2. The densest lattice packing with the double cone

Fig. 3. The densest lattice packing with the tetrahedron

The problem of the maximum density packing with congruent regular tetrahedra (allowing all isometries) remains open and appears to be extremely difficult. We report on the recent progress in Section 8, subsection 8.4.

8. Each of the results listed in the table is obtained "by hand," and, with the exception of Gauss, each of the authors uses Minkowski's method. The method often requires tedious computations with a large number of cases to analyze, which for some convex bodies becomes prohibitively complex. With the emergence of computer technology, however, it became possible to accomplish many such tasks in a very short

time. In an impressive article published in year 2000, Betke and Henk [2] present a fairly fast computer algorithm implementing Minkowski's method for finding lattice packing density of any 3-dimensional convex polytope. To show the algorithm's efficiency, the article lists lattice packing density of each of the regular and Archimedean polytopes, many of which would be practically impossible to handle without computers.

4. Packing Convex Bodies by Translations

Thus far no example of a convex body K has been found for which $\delta_T(K) > \delta_L(K)$. In fact, there are only a few types of convex bodies K whose packing density $\delta_T(K)$ is known, namely:

1. any convex polytope P that admits a tiling of space by its translates (it is known that each such polytope tiles space in a lattice-like manner, in every dimension, see Venkov [53] or McMullen [44]);

2. any cylinder $\mathbf{C}_s K$ with a convex base K, since obviously $\delta_T(\mathbf{C}_s K) = \delta_T(K)$;

3. any non-symmetric body K for which the packing density of the difference body $\delta_T(\mathbf{D}K)$ is known. For example, the difference body of a body K of constant width is a ball, hence the packing density of the ball can be used to find $\delta_T(K)$;

4. any convex body K such that $\mathbf{B}^3 \subset K \subset RhD$, where RhD denotes the rhombic dodecahedron circumscribing the unit ball $\mathbf{B}^3$, which is the Voronoi polytope associated with the densest lattice packing of $\mathbf{B}^3$.

The last two items are based on Hales' confirmation of the Kepler Conjecture, stating that $\delta(\mathbf{B}^3) = \delta_L(\mathbf{B}^3) = \mathrm{Vol}(\mathbf{B}^3)/\mathrm{Vol}(RhD)$.

In contrast, in $\mathbb{R}^2$ it is known that

$$(4.1) \qquad \delta(K) = \delta_L(K) \quad \text{for every centrally symmetric convex disk } K,$$

which implies that

$$(4.2) \qquad \delta_T(K) = \delta_L(K) \quad \text{for every convex disk } K,$$

see L. Fejes Tóth [17].

While equation (4.2) perhaps holds true for 3-dimensional convex bodies as well, equation (4.1) does not, as the following example shows.

Let P be the (slightly irregular) affine-regular octahedron in $\mathbb{R}^3$ with vertices of the form $(\pm 1, \pm 1, 0)$ and $(0, 0, \pm 1)$. It is easy to see that P cannot tile space by translates alone, hence $\delta_T(P) < 1$. On the other hand, P can tile space with translates of itself combined with translates of its copies rotated by 90° about the coordinate axes. Therefore $\delta(P) = 1$.

It should also be mentioned that already in dimension 2 the assumption of convexity is indispensable for equation (4.2). A. Bezdek and Kertész [5] constructed a non-convex polygon that allows a dense non-lattice packing of the plane by its translates, denser than any lattice packing, see Fig. 4. (The construction of Bezdek and Kertész was modified by Heppes [36] so as to obtain a starlike polygon with the same property.)

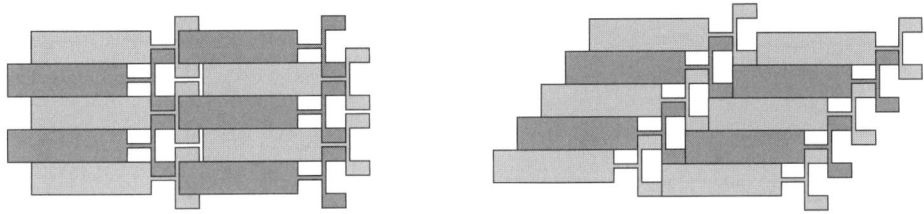

Fig. 4. An example of Bezdek and Kertész: a polygon whose translates can be packed more densely (left) than in its densest lattice packing (right)

The main question of this section remains open:

Is it true that the maximum density of a packing with translates of a convex body in $\mathbb{R}^3$ is attained in a lattice packing?

Similarly, the problem of whether or not $\delta_T^*(K) = \delta_L^*(K)$ holds for every 3-dimensional convex body K remains open.

5. AFFINE INVARIANCE AND COMPACTNESS

If $f : \mathbb{R}^d \to \mathbb{R}^d$ is an affine transformation, and if K_1 is a translate of a convex body K, then $f(K_1)$ is a translate of $f(K)$. Similarly, if K_1 is a translate of $-K$, then $f(K_1)$ is a translate of $-f(K)$. Therefore the affine image of a packing with translates of K is a packing with translates of the image of K, and these two packings have the same density. Moreover, the affine image of a lattice packing with copies of K is a lattice packing with the affine image of K. The same *affine invariance* holds true for any packing that combines translates of K and of $-K$. These simple facts imply that if convex bodies K and M are affine-equivalent, then:

$$\delta_T(K) = \delta_T(M), \quad \delta_{T^*}(K) = \delta_{T^*}(M), \quad \delta_L(K) = \delta_L(M),$$

and

$$\delta_{L^*}(K) = \delta_{L^*}(M).$$

Therefore we can say that the domain of each of the four density functions δ_T, δ_{T^*}, δ_L, and δ_{L^*} is the set of affine equivalence classes of convex bodies. Let $[K]$ denote the affine equivalence class of the convex body K. Following Macbeath [42], we supply the set of affine equivalence classes of convex bodies in $\mathbb{R}^d$ with the distance function d defined as follows: for every pair K, M of convex bodies, set

$$\rho(K, M) = \inf\{ \operatorname{Vol}(K')/\operatorname{Vol}(M) :$$

$$K' \text{ is affine equivalent to } K \text{ and } K' \supset M\}.$$

Since the function ρ is affine invariant, the function d given by

$$d([K], [M]) = \log \rho(K, M) + \log \rho(M, K)$$

is well-defined. It is easy to check that d is a metric on the set of all affine equivalence classes of convex bodies. The space of such classes supplied with this metric, denoted by $\mathcal{K}_a^d$, is compact (see Macbeath [42]), and each of the four packing density functions δ_T, δ_{T^*}, δ_L, and δ_{L^*} defined on $\mathcal{K}_a^d$ is continuous. Therefore each of them reaches its extreme values. Of course, the maximum value for each of them is 1, reached at any convex body that tiles $\mathbb{R}^d$ by its translates. However, none of the four minimum values is presently known.

Determining those minimum values and the convex bodies at which they are attained seems to be a very challenging problem, perhaps too difficult to expect to be solved in foreseeable future. Reasonably good estimates for these minimum values, however, should not be too hard to establish.

As for the maximum value of 1, attained at the corresponding space tiling bodies (polytopes), those that tile space by translations have been described in fairly simple terms by Venkov [53] and, independently, by McMullen [44]. However, the analogous question, asking which convex polytopes can tile space by their translates combined with translates of their negatives, still remains unanswered.

6. PACKING TRANSLATES OF CONES

We now turn our attention to the subspace $\mathcal{C}_a$ of $\mathcal{K}_a^3$ consisting of affine equivalence classes of cones, that is, affine classes of bodies of the form $\mathbf{C}_v(K)$, where K, the base, is a convex disk. Since $\mathcal{C}_a$ is a closed subset of $\mathcal{K}_a^3$, it is compact as well. Notice that the affine class of the cone $\mathbf{C}_v(K)$ is determined uniquely by the affine equivalence class of its base K. The affine class of a cone with base K will be denoted by $\mathbf{C}K$. Thus $\mathbf{C}K = \mathbf{C}M$ if and only if K and M are affinely equivalent convex disks.

Again, the problem of maximum and minimum values arises that each of the four packing density functions attains on the compact set $\mathcal{C}_a$. This time, however, the maximum value of each of them is strictly smaller than 1, since a cone cannot tile space, neither by its translates, nor by its translates combined with translates of its negative. Thus we face a set of eight questions:

Which convex disks produce cones of maximum and minimum packing density with respect to the four affine-invariant packing density functions?

The eight extremum density values over the set of cones will be denoted by $c^{\max}$ and $c^{\min}$ supplied with the corresponding subscripts T, T^*, L, and L^*. The case of cones with centrally symmetric bases is of special interest, raising another set of eight analogous questions.

We begin with a lower bound for the volume of the difference body of a cone, to be used in the inequality (1.1), producing an upper bound for the packing density δ_T for all cones. Figure 5 shows side-by-side two cones and their corresponding difference bodies.

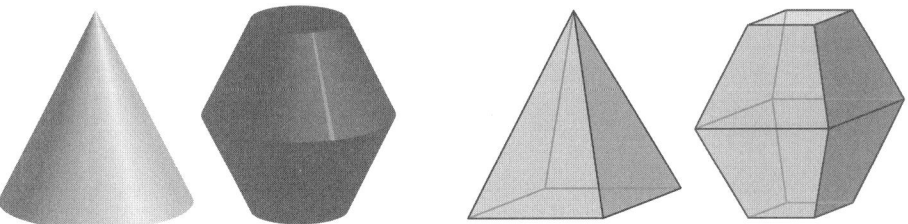

Fig. 5. Cones and their difference bodies: the circular cone and the square pyramid

For a cone with a centrally symmetric base, the volume ratio of the cone to its difference body is always $\frac{4}{7}$, which is easy to see. For a cone with non-symmetric base, the corresponding volume ratio is always smaller than $\frac{4}{7}$, which follows directly from the Brunn-Minkowski inequality (see *e.g.* [47])

in dimension 2. The minimum ratio $\frac{2}{5}$ occurs for the triangular cone (the tetrahedron) only. Thus we have the following upper bound:

$$\delta_T(\mathbf{C}K) < \frac{4}{7}$$

as equality cannot occur since the difference body of any cone cannot tile space by translations. Therefore

$$c_T^{\max} = \max\{\delta_T(\mathbf{C}K) : K \text{ is a convex disk}\} < \frac{4}{7}.$$

On the other hand, there is a lattice packing with translates of a square pyramid, of density $\frac{8}{15}$. The packing can be described as follows. Begin with a horizontal plane tiled by a lattice of "L"-shaped figures consisting of a unit square with a $\frac{1}{2} \times \frac{1}{2}$ square attached to it. Erect a square pyramid over each of the unit squares, get a layer of square pyramids, in which the small squares are vacant. Place upon the first layer its translate, shifted so that the peaks of the pyramids form the first layer plug the square holes in the second layer. The vertical shift from the first layer to the second one is equal to one-half of the pyramids' height. The two layers determine the entire lattice packing (see Fig. 6 for a top view of the two layers).

Fig. 6. A dense, though not the densest, lattice packing with the square pyramid

Thus $\delta_L(\mathbf{C}S) \geq \frac{8}{15}$, where S denotes the square. However, according to the information supplied in private communication by Betke and Henk, the authors of [2], the lattice packing density of the difference body of the square pyramid is $\frac{112}{117}$, therefore $\delta_L(\mathbf{CS}) \geq \frac{448}{819} = 0.547\ldots > \frac{8}{15} = 0.533\ldots$, and we get the bounds

$$0.547\ldots = \frac{448}{819} \leq c_T^{\max} < \frac{4}{7} = 0.571\ldots .$$

Remark 1. By request of the authors of the present article, Betke and Henk also computed the lattice packing densities of the difference bodies of the pyramids with a regular hexagonal and a regular octagonal bases. The results show that the lattice packing density of the square pyramid is greater than those of the other two. This seems to indicate that among all cones with centrally symmetric bases, the square-based cone has maximum lattice packing density.

Remark 2. The lattice packing density of the cone $\mathbf{CE}$ with a circular (elliptical) base E has not been computed yet. The best we know is

$$0.4469\ldots = \frac{2 + \sqrt{2}}{24}\pi \leq \delta_L(\mathbf{CE}) \leq \frac{\sqrt{2}}{9}\pi = 0.4936\ldots .$$

The upper bound is found by inscribing a maximum volume ellipsoid in the difference body of the circular cone (see Fig. 5) and using the lattice packing density of the ball. The lower bound is obtained by the construction shown in Fig. 7. Observe that the pattern is somewhat similar to that of the square pyramid seen in the previous figure. Neither of the two bounds seems best possible - improvements should not be hard to obtain.

Besides the tetrahedron T, we do not know of any examples of cones with non-symmetric bases whose lattice packing density has been computed. The lattice packing density of the tetrahedron is $\frac{18}{49} = 0.3673\ldots$ (see Section 3, Table 1), which is perhaps the value of $c_L^{\min}$.

Turning to cones with centrally symmetric bases, we obtain a common lower bound for their lattice packing density by a construction similar to that described for the square pyramid (see Fig. 6). First, observe that every centrally symmetric hexagon H is contained in a parallelogram whose sides are extensions of sides of H and of area at most $\frac{4}{3}$, maximum being reached by the regular hexagon. By an affine transformation we can assume that the parallelogram is a unit square, and H is obtained by cutting off two congruent right triangles at two of its opposite corners. Since the square is of minimum area among parallelograms containing H, the legs of the cut-off triangle cannot be longer than $\frac{1}{2}$. Therefore the arrangement shown in

Fig. 7. A dense, though not likely the densest, lattice packing with the circular cone

Fig. 8 is a lattice packing of the plane with the pair consisting of H and a translate of $\frac{1}{2}H$ attached to H, and the density of the collection of translates of H (the large hexagons) is at least $\frac{3}{4}$, minimum being reached when H is an affine regular hexagon.

In a similar way as in the construction for the square pyramid, treating the small hexagons as holes in one layer of hexagonal pyramids, this packing gives rise to a lattice packing of space with the cone $\mathbf{C}H$. The density of this packing is at least $\frac{1}{2}$.

Finally, by a theorem of Tammela [49], every centrally symmetric convex disk K of area 1 is contained in a centrally symmetric hexagon H of area at most $(3.570624)/4$, therefore, by the construction described above, we get the bound $\delta_L(\mathbf{C}K) \geq 0.446328\ldots$ for every cone with a centrally symmetric disk K. Therefore

$$c_T^{\min} \geq 0.446328\ldots.$$

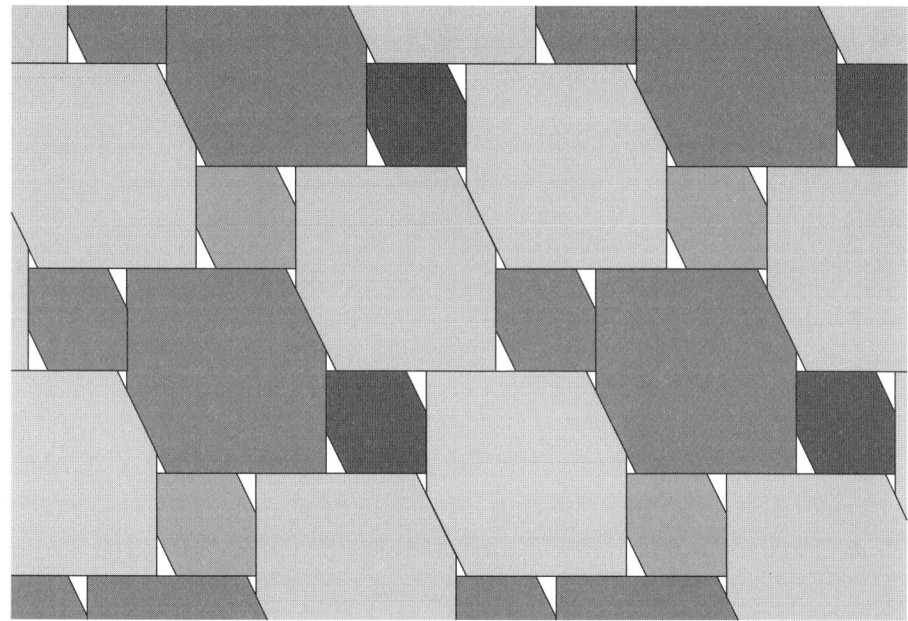

Fig. 8. A lattice packing with a pair of centrally symmetric hexagons H, $\frac{1}{2}H$. The large hexagons form a packingof density at least $\frac{3}{4}$

7. PACKING TRANSLATES OF CONES AND THEIR NEGATIVES

While we know that the value of $c_{T^*}^{\max}$ is smaller than 1, an explicit upper bound below 1 is not easy to produce. Bárány and Matoušek [1] found an explicit constant $\varepsilon > 0$ such that the density of every packing of space with translates of a cone and of its negative cannot exceed $1 - \varepsilon$. The value of ε produced by their proof is very small, about 10^{-42}, and there seems to be room for improvement.

The "best known" case is the densest lattice packing of regular octahedra, of density $\frac{18}{19}$ (see Table 1), showing that the constant ε cannot be greater than $\frac{1}{19}$, that is, $c_{L^*}^{\max} \geq \frac{18}{19}$, but $c_{L^*}^{\max}$ is very likely to be considerably greater than $\frac{18}{19}$. Namely, it is likely that in the densest packing with translates of a square pyramid combined with translates of its negative, the pyramids do not form pairs joined by their common base.

Perhaps it is true in general that the maximum density of a packing with translates of a double cone with a given centrally symmetric base is always smaller than some packing with translates of the cone and its negative. In other words, it seems likely that separating the two parts of the double cone from each other always allows them to reach higher density. It would be interesting to know at least whether or not it is so for the double cone with a circular base and for the double square pyramid.

Elaborating on the idea of packing translates of the cone and its negative in pairs joined by their common, centrally symmetric, base, we use a theorem of Petty [45], stating that every centrally symmetric disk of area 1 is contained in a parallelogram of area at most $\frac{4}{3}$, the bound being sharp only in case of an affine regular hexagon. This allows for enclosing such a pair of cones in an affine regular octahedron whose densest lattice packing produces a packing with a cone with an arbitrary centrally symmetric base. The density of so obtained packing is at least $\frac{3}{4} \times \frac{18}{19} = 0.7105\ldots$ A similar approach for cones with any convex base (the bases of the cone and its negative need not coincide) produces a much weaker lower bound of $\frac{1}{2} \times \frac{18}{19} = 0.47368\ldots$ (Here the factor of $\frac{1}{2}$ is reached only in the case of the triangular base, that is, when the cone is a tetrahedron.) Thus we have

$$(7.1) \qquad \delta_{T^*}(\mathbf{C}K) \geq \frac{27}{38} = 0.7105\ldots \text{ for all cones}$$

$$\text{with centrally symmetric convex bases } K,$$

and

$$(7.2) \quad \delta_{T^*}(\mathbf{C}K) \geq \frac{9}{19} = 0.47368\ldots \text{ for all cones with convex bases } K.$$

The bound in (7.1) is unlikely to be best possible, and the bound in (7.2) definitely is not, since the construction of the presently known densest packing with translates of the tetrahedron Δ and of $-\Delta$, recently found by Kallus, Elser and Gravel [38], is of density $(139 + 40\sqrt{10})/369 = 0.7194880\ldots$.

8. Packing Congruent Replicas of a Convex Body

Here we consider packing densities $\delta(K)$ of a convex body K in $\mathbb{R}^3$, with no restrictions on the nature of isometries used in packing. There are not many bodies K whose packing density is known. No good lower bound has been established for the packing density $\delta(K)$ valid for all convex 3-dimensional bodies K. A rather insignificant lower bound of $\frac{\sqrt{3}}{6} = 0.288\ldots$

is easy to prove based on the known result in the plane, namely that the packing density of every convex disk is at least $\frac{\sqrt{3}}{2}$ (see [40]).

For centrally symmetric convex bodies K, the best known bound of this type is due to E. H. Smith [48], who proved that $\delta(K) \geq 0.53835\ldots$ for every such body K in $\mathbb{R}^3$. The author indicates that the bound is not likely to be the best possible. No reasonable conjecture has been proposed, neither in the general case, nor under assumption of central symmetry, to point to a specific convex body whose packing density should be smallest among all convex bodies. It is not even certain that such a body exists.

Except for the trivial case of space-tiling polytopes, there are not many convex solids whose packing density δ (allowing all isometries) is known. In the following subsections we discuss known results for certain special cases.

8.1. The Kepler conjecture

The three-dimensional sphere packing problem in its general form, without restrictions on the structure of the spheres' arrangements is simple to state and easy to understand even for a non-expert. The conjecture states that the maximum density of a packing of $\mathbb{R}^3$ with congruent balls is $\frac{\pi}{\sqrt{18}} = 0.740480\ldots$, attained in the familiar lattice arrangement (see Fig. 1). The conjecture sounds very convincing to anyone who has ever seen spherical objects, such as oranges or apples, stacked in a pyramid, yet the proof eluded mathematicians for centuries. A problem so appealing attracts attention of experts and laymen alike, and a solution tends to instantly elevate its author to the status of celebrity. The Kepler conjecture, also known as the sphere packing conjecture, has a long and fascinating history, see [30]. The unsuccessful attempts at proof and the nature of the proof that was produced at last seem to indicate that this is one of those problems that cannot be resolved with a reasonably simple and reasonably short proof.

The proof is due to Thomas Hales, who announced it in [29], and then, during the past 13 years presented a series of articles on the subject (see [27, 28, 31, 32, 33, 34], see also [20] by Ferguson, a student of Hales). The description of the theoretical approach to the problem and results of the work of computer occupies nearly 300 pages in these articles. At the computational stage of the proof, computers examined some 5,000 computer-generated cases, each of the cases requiring optimization analysis of a system of non-linear inequalities with a large number of variables. Hales main approach follows a strategy suggested by L. Fejes Tóth in 1953 (see [18]) who anticipated a then insurmountable amount of computations needed for the case analysis.

As a corollary to Hales' result, the packing density of any convex body K such that $\mathbf{B}^3 \subset K \subset RhD$ is easily computed: $\delta(K) = \delta_T(K) = \frac{\text{Vol}(K)}{\text{Vol}(RhD)}$, where RhD denotes the rhombic dodecahedron circumscribed about the unit ball $\mathbf{B}^3$.

In 2003, Hales launched a project called FLYSPECK, designed for an automatic (computerized) formal verification of his proof. The project involves a number of experts in formal languages. They currently estimate that the project is about 65% complete. As a byproduct of the FLYSPECK project, Hales, jointly with five coauthors involved in the project, published recently another article [35] on the topic of the Kepler conjecture, revising the originally published text.

8.2. Packing space with congruent ellipsoids

The problem of packing space with ellipsoids is in sharp contrast with the analogous two-dimensional problem. In the plane, the density of any packing consisting of congruent ellipses, or even ellipses of equal areas (see L. Fejes Tóth [17], see also [18]), cannot exceed the circle's packing density $\frac{\pi}{\sqrt{12}}$. It has been noticed in [8] that ellipsoids E exist whose packing density is greater than that of a ball, that is, $\delta(E) > \frac{\pi}{\sqrt{18}}$. The first ellipsoid found that had this property was quite elongated, of a very high *aspect ratio*, that is, the ratio of its longest semiaxis to its shortest.

As an improvement of this construction, Wills [56] found a denser ellipsoid packing, with ellipsoids of a slightly smaller aspect ratio. However, a much more substantial improvement came about a few years ago. A. Donev, F.H. Stillinger, P. M. Chaikin, and S. Torquato [15] constructed a remarkably dense packing of congruent ellipsoids that do not differ from a sphere too much, namely with aspect ratio of $\sqrt{3}$ (or any greater than that). The packing they found using a computerized experimental simulation technique reaches density of 0.770732. This is the currently highest known density of a packing of space with congruent ellipsoids.

It is not known, however, whether or not there is an upper bound below 1 for such density. While no ellipsoid can tile space by its congruent replicas, thus the packing density of any ellipsoid is smaller than 1, it is conceivable that an ellipsoid with sufficiently high aspect ratio could have its packing density as close to 1 as desired.

8.3. Packing space with congruent cylinders

The first non-trivial case of a convex (though unbounded) solid whose packing density, allowing all isometries, was computed, was the circular cylinder, infinitely long in both directions, that is, the set $\{(x, y, z) \in \mathbb{R}^3 : x^2 + y^2 \leq 1\}$, see [6]. As expected, the maximum density is reached when all cylinders in the packing are parallel to each other and the plane cross-section of the packing perpendicular to the cylinders forms a densest circle packing in the plane. In other words, the packing density of the infinite circular cylinder is $\frac{\pi}{\sqrt{12}}$.

The first non-trivial case of a convex compact solid was resolved by A. Bezdek [4] who determined the exact value of the packing density of the rhombic dodecahedron slightly truncated at one of its trihedral vertices. Although the packing density of Bezdek's example can be derived from the now proven Kepler conjecture (the truncated rhombic dodecahedron contains the inscribed sphere), Bezdek's proof was published before the Kepler conjecture was settled and is independent from it.

The packing density of the circular cylinder $\{(x, y, z) \in \mathbb{R}^3 : x^2 + y^2 \leq 1, \ 0 \leq z \leq h\}$ of finite height $h > 0$, conjectured to be $\frac{\pi}{\sqrt{12}}$ as well, is not known for any value of h. The difficulty of this conjecture is indicated by an example of a certain elliptical cylinder that admits a packing of density greater than 0.99 (see [8]), while in any arrangement of the congruent copies of it such that all their generating segments are parallel to each other, the packing's density cannot exceed $\frac{\pi}{\sqrt{12}}$.

Related to the above problem is the following question about tiling space with congruent right cylinders (a cylinder is said to be *right* if its generating segment is perpendicular to the plane of its base):

If a right cylinder with a convex base admits a tiling of $\mathbb{R}^3$ with its congruent replicas, must its base admit a tiling of the plane?

The difficulty of this question is illustrated by two examples from [7]. First, there exists a space-tiling right cylinder with a non-convex polygonal base that cannot tile the plane (see Fig. 9). Second, there exists a skew prism (though as close to being right as we want) with a convex polygonal base that tiles space, but whose base cannot tile the plane (see Fig. 10).

To explain the construction shown in Fig. 10: (a) A regular hexagon is cut into three congruent, axially symmetric pentagons. (b) By an affine transormation, the pentagons are stretched slightly, each in the direction of its axis of symmetry, so that they cannot tile the plane. Then a skew pyramid is raised over each of the pentagons, so that when they are joined as shown in (c), they form a "hexagonal cup" whose projection to the plane of the original hexagon coincides with the hexagon. Such "cups" can be stacked, forming an infinite beam whose perpendicular cross-section is the original regular hexagon. Finally, such hexagonal parallel beams can fill space by the same pattern as the regular hexagon tiles the plane.

Fig. 9. A non-convex right prism that tiles space, with base that does not tile the plane

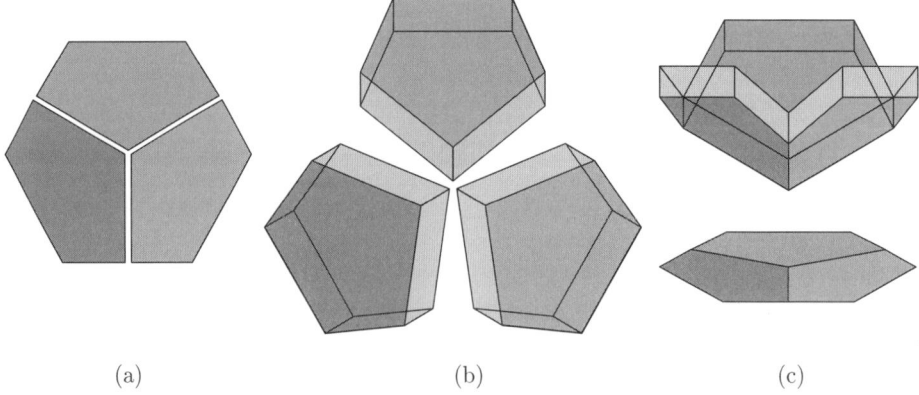

(a)   (b)   (c)

Fig. 10. A convex, slightly skew prism that tiles space, with base that does not tile the plane

The second example shows that the packing density of a cylinder $\mathbf{\Pi}_s(K)$ over a convex disk K could be greater than the packing density of K in the plane. In this example, however, the cylinder is skew. The same phenomenon, however, can occur with a right cylinder as well, as we saw it on the example of a right elliptical cylinder whose packing density is greater than 0.99. It seems natural to ask:

Which convex cylinders have their packing density in space the same as that of their bases in the plane, and which can be packed denser?

8.4. High density packing with congruent regular tetrahedra

Since no integer multiple of the dihedral angle $\varphi = \arccos\frac{1}{3} = 1.23\ldots$ formed by the faces of the regular tetrahedron Δ equals 2π ($5\varphi = 6.15\ldots$ is just slightly smaller than 2π), we know that $\delta(\Delta) < 1$. Then, how densely can space be packed with congruent regular tetrahedra? The question is of interest in areas other than mathematics as well, *e.g.* physics (compacting loose particles), chemistry (material design), etc. The past four years brought an exciting development: a series of articles appeared, each providing a surprisingly dense—denser than previously known—packing.

2006. Conway and Torquato [14] initiate the race by presenting a surprisingly dense packing with density $0.717455\ldots$, almost twice the lattice packing density of the tetrahedron (see Table 1). The packing is a lattice arrangement in which the "repeating unit" is a cluster of 17 congruent regular tetrahedra. Conway and Torquato also give a simple, uniform packing with density $\frac{2}{3}$ (here "uniform" means possessing a group of symmetry that acts transitively on the tetrahedra). This simple packing is a lattice arrangement in which the repeating unit consists of a pair of regular tetrahedra, one rotated by $\frac{\pi}{2}$ with respect to the other (see Fig. 11).

Same year, shortly after the appearance of Conway and Torquato's article, Chaikin, Jaoshvili, and Wang [11], a team composed of two physicists and a high-school student, announce results of an experiment with material tetrahedral dice, packing them tightly, but randomly in spherical and cylindrical containers. The experimental results indicate that the packing density of the regular tetrahedron should exceed 0.74, perhaps even 0.76.

2008. Elizabeth R. Chen [12], a graduate student at the University of Michigan, Ann Arbor, produces a packing reaching density 0.7786, well above the packing density of the ball.

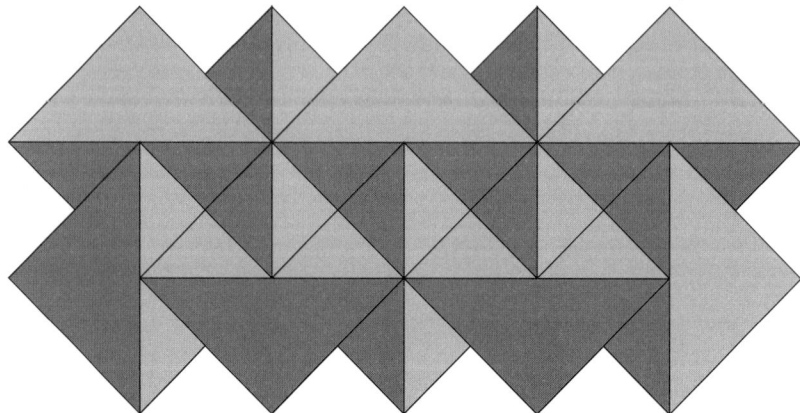

Fig. 11. A portion of the Conway and Torquato uniform packing of regular tetrahedra. Density: $\frac{2}{3}$

2009. Torquato and Jiao [50, 51], using computer simulation based on their "adaptive cell shrinking scheme" raise Chen's record first to $0.782\ldots$ and shortly thereafter to $0.823\ldots$.

At this point one could hardly expect or predict any significant improvements, but they kept coming without much delay.

2009. Haji-Akhbari *et al.* [26], using thermodynamic computer simulations that allow a system of particles to evolve naturally towards high-density states, find a packing whose density reaches 0.8324.

2009. Kallus, Elser, and Gravel [38] produce a surprisingly simple uniform one-parameter family of packings - a lattice arrangement of a repeating unit consisting of just four regular tetrahedra, one pair of tetrahedra joined by a common face and another pair a point-symmetric reflection of the first. New density record: $\frac{100}{117} = 0.85470\ldots$. The packings, though found with the aid of computer, are described analytically.

2010. Torquato and Jiao [52] produce an analytically described packing with regular tetrahedra bettering the density record of Kallus *et al.* Density: $\frac{12250}{14319} = 0.855506\ldots$.

2010. Chen, Engel, and Glotzer [13] set the most recent density record, reached by an analytically described packing. The currently highest known density is raised to $\frac{4000}{4671} = 0.856347\ldots$.

The last few density improvements seem to be inching towards its maximum value. Though it is difficult to conjecture what that value should be, any reasonable upper bound would be welcome as a valuable contribution.

Disappointingly, thus far no specific upper bound, not even by a miniscule amount below 1, has been established.

Added in proof. During the editorial process of publication of this article, S. Gravel, V. Elser and Y. Kallus [arXiv:1008.2830] obtained an upper bound for the packing density of the regular tetrahedron, around 2.6×10^{-24} below 1.

References

[1] I. Bárány and J. Matoušek, *Packing cones and their negatives in space,* Discrete Comput. Geom. **38** (2007), no. 2, 177–187.

[2] U. Betke and M. Henk, *Densest lattice packings of 3-polytopes,* Comput. Geom. **16** (2000), no. 3, 157–186.

[3] P. Brass, W. Moser and J. Pach, Research Problems in Discrete Geometry, Springer 2005.

[4] A. Bezdek, *A remark on the packing density in the 3-space,* in: Colloquia Mathematica Sociates János Bólyai, Intuitive Geometry, North Holland (edited by G. Fejes Tóth) **63** (1994), 17–22.

[5] A. Bezdek and G. Kertész, *Counter-examples to a packing problem of L. Fejes Tóth,* in: Intuitive Geometry (Siófok, 1985), 29–36, Colloq. Math. Soc. János Bolyai, **48**, North-Holland, Amsterdam, 1987.

[6] A. Bezdek and W. Kuperberg, *Maximum density space packing with congruent circular cylinders of infinite length,* Mathematika **37** (1990), 74–80.

[7] A. Bezdek and W. Kuperberg, *Examples of space–tiling polyhedra related to Hilbert's problem 18, question 2* in: Topics in Combinatorics and Graph Theory, Physica–Verlag Heidelberg, (edited by R. Bodendiek and R. Henn) (1990), 87–92.

[8] A. Bezdek, W. Kuperberg, *Packing Euclidean space with congruent cylinders and with congruent ellipsoids,* The Victor Klee Festschrift Applied Geometry and Discrete Mathematics, DIMACS Series in Discrete Mathematics and Theoretical Computer Science AMS ACM, (edited by P. Gritzman and B. Strumfels) **28** (1991), 71–80.

[9] J. H. H. Chalk, *On the frustrum of a sphere,* Ann. of Math. (2) **52**, (1950), 199–216.

[10] J. H. H. Chalk and C. A. Rogers, *The critical determinant of a convex cylinder,* J. London Math. Soc. 23, (1948), 178–187.

[11] P. Chaikin, S. Wang and A. Jaoshvili, *Packing of Tetrahedral and other Dice,* Abstract Submitted for the MAR07 Meeting of The American Physical Society (2007).

[12] E.R. Chen, *A dense packing of regular tetrahedra,* Discrete Comput. Geom. **40** (2008), no. 2, 214–240.

[13] E.R. Chen, M. Engel, and S.C. Glotzer, *Dense crystalline dimer packings of regular tetrahedra,* e-print arXiv:1001.0586.

[14] J. H. Conway and S. Torquato, *Packing, tiling, and covering with tetrahedra,* Proc. Natl. Acad. Sci. USA **103** (2006), no. 28, 10612–10617.

[15] A. Donev, F. H. Stillinger, P. M. Chaikin, and S. Torquato, *Unusually dense crystal packings of ellipsoids,* Phys. Rev. Lett. **92** (2004), 255506 1–4.

[16] P. Erdős, P. M. Gruber and J. Hammer, Lattice Points, Longman, Harlow and John Wiley, New York, 1989.

[17] L. Fejes Tóth, *Some packing and covering theorems,* Acta Sci. Math. Szeged **12**, (1950). Leopoldo Fejér et Frederico Riesz LXX annos natis dedicatus, Pars A, 62–67.

[18] L. Fejes Tóth, Lagerungen in der Ebene, auf der Kugel und im Raum, Springer Verl., Berlin/ Heidelberg/New York, 1972.

[19] G. Fejes Tóth and W. Kuperberg, *Packing and covering with convex sets,* Chapter 3.3 in Handbook of Convex Geometry, P. Gruber and J. Wills, Eds., North-Holland 1993, vol. B, 799–860.

[20] S. P. Ferguson, *Sphere packings. V. Pentahedral prisms,* Discrete Comput. Geom. **36** (2006), no. 1, 167–204.

[21] C. F. Gauss, *Untersuchungen über die Eigenschaften der positiven ternaren quadratischen Formen von Ludwig August Seber,* Göttingische gelehrte Anzeigen, 1831 Juli 9 = J. Reine Angew. Math. **20** (1840), 312–320 = Werke 11, 188–196.

[22] H. Groemer, *Über die dichteste gitterförmige Lagerung kongruenter Tetraeder. (German)* Monatsh. Math. **66** (1962) 12–15.

[23] H. Groemer, *Some basic properties of packing and covering constants,* Discrete Comput. Geom. **1** (1986), no. 2, 183–193.

[24] P. M. Gruber and C. G. Lekkerkerker, Geometry of Numbers, 2nd ed., North-Holland Math. Library 37, North-Holland, Amsterdam–New York, 1987.

[25] P. M. Gruber and J. M. Wills (eds.), Handbook of convex Geometry, Springer Verlag, New York 1991.

[26] A. Haji-Akbhari, M. Engel, A.S. Keys, X. Zheng, R. G. Petschek, P. Palffy-Muhoray, and S. C. Glotzer, *Disordered, quasicrystalline and crystalline phases of densely packed tetrahedra,* Nature **462**, 773 (2009).

[27] T. C. Hales, *Sphere packings. I,* Discrete Comput. Geom. **17** (1997), no. 1, 1–51.

[28] T. C. Hales, *Sphere packings. II,* Discrete Comput. Geom. **18** (1997), no. 2, 135–149.

[29] T. C. Hales, *The Kepler Conjecture – here it is at last!* www.math.pitt.edu/~thales/kepler98/

[30] T. C. Hales, *Cannonballs and honeycombs,* Notices Amer. Math. Soc. **47** (2000), no. 4, 440–449.

[31] T. C. Hales, *A proof of the Kepler conjecture,* Ann. of Math. (2) **162** (2005), no. 3, 1065–1185.

[32] T. C. Hales, *Sphere packings. III. Extremal cases,* Discrete Comput. Geom. **36** (2006), no. 1, 71–110.

[33] T. C. Hales, *Sphere packings. IV. Detailed bounds,* Discrete Comput. Geom. **36**, no. 1, 111–166.

[34] T. C. Hales, *Sphere packings. VI. Tame graphs and linear programs,* Discrete Comput. Geom. **36,** no. 1, 205–265.

[35] T. C. Hales, J. Harrison, S. McLaughlin, T. Nipkow, S. Obua, and R. Zumkeller, *A revision of the proof of the Kepler conjecture,* arXiv:0902.0350.

[36] A. Heppes, *On the packing density of translates of a domain,* Studia Sci. Math. Hungar. **25** (1990), no. 1–2, 117–120.

[37] D. J. Hoylman, *The densest lattice packing of tetrahedra,* Bull. Amer. Math. Soc. **76** (1970) 135–137.

[38] Y. Kallus, V. Elser and S. Gravel, *Dense periodic packings of tetrahedra with small repeating units,* Discrete Comput. Geom. DOI 10.1007/s00454-010-9254-3, published online March 03, 2010.

[39] J. Kepler, *The Six-Cornered Snowflake,* 1611; translated by L. L. Whyte, 1966 (Oxford Univ. Press).

[40] G. Kuperberg, and W. Kuperberg, *Double-lattice packings of convex bodies in the plane,* Discrete & Computational Geometry **5** (1990), 389–397.

[41] C. G. Lekkerkerker, Geometry of Numbers, Bibl. Math. **8**, Wolters-Noordhoff Publ., Amsterdam 1969.

[42] A. M. Macbeath, *A compactness theorem for affine equivalence-classes of convex regions,* Canadian J. Math. **3** (1951), 54–61.

[43] H. Minkowski, *Dichteste gitterförmige Lagerung kongruenter Körper,* Nachr. K. Ges. Wiss. Göttingen, Math.-Phys. KL (1904), 311–355 (see also: Gesammelte Abhandlungen vol. II, 3–42, Leipzig 1911).

[44] P. McMullen, *Convex bodies which tile space by translation,* Mathematika **27** (1980), no. 1, 113–121.

[45] C. M. Petty, *On the geometry of the Minkowski plane*, Riv. Mat. Univ. Parma **6** (1955), 269–292.

[46] C. A. Rogers, *The closest packing of convex two-dimensional domains*, Acta Math. **86** (1951), 309–321.

[47] R. Schneider, Convex bodies: the Brunn-Minkowski theory. Encyclopedia of Mathematics and its Applications, **44**. Cambridge University Press, Cambridge, 1993.

[48] E. H. Smith, *A new packing density bound in 3-space*, Discrete Comput. Geom. **34** (2005), no. 3, 537–544.

[49] P. Tammela, *An estimate of the critical determinant of a two-dimensional convex symmetric domain*, Izv. Vyss. Ucebn. Zaved. Mat. **12(103)** (1970), 103–107.

[50] S. Torquato and Y. Jiao, *Dense packings of the Platonic and Archimedean solids*, Nature **460**, 876 (2009).

[51] S. Torquato and Y. Jiao, *Dense packings of polyhedra: Platonic and Archimedean solids*, Phys. Rev. E **80**, 041104 (2009).

[52] S. Torquato and Y. Jiao, *Analytical constructions of a family of dense tetrahedron packings and the role of symmetry* e-print arXiv:0912.4210.

[53] B. A. Venkov, *On a class of Euclidean polyhedra*, Vestnik Leningrad. Univ. Ser. Mat. Fiz. Him. **9** (1954) no. 2 11–31.

[54] J. V. Whitworth, *On the densest packing of sections of a cube.* Ann. Mat. Pura Appl. **(4) 27**, (1948), 29–37.

[55] J. V. Whitworth, *The critical lattices of the double cone*, Proc. London Math. Soc. (2) **53**, (1951), 422–443.

[56] J. Wills, *An ellipsoid packing in E^3 of unexpected high density*, Mathematika **38** (1991), 318–320.

[57] Y. Yeh, *Lattice points in a cylinder over a convex domain*, J. London Math. Soc. **23**, (1948), 188–195.

András Bezdek

MTA Rényi Institute,
13-15 Réaltanoda u.,
Budapest, Hungary
also:
Mathematics & Statistics,
Auburn University,
Auburn, AL 36849-5310,
USA

e-mail: bezdean@auburn.edu

Włodzimierz Kuperberg

Mathematics & Statistics,
Auburn University,
Auburn, AL 36849-5310,
USA

e-mail: kuperwl@auburn.edu

BOLYAI SOCIETY
MATHEMATICAL STUDIES, 24

Geometry –
Intuitive, Discrete, and Convex
pp. 91–108.

GEOMETRIC PROBLEMS ON COVERAGE IN SENSOR NETWORKS

PETER BRASS*

In this article we survey results and state some open problems that are motivated by sensor networks applications.

1. INTRODUCTION

Sensors controlled by computers who communicate their acquired data to some central server by a network connection have been used for a fairly long time. *Sensor networks* now denote a more specific setting: a large number of sensors (nodes), each with an autonomous computer, wireless network connection, and power supply, distributed over a region of interest [74, 17]. The nodes form an *ad-hoc network* by maintaining a wireless network connection with some neighboring nodes, and distribute important sensor information by this network to all interested other nodes. The number of nodes is very large, thousands to possibly millions, the nodes are moderately cheap, and getting cheaper, so a much considered scenario is that they are just dropped from a plane in large numbers, and spontaneously organize and start sensing. Paradigmatic applications are environmental, e.g., detecting a forest fire, or presence of a dangerous chemical, or intrusive species; and military, detecting enemy forces, or hidden weapons. In this model, it is impossible to share all sensor information, since the number of sensors is potentially unlimited, but the network data rate is small. We assume there is one event somewhere, which should be detected by a sensor,

*Supported by NSF grant CCF-1017539.

and if it is detected, the communication of the event to some center is task of the network protocol. This divides the problem into the mostly geometric problem of finding the event, and the network protocol problem of spreading the news about the event. In this paper, we will concentrate on the first problem.

To further specify the model, we need to know when a sensor detects an event. The most widely used, and most easily accessible model is the *boolean sensing model*: there is a sensing radius r, and each sensor detects everything within that radius r, and nothing outside. In this case we arrive at a classical covering by discs setting. Alternatives are that the event needs to be discovered by several sensors, leading to multiple coverage problems, and that the discovery by a sensor is a random event whose probability depends on the distance from the sensor to the event.

Another important model parameter is whether we can freely choose the locations of the sensors, or the sensors are randomly distributed over the region of interest X. For random distribution of sensors, the most convenient model is a Poisson process of density λ, which distributes an expected number of $\lambda\,\mathrm{area}(X)$ sensors in X; an alternative, which is more difficult to analyze, is to drop a given number of sensors by independent uniform random choice from X. Many further model parameters are possible.

In the basic model, given a region X, a sensing radius r, and a number or density of available sensors $n = \lambda\,\mathrm{area}(X)$, we obtain versions of a classic geometric coverage problem, the optimum coverage of X by circles. The optimum coverage of the plane, i.e., $X = R^2$, was solved in [35, 29], and the more difficult case of a finite region X was essentially answered by Fejes Tóth in [25]. As example, we cite

Theorem 1. *To cover a region X by sensors with sensing radius r,*

$$\frac{2}{3\sqrt{3}}\,\frac{\mathrm{area}(X)}{r^2} + \frac{2}{\pi\sqrt{3}}\,\frac{\mathrm{perimeter}(X)}{r} + 1$$

sensors are sufficient.

If the number of available sensors is not enough to achieve complete coverage, but the area of interest X is large compared to a sensing disc, then, by the moment theorem, a triangular lattice arangement is essentially best possible. If X is the entire plane, it can be stated as follows.

Theorem 2. *If sensors with sensing radius r are placed with density λ in the plane, their coverage is at most the coverage reached by the triangular* *lattice arrangement of that density. So especially for $\lambda r^2 \leq \frac{1}{2\sqrt{3}}$ it is at most $\pi r^2 \lambda$, and for $\lambda r^2 \geq \frac{2}{3\sqrt{3}}$ it can reach 1.*

If the sensors are instead spread randomly in X by a Poisson process of density λ, there are never enough sensors to guarantee complete coverage. If we again disregard boundary effects by assuming X is the entire plane, it is easy to show [45]

Theorem 3. *If sensors with sensing radius r are placed with a Poisson process of density λ in the plane, then the expected coverage is $1 - e^{-\pi \lambda r^2}$.*

This follows, since a point will be covered if a sensor falls within distance r of it, so the probability of detecting the event by a sensor is the probability that at least one sensor falls into a disc of area πr^2. If the sensors are distributed in a finite region X, the expected coverage will be smaller, since for points p within distance r to the boundary, only the part of $D(p,r)$ inside X can possibly receive a sensor. So the probability of p being covered is only $1 - e^{-\lambda \operatorname{area}(D(p,r) \cap X)}$,

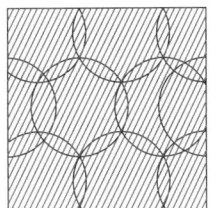 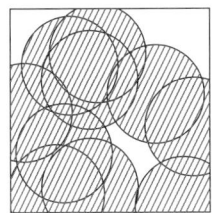 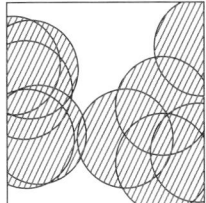

Ten discs covering a square, and two random arrangements of these discs

The density of the Poisson process necessary to give complete coverage with high probability has been studied in detail [56], one essentially needs an additional $\log n$-factor, where n is the number of necessary sensors under optimal placement. We cite a simple version.

Theorem 4. *Let sensors with sensing radius r be placed with a Poisson process of density λ in a sequence of scaled regions νX, $\nu \to \infty$, and let $\varepsilon > 0$.*

- *if $r = \sqrt{(1 - \varepsilon) \dfrac{\log \operatorname{area}(\nu X)}{\pi \lambda}}$, then $\lim\limits_{\nu \to \infty} \operatorname{Prob}(\nu X \text{ is covered}) = 0$*

- *if $r = \sqrt{(1 + \varepsilon) \dfrac{\log \operatorname{area}(\nu X)}{\pi \lambda}}$, then $\lim\limits_{\nu \to \infty} \operatorname{Prob}(\nu X \text{ is covered}) = 1$*

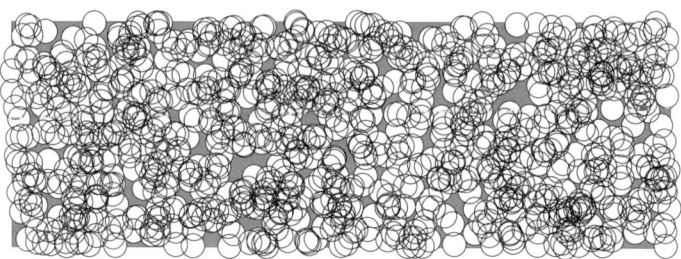

1000 random discs of combined area 300 in a rectangle of area 100

If we use as alternative randomization model that n sensors are independently placed according to a uniform distribution on X, the analysis gets more complicated but gives similar asymptotic behavior [39]. Yet another similar model is that the potential sensor locations are a sufficiently fine square grid in X, and each of these locations receives a sensor with some probability p. That model has the interpretation that sensors are placed on all grid points, but most are sleeping to conserve their battery, only with probability p they wake up [64, 61].

2. MOBILE TARGETS

If the event that the sensors should detect is not at a fixed point, but is a mobile target, this opens several further types of questions. If the sensors achieve complete coverage of X, mobility does not help the target since it will be caught immediately anyway. But if there are much less sensors, then the best we can do is catch the target in some coverage hole that is as small as possible. For randomly placed sensors, this becomes a question of percolation [49, 55, 3]. For chosen sensor positions this is the question of how to arrange n discs of radius r in X such that the largest component of the relative complement of the discs becomes as small as possible.

Problem 1. Given a region X and n discs of radius r, how can we find a placement of the discs that minimizes the maximum diameter of the components of the relative complement of the discs?

If the total area of the discs is small relative to X, then we divide X into cells by chains of sensors, so the problem becomes how to divide X into cells, using a given total length of cell-boundaries, such that the cells are as small as possible. If we wish to minimize the diameter of the cells, this becomes a known open isoperimetric-type problem [16], but the division into hexagonal cells is at least near-optimal.

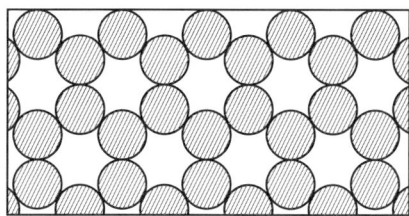

Arrangement of discs dividing the plane into small cells

Problem 2. Which placement of unit discs of density r in the plane minimizes the maximum diameter of the connected components of the complement of the discs?

For these and many further related settings, when the sensors or the target may be mobile, we refer to [14].

If something is known about the intended movement of the target, e.g., that it has to move from a position t_1 to t_2, the problem becomes that of barrier coverage. Several concepts have been proposed, best known are [51],

- the *maximal breach path*, that stays as far away as possible from the sensors (maximizing the minimum distance to the nearest sensor), and

- the *maximum support path*, that stays as close as possible to sensors (minimizing the maximum distance to the nearest sensor).

Both paths are easy to compute, e.g., the maximal breach path must be essentially following Voronoi edges to maximize the distance to the nearest sensor [42, 41, 50].

This definition considers only the maximum or minimum coverage of the path; it would be more interesting to consider the accumulated coverage, taking an integral of the coverage over the path length. This is the *path exposure* defined in [23]; it is there approximated by discretizing the region, computing the exposure of each cell, and computing the shortest path in this graph. Although this heuristic probably gives a constant-factor approximation, no bounds are known. The exposure of each cell could be defined in several ways, the simplest is to assume that within each sensing disc, there is at each moment a constant probability of discovery p, and

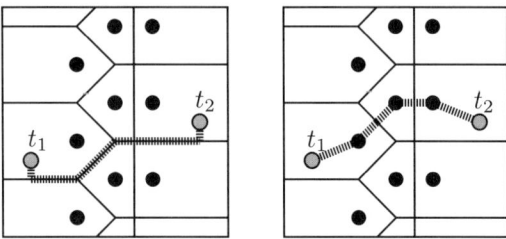

Maximal breach and maximum support path from t_1 to t_2

distinct sensors give independent events. Then the probability of discovery at point x is p to the number of discs containing x, and the path exposure is the integral over the path length of this probability.

Problem 3. Given n discs $D_1, \ldots, D_n$ and points t_1, t_2, how can we find or approximate the path from t_1 to t_2 that minimizes or maximizes exposure?

Two interpretations of this problem are: if the discs are cell phone transmitters, each having a probability of failure, how should we travel from t_1 to t_2 such that the expected time without connection is minimized? And: if the discs are guards, each with a constant probability of discovering the intruder while he is in the region of the guard, how should he move from t_1 to t_2 to minimize his probability of discovery? In general, these problems can be interpreted as a instances of the the shortest path in weighted regions problems, which is an apparently difficult problem in computational geometry [52].

A different formalization of barrier coverage is introduced in [37]; an arrangement of sensors forms a *k-barrier coverage* between t_1 and t_2 if any path in the region from t_1 to t_2 crosses at least k sensing discs. If we can choose the positions of the sensors, the best k-barrier separating opposite sides of a rectangle consists of k chains of sensors parallel to those sides. In [37], the probability that sensors placed by a Poisson process of density λ form a k-barrier coverage is analyzed. Several other measures are analyzed in this random placement setting in [58].

If the target is known to move on a straight line, we obtain the model of *track coverage* proposed in [12]; they try to place the sensors as to maximize the measure of the set of lines crossing the region of interest or barrier that intersect at k sensing discs.

Problem 4. How should n unit discs be arranged in $a \times b$-rectangle such that the measure of lines that cross the rectangle, intersecting both a-sides, and that intersect at least k discs, is maximized?

The problem of intercepting a mobile target by sensors might also be viewed as a geometric game, with randomized strategies on either side [21]; geometric games are analyzed in [60].

3. MOBILE SENSORS

If the sensors are mobile, we can actively search for targets, or improve the coverage properties of our arrangement. Here is an overlap between mobile sensors and robots; sensors are typically assumed to be very energy-constrained and cannot move much, if at all, whereas no such restriction exists for robots.

The improvement of an initial random sensor arrangement by moving sensors is a special case of the problem of sensors spreading out studied in the next section; if the initial density of the sensor network is high enough, then only a small fraction of the sensors have to be moved a moderately short distance to get coverage, or even k-fold coverage, with a high probability [70].

If movement is less restricted, and a sensor is sensing everything within distance r while it is traveling along a path γ of length l, then this sensor covers the set $\{p \mid \exists q \in \gamma \colon d(p,q) \leq r\}$, that is the Minkowski sum of the disc around 0 of radius r and the curve γ. This set has area at most $\pi r^2 + 2rl$. If we can freely choose the initial placement and the paths taken, it is nearly optimal to arrange the sensors in columns of searchers, with distance $2r$ from one searcher to the next, and as many columns as the available sensor density allows [14]. If the initial position is randomly distributed, then a simple strategy is to choose some direction and move along it; this model is again easy to analyze [43]. Such random direction models were further studied in [53, 47].

A natural problem which up to now did not receive any study is to arrange periodic patrol paths in a region. The model here is that we have k mobile guards that move at a fixed speed through the region of interest X, and at each moment see everything within distance r to their current position. At any time, a spontaneous event might occur anywhere in X, and once it occured, it stays visible. We want to arrange the patrol paths in such a way that the time to discovery is minimized.

Problem 5. How should k mobile guards patrol a region X as to minimize the time to discovery of an event?

For $k = 1$ this problem was discussed as lawn mowing problem: a classical rotating-blade lawnmower mows a disc, and one wants to find the shortest path to mow the entire lawn X. Even there, only a 3-approximation for the optimum path length is known [4]. We do not know any way to compute the optimum patrol path of a given set X and radius r, e.g., what is the optimum patrol path for a square?

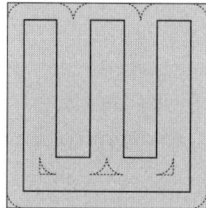 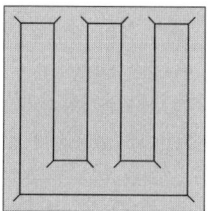

The trivial patrol path in a square leaves many corners uncovered

If both target and sensors are mobile, we get several types of searching problems, depending on the mutual information of searcher and target, and their speed. The most natural in our context is that the searchers see everything within their sensing radius r, whereas the target sees the position of all the searchers. This setting has been studied in [24] under the title 'offline variants of lion and man', referring to the unrelated 'Lion and Man' problem studied by Besicovitch, and others, in which the lion sees the position of the man and has to intercept him. The problem has been studied mainly in a graph variant, where there are k searchers on an $n \times n$-grid graph, and one target; the searchers see only their current vertex, and have to catch the target [24, 15, 13]. Using a discretization of the unit square, there is some asymptotic relation between the graph setting and the geometric setting, although not enough to give the correct multiplicative constants. An interesting problem here is the following

Problem 6. What is the largest square that can be searched by k unit-radius searchers against a mobile target with the same speed as the searchers?

Some small cases have been studied in [1]. The obvious answer, that the k searchers can form a row and sweep a square of sidelength $2k$ is not optimal. A unit radius disc wobbling between two unit radius discs slightly more than 4 units apart can close both gaps alternatingly, by overlapping the outer discs at a length which is much larger than the necessary movement of the middle disc, so the target would be caught if it entered that gap. This has been worked out and extended to 3d in [2]; the 2d movement scheme proposed there might be optimal, the 3d version almost certainly is not.

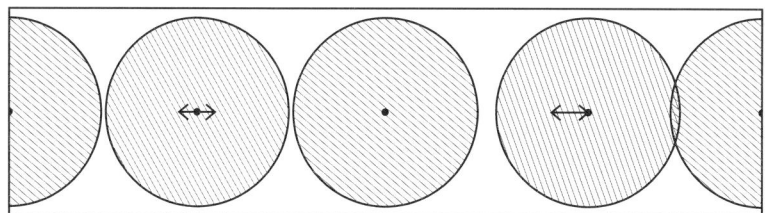

Discs blocking slightly more than their diameter, alternatingly closing the gaps

4. Mobile Sensors Spreading Out

A problem that has been studied in a number of papers without reaching a satisfactory theoretical analysis is that of mobile sensors spreading out from some starting positions to cover a region X. The idea is that each sensor looks at his own position and that of neigboring sensors, and moves to a new position to locally improve the covering quality. By repeating these independent local improvements, one hopes to arrive at a global near-optimal solution. Such a motion strategy would allow autonomous spreading out to cover some unknown regions, as well as automatic filling in of coverage holes if some sensors fail.

If the current sensor positions are $p_1, \ldots, p_n$, then each sensor p_i covers some part of his Voronoi-cell $\mathrm{Vor}(p_i) \cap X$, so natural movement strategies that were suggested is to move p_i so that it best covers its own Voronoi-cell [66, 63, 30, 38], or to sct up some virtual forces against the neighboring robots and try to balance these forces [75, 57, 28, 54, 46, 19, 40, 20]. Since the actual movement of sensors is slow compared to communication and computation, one can run the algorithm on the intended positions instead of the real positions, and start moving only when the intended positions for each sensor have stabilized [67]. But up to now it has not been proved for any algorithm of this type that it will converge to a near-optimal

Points moving away from each other, with increasing number of steps.
Discs show half the distance to nearest neighbor

covering, and neither any bound for the expected pathlength. A square lattice arrangement would give a suboptimal stable arrangement under this motion rule. Indeed, if we disregard the boundary effects and fill the entire plane, there are arbitrary bad arrangements which are necessarily stable under any local motion rule since each sensor has a symmetric neighborhood. These are, however, not stable under small perturbations.

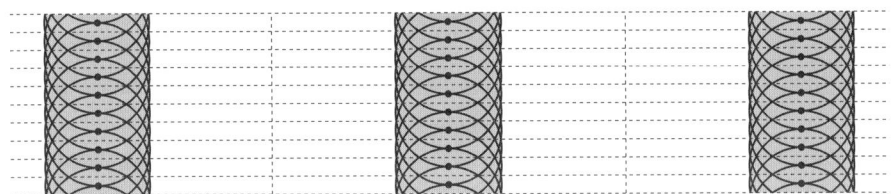

Bad arrangement of discs in widely spaced dense rows, unstable under
perturbations

Problem 7. Show for some local motion rule that any stable arrangement in the entire plane which is also stable under infinitesimal perturbations will have a covering density at most a constant factor worse than the optimal covering density.

A more fundamental question is to demand both travel distance and covering quality to be near-optimal.

Problem 8. Show for some motion rule that is based only on local information that the positions of the sensors converge to a solution for which both the covering quality is a constant-factor approximation of the optimum covering quality, and the total travelled path length is a constant-factor approximation of the optimal path length.

A different type of strategy was proposed in [8, 9]; they force the tri-angular lattice structure, first choosing the right edgelength for it from the knowledge of X, and then moving sensors to these lattice positions along a growing patch. Every sensor who reaches a new lattice position on the boundary of the patch becomes fixed and extends the boundary. This solution necessarily converges to a near-optimum, but it needs a-priori knowledge of the region X and the number of available sensors, whereas the local rules can be used to fill an unknown region, and deal with sensor failures and coverage obstacles. Further algorithms with more global planning to match redundant sensors to coverage holes with small movement distances were proposed in [68, 73, 18].

The same problem of sensors spreading out through an unknown region has also been studied for more complicated sensing models [10, 31].

5. COVERAGE AND CONNECTIVITY

In the previous models, we described only coverage; if we also take the communication in the network into account, we have a set of sensors with their sensing discs, and a graph of sensors which are connected by communication links. Typically, the graph is given by a distance constraint on the sensor locations: there is a communication distance d, and the communication network consists of those sensors of distance at most d.

A minimal condition on the communication network is that it must be connected. If the sensors cover the region, then it is easy to see that $d \geq 2r$ guarantees that the network is connected. If the region is just a long, narrow strip, this is clearly best possible, connectivity does not follow from coverage if $d < 2r$. But even if we cover the plane, an arrangement of two halfplanes each densely filled with sensors, but separated by an empty strip of width $2r - \varepsilon$, show that connectivity does not follow from coverage for any $d < 2r$.

To introduce redundancy in the communication network, which typically suffers from many link failures due to transmission obstacles, one frequently aims for k-edge-connected communication networks. Node failures can be modelled by requiring higher node connectivity.

Problem 9. What is the smallest c_k such that for any arrangement of sensors whose sensing discs of radius r cover the plane, the graph of distances less than $c_k r$ is k-edge-connected?

A problem which has already received partial study is the minimum density of a sensor network, given d and r, which gives complete coverage and k-connectivity. Since minimum density for complete coverage is achieved by the triangular lattice arrangement, which for $d \geq \sqrt{3}r$ has a communication network which is at least 6-connected, this question is interesting only for $d < \sqrt{3}$ or large k. For 1- and 2-connectivity, it was proved [5, 33, 69] that the optimum arrangement, for any $d < \sqrt{3}r$, consists of rows of sensors at distance d, which are connected by one column for 1-connectivity, or by columns spaced arbitrarily far (e.g., at exponentially increasing distances) for 2-connectivity.

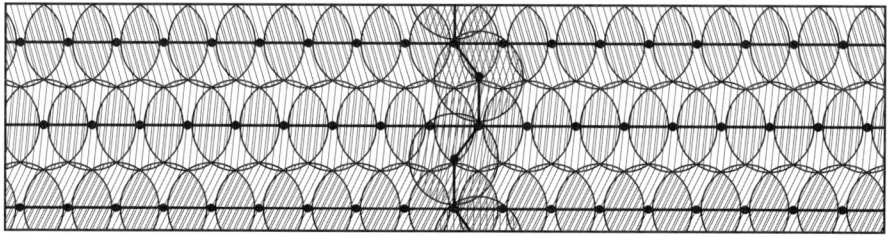

Thinnest covering arrangement with 1-connected communication network

The optimum arrangement for 4-connectivity was determined for the interval $\sqrt{3}r \geq d \geq \sqrt{2}r$ in [6]; beyond $\sqrt{3}r$, the triangular lattice arrangement is extremal, and for $d < \sqrt{2}r$, probably again a row-based network, with 4-connected rows, will be optimal. There are no results for 3-connected arrangements, but it seems that 3- and 4-connectivity give the same extremal arrangements, as it is true for 1- and 2-connectivity.

Problem 10. What is the arrangement of minimum density whose sensing discs of radius r cover the plane, and whose graph of distances less than d is k-edge-connected?

Similar questions for networks in three-dimensional space, e.g., floating in the sea, were studied in [59, 7].

REFERENCES

[1] Alonso, L., Reingold, E. M., Bounds for cops and robber pursuit, Computational Geometry—Theory and Applications **43** (2010) 749–766.

[2] Alonso, L., Reingold, E. M., Improved bounds for cops-and-robber pursuit, Computational Geometry—Theory and Applications **44** (2011) 365–369.

[3] Ammari, H. M. and Das, S. K., Integrated coverage and connectivity in wireless sensor networks: a two-dimensional percolation problem, *IEEE Transactions on Computers* **57** (2008) 1423–1434.

[4] Arkin, E. M., Fekete, S. P. and Mitchell, J. S. B., Approximation algorithms for lawn mowing and milling, *Computational Geometry—Theory and Applications* **17** (2000) 25–50.

[5] Bai, X., Kumer, S., Xuan, D., Yun, Z. and Lai, T. H., Deploying wireless sensor networks to achieve both coverage and connectivity, in: *MobiHoc '06* (Proc. ACM International Symposium on Mobile Ad-Hoc Networking and Computing) 2006, 131–142.

[6] Bai, X., Yun, Z., Xuan, D., Lai, T. H. and Jia, W., Deploying four-connectivity and full-coverage wireless sensor networks, in: *Proc. IEEE INFOCOM* 2008, 906–914.

[7] Bai, X., Zhang, C., Xuan, D., Jia, W., Full coverage and k-connectivity ($k = 14, 6$) in three-dimensional networks, in: *Proc. IEEE INFOCOM* 2009, 388–396.

[8] Bartolini, N., Calamoneri, T., Fusco, E. G., Massini, A. and Silvestri, S., Snap and Spread: a self-deployment algorithm for mobile sensor networks, in: *DCOSS '08* (Proc. IEEE Conference on Distributed Computing in Sensor Systems) 2008, 451–456.

[9] Bartolini, N., Calamoneri, T., Fusco, E. G., Massini, A. and Silvestri, S., Autonomous deployment of self-organizing mobile sensors foor complete coverage, in: *IWSOS '08* (Proc. International Workshop on Self-Organizing Systems) Springer LNCS **5343** (2008) 194–205.

[10] Batalin, M. and Sukhatme, G. S., Spreading out: a local approach to multi-robot coverage, in: *Proc. Symp. Distributed Autonomous Robotic Systems* 2002, 373–382.

[11] Batalin, M. and Sukhatme, G. S., Coverage, exploration and deployment by a mobile robot and communication network, *Telecommunication Systems Journal* **26** (2004) 181–196.

[12] Baumgartner, K. and Ferrari, S., A geometric transversal approach to analyzing track coverage in sensor networks, *IEEE Transactions on Computers* **57** (2008) 1113–1128.

[13] Berger, F., Gilbers, A., Grüne, A. and Klein, R., How many lions are needed to clear a grid?, *Algorithms* **2** (2009) 1069–1086.

[14] Brass, P., Bounds on coverage and target detection capabilities for models of networks of mobile sensors, *ACM Transactions on Sensor Networks* **3** (2007) Article 9.

[15] Brass, P., Kim, K. D., Na, H.-S. and Shin, C.-S., Escaping off-line searchers and a discrete isoperimetric theorem, in: *ISAAC '07* (Proc. International Symposium on Algorithms and Computation), Springer LNCS **4835** (2007) 65–74, see also *Computational Geometry—Theory and Applications* **42** (2009) 119–126.

[16] Brass, P., Moser, W. O. J. and Pach, J., *Research Problems in Discrete Geometry*, Springer 2005.

[17] Cardei, M., Wu, J., Coverage in wireless sensor networks, in: *Handbook of Sensor Networks*, M. Ilyas, I. Mahgoub, eds., CRC Press 2004.

[18] Chellappan, S., Gu, W., Bai, X., Xuan, D., Ma, B. and Zhang, K., Deploying wireless sensor networks under limited mobility constraints, *IEEE Transactions on Mobile Computing* **6** (2007) 1142–1157.

[19] Chen, J., Li, S. and Sun, Y., Novel deployment schemes for mobile sensor networks, *Sensors* **7** (2007) 2907–2919.

[20] Chen, J., Shen, E. and Sun, Y., The deployment algorithms in wireless sensor networks: a survey, *Information Technology Journal* **8** (2009) 293–301.

[21] Chin, J.-C., Dong, Y., Hon, W.-K. and Yau, D. K. Y., On intelligent mobile target detection in a mobile sensor network, in: *MASS '07* (Proc. IEEE Conference on Mobile Ad-Hoc and Sensor Systems) 2007, 1–9.

[22] Chin, T.-L., Ramanathan, P., Saluja, K. K. and Wang, K.-C., Exposure for collaborative detection using mobile sensor networks, in: *MASS '05* (Proc. IEEE Conference on Mobile Ad-Hoc and Sensor Systems) 2005, 743–750.

[23] Clouqueur, T., Phipatanasuphorn, V., Ramanathan, P. and Saluja, K. K., Sensor deployment strategy for target detection, in: *WSNA '02* (Proc. ACM Workshop Wireless Sensor Networks Applications) 2002, 42–48.

[24] Dumitrescu, A., Suzuki, I. and Zylinski, P., Offline variants of the "lion and man" problem, in: *SoCG '07* (Proc. ACM Symposium on Computational Geometry) 2007, 102–111, see also *Theoretical Computer Science* **399** (2008) 220–235.

[25] Fejes Tóth, L., *Lagerungen in der Ebene, auf der Kugel, und im Raum* (2. Auflage), Springer-Verlag 1972.

[26] Gao, Y., Wu, K. and Li, F., Analysis of the redundancy or wireless sensor networks, in: *WSNA '03* (Proc. ACM Workshop Wireless Sensor Networks Applications) 2003, 108–114.

[27] Gui, C. and Mohapatra, P., Power convervation and quality of surveillance in target tracking application, in: *MobiCom '04* (Proc. ACM Conference on Mobile Computing and Networking) 2004, 129–143.

[28] Guo, P., Zhu, G. and Fang, L., An adaptive coverage algorithm for large-scale mobile sensor networks, in: *UIC '06* (Proc. Conference on Ubiquitous Intelligence and Computing) Springer LNCS **4159** (2006) 468–477.

[29] Hadwiger, H., Überdeckungen ebener Bereiche durch Kreise und Quadrate, *Commentarii Math. Helvetici* **13** (1940/41) 195–200.

[30] Heo, N. and Varshney, P., Energy-efficient deployment of intelligent mobile sensor networks, *IEEE Transactions on Systems, Man, and Cybernetics* **35** (2005) 78 92.

[31] Howard, A., Matarić, M. J. and Sukhatme, G. S., An incremental self-deployment algorithm for mobile sensor networks, *Autonomous Robots* **13** (2002) 113–126.

[32] Huang, C.-F. and Tseng, Y.-C., The coverage problem in a wireless sensor network, in: *WSNA '03* (Proc. ACM Workshop Wireless Sensor Networks Applications) 2003, 115–121.

[33] Iyengar, R., Kar, K. and Banerjee, S., Low-coordination topologies for redundancy in sensor networks, in: *MobiHoc '05* (Proc. ACM International Symposium on Mobile Ad-Hoc Networking and Computing) 2005, 332–342.

[34] Iyengar, S. S., Tandon, A., Wu, Q., Cho, E., Rao, N. S. V. and Vaishnavi, V. K., Deployment of sensors: an overview, in: *Distributed Sensor Networks,* S. S. Iyengar, R. R. Brooks, eds., CRC Press 2005, 483–504.

[35] Kershner, R. B., The number of circles covering a set, *Amer. J. Math.* **61** (1939) 665-671.

[36] Kumar, S., Lai, T. H. and Balogh, J., On *k*-coverage in a mostly sleeping network, in: *MobiCom '04* (Proc. ACM Conference on Mobile Computing and Networking) 2004, 144–158.

[37] Kumar, S., Lai, T. H. and Arora, A., Barrier coverage with wireless sensors, in: *MobiCom '05* (Proc. ACM Conference on Mobile Computing and Networking) 2005, 284–298.

[38] Kwok, A. and Martinez, M., Deployment algorithms for a power-constrained sensor network, in: *ICRA '08* (Proc. IEEE Conference on Robotics and Automation) 2008, 140–145.

[39] Lan, G.-L., Ma, Z.-M. and Sun, S.-Y., Coverage of wireless sensor networks, in: *CJCDGCGT '05* (Proc. China-Japan Conference on Discrete Geometry, Combinatorics and Graph Theory) Springer LNCS **4381** (2007) 88–100.

[40] Lee, G. and Chong, N. Y., Self-configurable mobile robot swarms with hole repair capability, in: *IROS '08* (Proc. IEEE Conference on Intelligent Robots and Systems) 2008, 1403–1408.

[41] Li, Q. and Rus, D., Navigation protocols in sensor networks, *ACM Transactions on Sensor Networks* **1** (2005) 3–35.

[42] Li, X. Y., Wan, P. J. and Frieder, O., Coverage in wireless ad-hoc sensor networks, *IEEE Transactions on Computers* **52**(6) 2003, 753–763.

[43] Liu, B., Brass, P., Dousse, O., Nain, P. and Towsley, D., Mobility improves coverage of sensor networks, in: *MobiHoc '05* (Proc. ACM Conference on Mobile Ad-Hoc Networks and Computing) 2005, 300–308.

[44] Liu, B. and Towsley, D., On the coverage and detectability of wireless sensor networks, in: *Proc. WiOpt: Modeling Optimization Mobile Ad-hoc Wireless Networks*, 2003, 201–204.

[45] Liu, B. and Towsley, D., A study on the coverage of large-scale sensor networks, in: *MASS 2004* (Proc. IEEE Conference on Mobile Ad-hoc and Sensor Systems), 2004, 475–483.

[46] Ma, M. and Yang, Y., Adaptive triangular deployment algorithm for unattended mobile sensor networks, *IEEE Transactions on Computers* **56** (2007) 946–958.

[47] McGuire, M., Stationary distributions of random walk mobility models for wireless ad hoc networks, in: *MobiHoc '05* (Proc. ACM Conference on Mobile and Ad-Hoc Networking and Computing) 2005, 90–98.

[48] Mecke, J., Schneider, R. G., Stoyan, D. and Weil, W. R. R., *Stochastische Geometrie*, DMV Seminar Ser. 16, Birkhäuser 1990.

[49] Meester, R. and Roy, R., *Continuum Percolation,* Cambridge University Press 1996.

[50] Megerian, S., Koushanfar, F., Potkonjak, M. and Srivastava, M. B., Worst and best-case coverage in sensor networks, *IEEE Transactions on Mobile Computing* **4** (2005) 84–92.

[51] Meguerdichian, S., Koushanfar, F., Potkonjak, M. and Srivastava, M. B., Coverage problems in wireless ad-hoc sensor networks, in: *Proc. IEEE INFOCOM* 2001, 1380–1387.

[52] Mitchell, J. S. B. and Papadimitriou C. H., The Weighted Region Problem: Finding Shortest Paths Through a Weighted Planar Subdivision, *Journal of the ACM* **38** (1991) 18–73.

[53] Nain, P., Towsley, D., Liu, B. and Liu, Z., Properties of random direction models, in: *Proc. IEEE INFOCOM* 2005, 1897–1907.

[54] Pac, M. R., Erkmen, A. M. and Erkmen, I., Scalable self-deployment of mobile sensor networks: a fluid dynamics approach, in: *IROS '06* (Proc. IEEE Conference on Intelligent Robots and Systems) 2006, 1446–1451.

[55] Penrose, M., *Random Geometric Graphs,* Oxford University Press 2003.

[56] Philips, T. K., Panwar, S. S. and Tantawi, A. N., Connectivity properties of a packet radio network model, *IEEE Transactions on Information Theory* **35**(5) (1989), 1044–1047.

[57] Poduri, S. and Sukhatme, G. S., Constrained coverage for mobile sensor networks, in: *ICRA '04* (Proc. IEEE International Conference on Robotics and Automation) 2004, 165–171.

[58] Ram, S. S., Manjunath, D., Iyer, S. K. and Yogeshwaran, D., On path coverage properties of random sensor networks, *IEEE Transactions on Mobile Computing* **6** (2007) 494–506.

[59] Ravelomanana, V., Extremal properties of three-dimensional sensor networks with applications, *IEEE Transactions on Mobile Computing* **3** (2004) 246–257.

[60] Ruckle, W. H., *Geometric Games and their Applications,* Pitman Publishing 1983.

[61] Shakkottai, S., Srikant, R. and Shroff, N., Unreliable sensor grids: coverage, connectivity, and diameter, in: *Proc. IEEE INFOCOM* 2003, 1073–1083.

[62] Stoyan, D., Kendall, W. S. and Mecke, J., *Stochastic Geometry and its Applications,* second edition, John Wiley & Sons 1995.

[63] Tan, J., Xi, N., Sheng, W. and Xiao, J., Modeling multiple robot systems for area coverage and cooperation, in: *Proc. IEEE Conference on Robotics and Automation* 2004, 2568–2573.

[64] Tian, D. and Georganas, N. D., A coverage-preserving node scheduling scheme for large wireless sensor networks, in: *WSNA '02* (ACM Workshop on Wireless Sensor Networks and Applications) 2002, 32–41.

[65] Wan, P.-J. and Yi, C.-W., Coverage by randomly deployed wireless sensor networks, *IEEE Transactions on Information Theory* **52**(6) (2006) 2658–2669.

[66] Wang, G., Cao, G., La Porta, T., Movement-assisted sensor deployment, in: *Proc. IEEE INFOCOM 2004,* 2469–2479, see also *IEEE Transactions on Mobile Computing* **6** (2006) 640–652.

[67] Wang, G., Cao, G. and La Porta, T., Proxy-based sensor deployment for mobile sensor networks, in: *MASS 2004* (Proc. IEEE Conference on Mobile Ad-hoc and Sensor Systems), 2004, 493–502.

[68] Wang, G., Cao, G., La Porta, T., Zhang, W., Sensor relocation in mobile sensor networks, in: *Proc. IEEE INFOCOM 2005,* 2302–2312.

[69] Wang, Y., Hu, C. and Tseng, Y., Efficient deployment algorihms for ensuring coverage and connectivity in wireless sensor networks, in: *WICON '05* (Proc. International Conference on Wireless Internet) 2005, 114–121.

[70] Wang, W., Srinivasan, V. and Chua, K.-C., Trade-offs between mobility and density for coverage in wireless sensor networks, in: *MobiCom '07* (Proc. ACM Conference on Mobile Computing) 2007, 39–48.

[71] Xing, G., Wang, X., Zhang, Y., Lu, C., Pless, R. and Gill, C., Integrated coverage and connectivity configuration for energy conservation in sensor networks, *ACM Transactions on Sensor Networks* **1** (2005) 36–72.

[72] Zhang, H. and Hou, J., On deriving the upper bound for the α-lifetime for large sensor networks, in: *MobiHoc '04* (Proc. ACM Conference on Mobile and Ad-Hoc Networking and Computing) 2004, 121–132.

[73] Zhang, M., Du, X. and Nygard, K., Improving coverage performance in sensor networks by using mobile sensors, in: *Proc. IEEE MILCOM 2005,* 7p.

[74] Zhao, F. and Guibas, L., Wireless Sensor Networks: An Information Processing Approach, Morgan Kaufmann 2004.

[75] Zou, Y. and Chakrabarty, K., Sensor deployment and target localization based on virtual forces, in: *Proc. IEEE INFOCOM 2003,* 1293–1303.

[76] Zou, Y. and Chakrabarty, K., Coverage-oriented sensor deployment, in: *Distributed Sensor Networks,* S. S. Iyengar, R. R. Brooks, eds., CRC Press 2005, 483–504.

Peter Brass

Department of Computer Science,
The City College of New York,
138th Street at Convent Avenue,
New York, NY 10031

e-mail: peter@cs.ccny.cuny.edu

BOLYAI SOCIETY
MATHEMATICAL STUDIES, 24

Geometry –
Intuitive, Discrete, and Convex
pp. 109–157.

APPLICATIONS OF AN IDEA OF VORONOĬ,
A REPORT

PETER M. GRUBER

In memoriam László Fejes Tóth (1915–2005)

The idea of Voronoĭ's proof of his well-known criterion that a positive definite quadratic form is extreme if and only if it is eutactic and perfect, is as follows: Identify positive definite quadratic forms on $\mathbb{E}^d$ with their coefficient vectors in $\mathbb{E}^{\frac{1}{d}d(d+1)}$. This translates certain problems on quadratic forms into more transparent geometric problems in $\mathbb{E}^{\frac{1}{2}d(d+1)}$ which, sometimes, are easier to solve. Since the 1960s this idea has been applied successfully to various problems of quadratic forms, lattice packing and covering of balls, the Epstein zeta function, closed geodesics on the Riemannian manifolds of a Teichmüller space, and other problems.

This report deals with recent applications of Voronoĭ's idea. It begins with geometric properties of the convex cone of positive definite quadratic forms and a finiteness theorem. Then we describe applications to lattice packings of balls and smooth convex bodies, to the Epstein zeta function and a generalization of it and, finally, to John type and minimum position problems.

1. INTRODUCTION

A classical criterion of Voronoĭ [80, 81, 82] says that a positive definite quadratic form on Euclidean d-space $\mathbb{E}^d$ is (locally) extreme if and only if it is eutactic and perfect. Equivalently, a lattice packing of balls has (locally) maximum density, if it is eutactic and perfect. To prove this result, Voronoĭ identified the positive definite quadratic forms on $\mathbb{E}^d$ with their coefficient vectors in $\mathbb{E}^{\frac{1}{2}d(d+1)}$. Slightly earlier Plücker [59] and Klein [51] used a similar idea in the context of line geometry. By Voronoĭ's method, certain problems on positive definite quadratic forms, resp. on lattice packing of balls in $\mathbb{E}^d$, are translated into more transparent geometric problems in $\mathbb{E}^{\frac{1}{2}d(d+1)}$.

While Voronoï's criterion won immediate recognition and was widely acclaimed, the idea of his proof was ignored for decades. It drew attention only since about 1960. It was applied systematically to the following areas: Barnes, Dickson and the Russian school of the geometry of numbers led by Delone and Ryshkov and their collaborators Stogrin and Dolbilin used it for lattice packing and covering problems. For coverings we add Bambah, Schürmann, Vallentin and the author and refer to [6, 7, 22, 41, 75, 77]. The minimization problem for the Epstein zeta function was at first investigated by the British school of the geometry of numbers (Rankin [62], Cassels [16], Ennola [24] and Montgomery [57]). Later, following a suggestion of Sobolev, who re-discovered the zeta function in the context of numerical integration, this problem was studied by the Russian school, see [21, 64]. Related recent results are due to Sarnak and Strömbergsson [68], Coulangeon [19] and the author [45]. General properties of the density of lattice packings of balls, considered as a function on the space of lattices, were studied by Ash [1], who showed that the density is a Morse function. His work was continued by Bergé and Martinet [12]. Extensions and refinements of Voronoï's results on extremum properties of quadratic forms are due to the school on quadratic forms in Bordeaux. It includes Martinet, Bergé, Bachoc, Nebe and Coulangeon, see the monograph [54] of Martinet. We mention also the contributions of Barnes, Sloane and Conway for which we refer to the comprehensive volume [18]. Extensions to periodic sets are due to Schürmann [75]. The kissing number of a lattice packing is related to the number of closed geodesics on the fundamental torus of the lattice. This observation led Bavard [10] and Schmutz Schaller [69, 70, 71] to investigate the closed geodesics on the Riemannian manifolds of a Teichmüller space.

This article gives an overview of the pertinent work of the author. See, in particular the papers [34, 38, 39, 44, 45] and the joint article with Schuster [47]. We have included also work of other authors. A few results are new. The section headings give a first idea of the results that will be presented:

> The cone of positive definite quadratic forms,
> Weakly eutactic lattices,
> Extremum properties of the lattice packing density,
> Extremum properties of the product of the lattice packing density and its polar,
> Extremum properties of zeta functions,

> Extremum properties of the product of zeta functions and their polars,
> John type results and minimum ellipsoidal shells,
> Minimum position problems.

These results belong to the geometry of numbers, to convex geometry and to the asymptotic theory of normed spaces. In many cases, similar results hold both for Euclidean balls and o-symmetric, smooth convex bodies. If so, the results for balls are presented in more detail since in some cases they are more far reaching and have classical arithmetic interpretations in terms of positive definite quadratic forms. While the results in different areas seem to be unrelated, they are bound together by their outlook and the method of proof. One may speculate, whether they are related in a deeper sense. For one such relation see Corollary 12. A few proofs have been included. This was done in case of new results, or to illustrate the technique of proof.

Since this is a report, the material is organized as follows: For each topic the definitions, the results and the comments are put together, while the proofs are presented later and may well be skipped.

For general information on the geometry of numbers, on positive quadratic forms and on convex geometry we refer to the author and Lekkerkerker [46], Conway and Sloane [18], Zong [84], Martinet [54], the author [37], and Schürmann [75].

Let the symbols tr, dim, bd, relint, $\text{relint}_\mathcal{S}$, pos, lin, conv, $\|\cdot\|$, $\cdot$, V, B^d, S^{d-1}, ^{T}, $^\perp$ stand for trace, dimension, boundary, interior relative to the affine hull, interior relative to the linear subspace $\mathcal{S}$, positive(=non-negative), linear and convex hull, Euclidean norm, inner product, volume, unit ball and unit sphere in Euclidean d-space $\mathbb{E}^d$, transposition, and orthogonal complement.

2. THE CONE OF POSITIVE DEFINITE QUADRATIC FORMS

Most results in this report deal in one way or another with positive definite quadratic forms. In some cases geometric properties of the cone $\mathcal{P}^d$ of positive definite quadratic forms or of certain subsets of it are indispensable tools for the proofs. It thus seems justified to begin this overview with an investigation of geometric properties of the cone $\mathcal{P}^d$.

A (real) quadratic form on $\mathbb{E}^d$,

$$q(x) = \sum a_{ik} x_i x_k, \quad x \in \mathbb{E}^d,$$

its (real) symmetric $d \times d$ coefficient matrix

$$A = (a_{ik})$$

and its (real) coefficient vector

$$\left(a_{11}, \ldots, a_{1d}, a_{22}, \ldots, a_{2d}, \ldots, a_{dd}\right)^{T}$$

in $\mathbb{E}^{\frac{1}{2}d(d+1)}$ may be identified. The family of all positive definite quadratic forms on $\mathbb{E}^{d}$ then corresponds to an open convex cone $\mathcal{P}^{d}$ in $\mathbb{E}^{\frac{1}{2}d(d+1)}$ with apex at the origin O, the *cone of positive definite quadratic forms*. The closure $\mathcal{Q}^{d}$ of $\mathcal{P}^{d}$ is the *cone of positive semi-definite quadratic forms* on $\mathbb{E}^{d}$. The cones $\mathcal{P}^{d}$ and $\mathcal{Q}^{d}$, certain polyhedra and unbounded convex bodies in $\mathcal{P}^{d}$, as well as polyhedral subdivisions of $\mathcal{P}^{d}$ play an important role in the geometric theory of positive definite quadratic forms, including reduction theory.

Thus, $\mathcal{P}^{d}$ and $\mathcal{Q}^{d}$ appear as natural objects of investigation. To our surprise, we found only a few pertinent results, due to Ryshkov and Baranovskiĭ [67], Ryshkov [66], Bertraneu and Fichet [14], Barvinok [8], Wickelgren [83] and the author [39]. Ryshkov and Baranovskiĭ showed that the group of linear automorphisms of $\mathcal{P}^{d}$ is transitive on $\mathcal{P}^{d}$. Ryshkov seems to have proved that each linear automorphism of $\mathcal{P}^{d}$ is of a particularly simple form. Wickelgren characterized the linear automorphisms of the Ryshkov polyhedron $\mathcal{R}^{d}$ in $\mathcal{P}^{d}$ and Bertraneu and Fichet gave a description of the extreme faces of $\mathcal{Q}^{d}$ and, as a consequence, showed that the lattice of extreme faces of $\mathcal{Q}^{d}$ is isomorphic to the lattice of linear subspaces of $\mathbb{E}^{d}$ and, thus, modular.

In this section, we report on the results of the author [39], beginning with an analog for exposed faces of the result of Bertraneu and Fichet. The next result says that the exposed and the extreme faces of $\mathcal{Q}^{d}$ coincide. These results then are used as tools for the proofs of all further results: First, extending well-known notions for polytopes, flag transitivity of the group of all orthogonal transformations and neighborliness properties of the convex cone $\mathcal{Q}^{d}$ are studied. Then we investigate singularity properties of boundary points and faces of $\mathcal{Q}^{d}$, and show the simple fact that $\mathcal{Q}^{d}$ is self-dual. Finally, the group of isometries of $\mathcal{Q}^{d}$ will be described. Each isometry is generated by an orthogonal transformation of $\mathbb{E}^{d}$.

Extreme and Exposed Faces of $\mathcal{Q}^d$

An *extreme face* or *face* $\mathcal{F}$ of the cone $\mathcal{Q}^d$ is a subset of $\mathcal{Q}^d$ with the following property: If a relative interior point of a line segment in $\mathcal{Q}^d$ is contained in $\mathcal{F}$, then the whole line segment is contained in $\mathcal{F}$. The empty set $\emptyset$ and the cone $\mathcal{Q}^d$ are faces of $\mathcal{Q}^d$. Each extreme face of $\mathcal{Q}^d$ is itself a closed convex cone. A special face is an *exposed face*, i.e. the intersection of $\mathcal{Q}^d$ with a support hyperplane. To simplify, also $\emptyset$ and $\mathcal{Q}^d$ are said to be exposed.

For $u \in \mathbb{E}^d$ define the *tensor product* $u \otimes u$ to be the symmetric $d \times d$ matrix $u\,u^T \in \mathbb{E}^{\frac{1}{2}d(d+1)}$. (The linear mapping $x \to u \otimes u\, x$ maps $x \in \mathbb{E}^d$ onto the point $(u \cdot x)u$ and, if u is a unit vector, this is the orthogonal projection of x onto the line $\lim\{u\}$.)

Theorem 1. *Let $\mathcal{F} \subseteq \mathcal{Q}^d$. Then the following properties (i) and (ii) of the set $\mathcal{F}$ are equivalent:*

  (i) $\mathcal{F}$ *is an exposed face of* $\mathcal{Q}^d$.
  (ii) *There is a linear subspace S of $\mathbb{E}^d$ such that* $\mathcal{F} = \mathrm{pos}\,\{u \otimes u : u \in S\}$.

Moreover,

  (iii) *if (ii) holds, then* $\dim \mathcal{F} = \frac{1}{2}c(c+1)$, *where* $c = \dim S$.

Theorem 2. *Each extreme face of $\mathcal{Q}^d$ is exposed.*

Since by Theorem 2, extreme and exposed faces coincide, from now on we will speak simply of *faces* of $\mathcal{Q}^d$.

The Face Lattice of $\mathcal{Q}^d$

The above results, which show that the faces of $\mathcal{Q}^d$ can be represented in a particularly simple way, lead to a series of properties of $\mathcal{Q}^d$.

An (algebraic) lattice $\langle L, \vee, \wedge \rangle$ is *modular* if it satisfies the *modular law*,

$$(l \wedge m) \vee n = l \wedge (m \vee n) \quad \text{for} \quad l, m, n \in L.$$

It is *orthomodular,* if it has 0 and 1 and for each $l \in L$ there is an *ortho-complement,* i.e. an element $l^\perp \in L$ such that

$$l \vee l^\perp = 1, \quad l \wedge l^\perp = 0, \quad (l^\perp)^\perp = l \quad \text{and} \quad l \le m \Rightarrow l^\perp \ge m^\perp,$$

and satisfies the *orthomodular law,*

$$l \leq m \ \Rightarrow \ m = l \vee (m \wedge l^{\perp}) \quad \text{for} \quad l, m \in L.$$

It is well-known that the family of all linear subspaces of $\mathbb{E}^d$, including $\emptyset = 0$ and $\mathbb{E}^d = 1$, with the following definitions of $\wedge$, $\vee$ is a lattice with 0 and 1:

$$S \wedge T = S \cap T,$$

$$S \vee T = \bigcap \{U \ : \ U \text{ linear subspace of } \mathbb{E}^d \text{ with } S, T \subseteq U\} = S + T$$

for linear subspaces S, T of $\mathbb{E}^d$.

This lattice is both modular and orthomodular. The family of all faces of $\mathcal{Q}^d$, including $\emptyset$ and $\mathcal{Q}^d$, is a lattice with respect to the following lattice operations $\wedge$, $\vee$:

$$\mathcal{F} \wedge \mathcal{G} = \mathcal{F} \cap \mathcal{G},$$

$$\mathcal{F} \vee \mathcal{G} = \bigcap \{\mathcal{H} \ : \ \mathcal{H} \text{ face of } \mathcal{Q}^d \text{ with } \mathcal{F}, \mathcal{G} \subseteq \mathcal{H}\}$$

for faces $\mathcal{F}$, $\mathcal{G}$ of $\mathcal{Q}^d$.

Since by Theorem 1 and 2, these lattices are isomorphic, we get the following result:

Corollary 1. *The lattice of all faces of $\mathcal{Q}^d$ is modular and orthomodular.*

The Flag Transitivity of $\mathcal{Q}^d$

For $d \times d$ matrices $A = (a_{ik})$ and $B = (b_{ik})$ in $\mathbb{E}^{\frac{1}{2}d(d+1)}$ or E^{d^2} define an inner product and a norm by $A \cdot B = \sum a_{ik} b_{ik}$ and $\|A\| = \left(\sum a_{ik}^2 \right)^{\frac{1}{2}}$. A group of transformations which map $\mathcal{P}^d$ or $\mathcal{Q}^d$ onto itself is called a *group of automorphisms* or *symmetries* of $\mathcal{P}^d$ or $\mathcal{Q}^d$. If the transformations are linear, orthogonal or isometric we speak of *linear* or *orthogonal automorphisms,* or, of a *group of isometries* with respect to the norm just defined. Then the following holds: Let U be an orthogonal transformation of $\mathbb{E}^d$. Then the transformation

$$\mathcal{U} \ : \ A \to UAU^T \text{ for } A \in \mathcal{P}^d \text{ or } \mathcal{Q}^d, \text{ respectively,}$$

is an orthogonal automorphism of $\mathcal{P}^d$ or $\mathcal{Q}^d$, respectively.

Extending the definition for convex polytopes, a sequence $\mathcal{F}_1, \mathcal{F}_2, \ldots,$ $\mathcal{F}_{d-1}$ of faces of $\mathcal{Q}^d$ is called a *flag* or a *tower* of $\mathcal{Q}^d$, if

$$\mathcal{F}_1 \subseteq \mathcal{F}_2 \subseteq \cdots \subseteq \mathcal{F}_{d-1} \quad \text{and} \quad \dim \mathcal{F}_c = \frac{1}{2}c(c+1) \quad \text{for} \quad c = 1, 2, \ldots, d-1.$$

A group of automorphisms of $\mathcal{Q}^d$ is *flag transitive* if for any two flags

$$\mathcal{F}_1, \mathcal{F}_2, \ldots, \mathcal{F}_{d-1} \quad \text{and} \quad \mathcal{G}_1, \mathcal{G}_2, \ldots, \mathcal{G}_{d-1}$$

there is an automorphism $\mathcal{U}$ in the group such that $\mathcal{U}\mathcal{F}_i = \mathcal{G}_i$ for $i = 1, 2, \ldots, d-1$.

Corollary 2. *The group of orthogonal automorphisms of $\mathcal{Q}^d$ is flag transitive.*

The Neighborliness of $\mathcal{Q}^d$

The notion of neighborliness for a convex polytope, see Grünbaum [48], Ch. 7, can be adapted to the present situation as follows: For $k = 1, 2, \ldots,$ the convex cone $\mathcal{Q}^d$ is said to be *k-almost neighborly*, if the positive (=non-negative) hull of any k extreme rays of $\mathcal{Q}^d$ with endpoint O is contained in a proper face of $\mathcal{Q}^d$.

Corollary 3. $\mathcal{Q}^d$ *is $(d-1)$-, but not d-almost neighborly.*

Polarity and Self-Polarity of $\mathcal{Q}^d$

The *dual* or *polar cone* of the convex cone $\mathcal{Q}^d$ with apex O is the convex cone

$$\mathcal{Q}^{d*} = \left\{ N \in \mathbb{E}^{\frac{1}{2}d(d+1)} : A \cdot N \geq 0 \text{ for } A \in \mathcal{Q}^d \right\},$$

i.e. the (interior) normal cone of $\mathcal{Q}^d$ at its apex O.

Corollary 4. $\mathcal{Q}^d = \mathcal{Q}^{d*}$, *i.e., $\mathcal{Q}^d$ is self-polar.*

How Singular are the Faces of $\mathcal{Q}^d$?

A face $\mathcal{F}$ of $\mathcal{Q}^d$ is *k-singular,* if k is the dimension of the (interior) normal cone of the convex cone $\mathcal{Q}^d$ at (a relative interior point of) $\mathcal{F}$.

Corollary 5. *Let $\mathcal{F}$ be a non-empty proper face of $\mathcal{Q}^d$ with* $\dim \mathcal{F} = \frac{1}{2}c(c+1)$. *Then $\mathcal{F}$ is $\frac{1}{2}(d-c)(d-c+1)$-singular.*

The Isometries of $\mathcal{Q}^d$

The following result shows that the isometries of $\mathcal{Q}^d$ are orthogonal automorphisms and thus are determined by orthogonal transformations of the underlying space $\mathbb{E}^d$. This is a phenomenon which appears also in several other instances in convex geometry, see the survey [35], to which we add Böröczky and Schneider [15] and Schneider [74]. This result shows that, in particular, the space of isometries of $\mathcal{Q}^d$ is rather small.

Theorem 3. *Let $\mathcal{U}$ be a mapping of $\mathcal{Q}^d$ onto itself. Then the following properties are equivalent:*

   (i) *$\mathcal{U}$ is an isometry.*
   (ii) *There is an orthogonal $d \times d$ matrix $U \in \mathbb{E}^{d^2}$ such that $\mathcal{U}A = UAU^T$ for $A \in \mathcal{Q}^d$.*

Conclusion

Remark 1. While $\mathcal{Q}^d$ is far from being a polyhdral cone, it shares many properties with highly symmetric, neighborly, and self dual polyhedral convex cones.

3. Weakly Eutactic Lattices

In later sections we will frequently encounter (geometric) lattices which are eutactic, possibly in a weaker or stronger sense. Thus it is appropriate to give some information on such lattices. Bergé and Martinet [13] and

Bavard [11] gave descriptions of the weakly eutactic lattices in $\mathbb{E}^2$, $\mathbb{E}^3$, $\mathbb{E}^4$. We extract from their results the semi-eutactic lattices in $\mathbb{E}^2$, $\mathbb{E}^3$ and point out their relationship to the Bravais classification of lattices in crystallography. The aforementioned authors, with Ash [2] as a forerunner, showed that in general dimensions there are only finitely many similarity classes of weakly eutactic lattices. We outline a new geometric proof of this result, using the Ryshkov polyhedron.

Eutactic Lattices and the Bravais Classification

Let L be a (geometric) *lattice* in $\mathbb{E}^d$, that is the set of all integer linear combinations of d linearly independent vectors. The volume of the parallelepiped generated by these vectors is the *determinant* $d(L)$ of L. The *set M_L of minimum points*, or the *first layer* of L, consists of all points $l \in L\setminus\{o\}$ with minimum Euclidean norm. The lattice L is called *weakly eutactic, semi-eutactic, eutactic, strongly eutactic,* or *perfect* with respect to B^d, or $\|\cdot\|$, if

$$I = \sum_{l \in M_L} \lambda_l\, l \otimes l \text{ with suitable } \lambda_l \begin{cases} \text{real} \\ \geq 0 \\ > 0 \\ = \text{const} \end{cases}, \quad \text{resp.}$$

$$\mathbb{E}^{\frac{1}{2}d(d+1)} = \lin\{u \otimes u \,:\, u \in M_L\}.$$

Note that any perfect lattice is weakly eutactic.

The *Bravais classification* of lattices is used in crystallography and classifies lattices by their groups of orthogonal automorphisms which keep the origin o fixed. In dimensions 2 and 3 there are 5, resp. 14 Bravais classes of lattices. For more information see Erdös, Gruber and Hammer [25] and Engel [23]. The *kissing number* of L is the number of minimum points.

Theorem 4. *The following is a list of the similarity classes of the semi-eutactic lattices in $\mathbb{E}^2$ and $\mathbb{E}^3$, containing the symbols of their Bravais classes, their usual names, their eutaxy type, a remark whether they are* *perfect, and their kissing number.*

$d = 2$:	*tp*	*square*	*strongly cutactic*	4
	hp	*hexagonal*	*strongly eutactic perfect*	6
$d = 3$:	*cP*	*cubic primitive*	*strongly eutactic*	6
	hP	*special hexagonal primitive*	*eutactic*	8
	cF	*cubic face centered*	*strongly eutactic perfect*	12
	cI	*cubic body centered*	*strongly eutactic*	8
	tI	*special tetragonal body centered*	*eutactic*	8

These lattices in $\mathbb{E}^2$ and $\mathbb{E}^3$ make up certain Bravais types (tp, hp, cP, cF, cI), or form a subset of a Bravais type (hP, tI). In the latter case we have added the adjective 'special'.

Theorem 5. *There are only finitely many similarity classes of weakly eutactic, resp. perfect lattices in $\mathbb{E}^d$.*

Since each perfect lattice is weakly eutactic, the result for perfect lattices follows from that for weakly eutactic lattices.

Open Problems

In the later Corollaries 6 and 11 there are specified the families of Bravais classes corresponding to those lattices in $\mathbb{E}^2$ and $\mathbb{E}^3$ which have particular extremum properties. It would be of interest, to know whether there are other properties of lattices which lead to the same families of Bravais classes.

Problem 1. Specify geometric properties of lattices which single out the Bravais classes

$$\{hp\}, \ \{tp, hp\}, \ \{cF\}, \ \{cP, cF, cI\}, \ \{hP, cP, cI, cF, tI\}$$

among the 5 Bravais classes for $d = 2$, and the 14 Bravais classes for $d = 3$, respectively. Is there a connection between such properties and extremum properties of the density of lattice packings of balls or the Epstein zeta function?

Problem 2. What is the precise relation of similarity classes of weakly eutactic lattices and Bravais classes of lattices in $\mathbb{E}^d$ for general d?

Outline of the Proof of Theorem 5

We begin with some preparations. A lattice L may be represented in the form $L = B\mathbb{Z}^d$, where B is a non-singular $d \times d$-matrix and $\mathbb{Z}^d$ is the integer lattice. The columns of the matrix B then form a basis of L. The positive definite quadratic form $q(x) = (Bx)^2 = B^T B \cdot x \otimes x$ then is the *metric form* (of the basis matrix B) of L. The metric forms of L are unique up to equivalence. Conversely, a quadratic form q in $\mathcal{P}^d$ can be written in the form $q(x) = A \cdot x \otimes x$, where A is a symmetric $d \times d$ matrix. Then q is the metric form of all lattices of the form $RA^{\frac{1}{2}}\mathbb{Z}^d$, where R is orthogonal. The set M_q of *minimum points* of q consists of all $u \in \mathbb{Z}^d \setminus \{o\}$ such that $q(u)$ is minimum. We have the following dictionary:

$L = B\mathbb{Z}^d$	$q(x) = B^T B \cdot x \otimes x = A \cdot x \otimes x$
class of all lattices L	class of all quadratic forms in $\mathcal{P}^d$
similar to L	equivalent to a multiple of q
$M_L = BM_q$	$M_q = B^{-1}M_L$
$I = \displaystyle\sum_{l \in M_L} \lambda_l \, l \otimes l$	$A^{-1} = \displaystyle\sum_{u \in M_q} \lambda_u \, u \otimes u$
$\mathbb{E}^{\frac{1}{2}d(d+1)} = \operatorname{lin}\{l \otimes l : l \in M_L\}$	$\mathbb{E}^{\frac{1}{2}d(d+1)} = \operatorname{lin}\{u \otimes u : u \in M_q\}$

Call the positive definite quadratic form q *weakly eutactic*, resp. *perfect* if the corresponding lattice L satisfies this condition.

Let $m > 0$. The *Ryshkov polyhedron* $\mathcal{R}^d(m)$ is defined by

$$\mathcal{R}^d(m) = \left\{ A \in \mathcal{P}^d : A \cdot u \otimes u \geq m \text{ for } u \in \mathbb{Z}^d \setminus \{o\} \right\}$$

$$= \bigcap_{\substack{u \in \mathbb{Z}^d \setminus \{o\} \\ \text{primitive}}} \left\{ A \in \mathbb{E}^{\frac{1}{2}d(d+1)} : A \cdot u \otimes u \geq m \right\} \cap \mathcal{P}^d,$$

where u is *primitive* if the points o and u are the only points of $\mathbb{Z}^d$ on the line segment $[o, u]$. The following properties of $\mathcal{R}^d(m)$ can easily be verified, see [37]:

$\mathcal{R}^d(m)$ is a generalized polyhedron, i.e. its intersection with any convex polytope, is a convex polytope.

The facets of $\mathcal{R}^d(m)$ are precisely the sets

$$\mathcal{R}^d(m) \cap \left\{A \in \mathbb{E}^{\frac{1}{2}d(d+1)} : A \cdot u \otimes u = m\right\}, \quad u \in \mathbb{Z}^d\backslash\{o\} \text{ primitive.}$$

bd $\mathcal{R}^d(m)$ is the set of all (coefficient matrices of positive definite quadratic forms) $q \in \mathcal{P}^d$ with (homogeneous) *minimum* $\min\{q(u) : u \in \mathbb{Z}^d\backslash\{o\}\} = m$.

bd $\mathcal{R}^d(m)$ is the disjoint union of the relative interiors of its faces.

The mappings $\mathcal{U} : A \to U^T A U$ for $A \in \mathbb{E}^{\frac{1}{2}d(d+1)}$, where U is an integer $d \times d$ matrix with determinant ± 1, map $\mathcal{R}^d(m)$ onto itself. Two faces $\mathcal{F}$, $\mathcal{G}$ of $\mathcal{R}^d(m)$ are *equivalent* if there is such a mapping $\mathcal{U}$ with $\mathcal{G} = \mathcal{U}\mathcal{F}$.

There are pairwise non-equivalent vertices $V_1, \dots, V_k$ of $\mathcal{R}^d(m)$, such that any other vertex is equivalent to one of these.

The latter result can be used to show that

(1)   there are pairwise non-equivalent faces $\mathcal{F}_1, \dots, \mathcal{F}_p$ of $\mathcal{R}^d(m)$, such that any other face is equivalent to one of these.

Finally, the following hold:

(2)   Let $A \in \operatorname{relint} \mathcal{F}$, where $\mathcal{F}$ is a face of $\mathcal{R}^d(m)$. Then the (interior) normal cone of $\mathcal{R}^d(m)$ at A resp. $\mathcal{F}$, equals $\operatorname{pos}\{u_1 \otimes u_1, \dots, u_j \otimes u_j\}$ where $u_1 \otimes u_1, \dots, u_j \otimes u_j$ are (interior) normal vectors of those facets of $\mathcal{R}^d(m)$ which contain A and thus $\mathcal{F}$.

(3) $q = A \cdot x \otimes x$ is contained precisely in those facets of $\mathcal{R}^d(m)$ with normal vectors $u \otimes u : u \in M_q$.

Let $\delta > 0$. The *discriminant body* $\mathcal{D}^d(\delta)$, is given by

$$\mathcal{D}^d(\delta) = \{A \in \mathcal{P}^d : \det A \geq \delta\}.$$

It is well-known that

(4) $\mathcal{D}^d(\delta)$ is an unbounded, smooth, and strictly convex set in $\mathcal{P}^d$ with non-empty interior. For A in bd $\mathcal{D}^d(\delta)$ the vector A^{-1} is an interior normal vector of bd $\mathcal{D}^d(\delta)$ at A.

After these preparations, the first step of the proof is to show the following

(5)   Let the positive definite quadratic form $q = A \cdot x \otimes x$ have minimum m and determinant δ. Assume that $A \in \operatorname{relint} \mathcal{F}$, where $\mathcal{F}$ is a face of $\mathcal{P}^d$, and that q is weakly eutactic. Then $\mathcal{D}(\delta)$ touches $\mathcal{F}$ at A.

The normal cone of $\mathcal{R}^d(m)$ at q (or $\mathcal{F}$) is $\operatorname{pos} \{u_1 \otimes u_1, \ldots, u_j \otimes u_j\}$, where $u_1 \otimes u_1, \ldots, u_j \otimes u_j$ are normal vectors of those facets of $\mathcal{R}^d(m)$ which contain q (or $\mathcal{F}$), see (2). Hence,

$$\mathcal{F} - A \subseteq \operatorname{lin} \{u_1 \otimes u_1, \ldots, u_j \otimes u_j\}^\perp.$$

By (3), the vectors $u_1, \ldots, u_j$ are the minimum vectors of the quadratic form $q = A \cdot x \otimes x$. The weak eutaxy of q then shows that

$$A^{-1} \in \operatorname{lin} \{u_1 \otimes u_1, \ldots, u_j \otimes u_j\}, \quad \text{or} \quad A^{-1\perp} \supseteq \operatorname{lin} \{u_1 \otimes u_1, \ldots, u_j \otimes u_j\}^\perp.$$

By (4), the matrix A^{-1} is an interior normal vector of $\operatorname{bd} \mathcal{D}^d(\delta)$ at A. Hence,

$$A + A^{-1\perp}$$

is the tangent hyperplane of $\mathcal{D}^d(\delta)$ at A (or q). This together with the earlier inclusions finally yields,

$$\mathcal{F} \subseteq A + \operatorname{lin} \{u_1 \otimes u_1, \ldots, u_j \otimes u_j\}^\perp \subseteq A + A^{-1\perp},$$

concluding the proof of (5).

Since, by (4), $\mathcal{D}^d(\delta)$ is smooth and strictly convex and the surfaces $\operatorname{bd} \mathcal{D}^d(\delta)$, $\delta > 0$, are strictly convex and their union is $\mathcal{P}^d$,

> there is for each face $\mathcal{F}$ of $\mathcal{R}^d(m)$ at most one value of $\delta > 0$ such that $\mathcal{D}(\delta)$ touches $\mathcal{F}$ at a relative interior point.

Thus, by (5), there are at most n weakly eutactic forms contained in the facets $\mathcal{F}_1, \ldots, \mathcal{F}_p$, say $q_1, \ldots, q_n$. Since weak eutaxy is invariant with respect to equivalence and multiplication with positive integers, we see, by (1), that

> the weakly eutactic forms are precisely the forms in $\mathcal{P}^d$ which are equivalent to positive multiples of the forms $q_1, \ldots, q_n$.

Taking into account the above dictionary, the proof is complete.   ∎

4. MAXIMUM PROPERTIES OF THE LATTICE PACKING DENSITY

In this section refined maximum properties of the density of lattice packings of balls and convex bodies are studied. The results obtained are due to the author [44] and refine and extend the classical criterion of Voronoĭ. The notions of semi-eutaxy, eutaxy and perfection are used to characterize lattices which provide lattice packings of balls, resp. of smooth convex bodies with semi-stationary, maximum and ultra-maximum density. It is surprising to observe that maximum and ultra-maximum lattice packings of balls coincide, and that the proof is simple. Relations to the Bravais classification of lattices are specified.

Let C be a *convex body*, i.e. a compact convex subset of $\mathbb{E}^d$ with non-empty interior. We assume that C is o-symmetric and smooth that is, the boundary is of class C^1. Note that for lattice packing problems the assumption of central symmetry of C is not an essential restriction. Let $\|l\|_C$ be the norm on $\mathbb{E}^d$ for which C is the unit ball. Let L be a lattice. The *homogeneous* or *first successive minimum* of L with respect to C is defined by

$$\lambda = \lambda(C, L) = \min \left\{ \|l\|_C : l \in L \backslash \{o\} \right\}.$$

Then the convex bodies $\frac{\lambda}{2}C + l : l \in L$ do not overlap, and thus form a *lattice packing* with *packing lattice* L. This lattice packing is said to be *provided* by L. In the following the *density*

$$\delta(C, L) = \frac{V\left(\frac{\lambda}{2}C\right)}{d(L)} = \frac{\lambda^d V(C)}{2^d d(L)}$$

of this lattice packing will be investigated for given C, as L ranges over the space of all lattices in $\mathbb{E}^d$. Let $M_L = \left\{ l \in L : \|l\|_C = \lambda(C, L) \right\}$ be the set of minimum points or the first layer of L with respect to C.

The connection between Voronoĭ type and maximum properties of a lattice L can roughly be described as follows: The different Voronoĭ type properties of L are equivalent to different positions of a certain convex polytope in $\mathbb{E}^{d^2}$ relative to the origin. (The origin is an exterior point, a point on the relative boundary, or in the relative interior.) These simple geometric properties turn out to be equivalent to different maximum properties of L.

Extremum and Voronoĭ Type Properties

The lattice L is (upper) *semi-stationary, stationary, maximum* or *ultra-maximum* with respect to $\delta(B^d, \cdot)$, if

$$\frac{\delta\big(B^d, (I+A)L\big)}{\delta(B^d, L)} \begin{cases} \leq 1 + o\big(\|A\|\big) \\ = 1 + o\big(\|A\|\big) \\ \leq 1 \\ \leq 1 - \text{const}\|A\| \end{cases} \quad \text{as } A \to O, \ A \in \mathcal{T},$$

where $\mathcal{T}$ is the subspace

$$\mathcal{T} = \left\{A \in \mathbb{E}^{\frac{1}{2}d(d+1)} : \operatorname{tr} A = A \cdot I = 0\right\} = I^{\perp}$$

of $\mathbb{E}^{\frac{1}{2}d(d+1)}$ of codimension 1 with normal vector I. The restriction of A to $\mathcal{T}$ is not essential. It helps to avoid clumsy formulations of our results. An inequality holds as $A \to O$, $A \in \mathcal{T}$, if it holds for all $A \in \mathcal{T}$ with sufficiently small norm. The symbols $o(\cdot)$ and const > 0 depend only on B^d and L.

In order to characterize these maximum properties, we need the Voronoĭ type notions of *semi-eutactic, eutactic,* and *perfect* lattice (or first layer) with respect to B^d, see Sect. 3.

Characterization of Semi-Stationary and Ultra-Extreme Lattices

Theorem 6. *The following properties (i) and (ii) of $\delta(B^d, \cdot)$ and L are equivalent:*

(i) *L is semi-stationary.*
(ii) *L is semi-eutactic.*

Moreover,

(iii) *there is no stationary lattice.*

This result implies, in particular, that $\delta(B^d, \cdot)$, considered as a function on the space of lattices in $\mathbb{E}^d$, is 'not differentiable'. More surprising is the next result, the main result of this section. It shows that maximality and ultra-maximality with respect to $\delta(B^d, \cdot)$ coincide.

Theorem 7. *The following properties of $\delta(B^d, \cdot)$ and L are equivalent:*

(i) *L is ultra-maximum.*

(ii) *L is maximum.*

(iii) *L is perfect and eutactic.*

The equivalence of (ii) and (iii) is Voronoĭ's criterion.

Bravais Types of Lattices with Maximum Properties

The following result is a consequence of Theorems 4, 6 and 7:

Corollary 6. *In $\mathbb{E}^2$ and $\mathbb{E}^3$ it is the following lattices of determinant 1 which are semi-stationary, resp. ultra-maximum with respect to $\delta(B^d, \cdot)$:*

$d = 2:$	*tp*	*square*	$d = 3:$	*cP*	*cubic primitive*
	hp	*hexagonal*		*cF*	*cubic face centered*
				cI	*cubic body centered*
				hP	*special hexagonal primitive*
				tI	*special tetragonal body centered*
$d = 2:$	*hp*	*hexagonal*	$d = 3:$	*cF*	*cubic face centered*

For general d, there are, up to orthogonal transformations, only finitely many lattices of determinant 1 which are semi-stationary, resp. ultra-maximum.

Extremum and Voronoĭ Type Properties

In the following, the results for balls will be extended to convex bodies. Let C be an o-symmetric, smooth convex body. Replace $\mathbb{E}^{\frac{1}{2}d(d+1)}$, $\mathcal{T}$, B^d and $\|\cdot\|$ by $\mathbb{E}^{d^2}$, $\mathcal{S} = \{A \in \mathbb{E}^{d^2}, \text{tr } A = A \cdot I = 0\}$, C and $\|\cdot\|_C$. The density $\delta(C, L)$ and the notions of semi-stationary, etc. lattice with respect to $\delta(C, \cdot)$ are defined as earlier.

In order to specify the versions of eutaxy and perfection which are needed to characterize the maximum properties of $\delta(C, \cdot)$, we proceed as follows: For $l \in \mathbb{E}^d \backslash \{o\}$ let u be the exterior unit normal vector of the smooth convex body $\|l\|_C C$ at its boundary point l, and put $n = l/l \cdot u$. Then the lattice L, or the set M_L of its minimum points with respect to C, is *semi-eutactic, eutactic, strongly eutactic,* or *perfect* with respect to C, if

$$I = \sum_{l \in M_L} \lambda_l \, u \otimes n \quad \text{with suitable} \quad \lambda_l \begin{cases} \geq 0 \\ > 0 \\ = \text{const} \end{cases}, \quad \text{resp.}$$

$$\mathbb{E}^{d^2} = \lim \{ u \otimes n \, : \, l \in M_L \}.$$

Characterization of Semi-Stationary and Ultra-Maximum Lattices

Theorems 6 and 7 now assume the following form.

Theorem 8. *The following properties* (i) *and* (ii) *of* $\delta(C, \cdot)$ *and* L *are equivalent:*

(i) L *is semi-stationary.*

(ii) L *is semi-eutactic.*

Moreover,

(iii) *there is no stationary lattice.*

Theorem 9. *The following properties of* $\delta(C, \cdot)$ *and* L *are equivalent:*

(i) L *is ultra-maximum.*

(ii) L *is perfect and eutactic.*

While there are many semi-stationary lattices for $\delta(C, \cdot)$, for example all lattices which provide lattice packings of C of maximum density, this is not clear for ultra-maximum lattices. See the later discussion.

The *kissing number* $k(C, L)$ of the lattice L with respect to the convex body C or the norm $\| \cdot \|_C$ is $\# M_L$, the number of minimum points. Equivalently, let $\lambda = \lambda(C, L)$. Then $k(C, L)$ is the number of bodies of the lattice packing $\{ \frac{\lambda}{2} C + l \, : \, l \in L \}$, which touch the body $\frac{\lambda}{2} C$. The next estimate is an immediate consequence of Theorem 9.

Corollary 7. *Let* L *be an ultra-maximum lattice for* $\delta(C, \cdot)$. *Then the kissing number satisfies the inequality* $k(C, L) \geq 2d^2$.

Remark 2. If L is ultra-maximum then $k(C, L) \geq 2d^2$ by Corollary 7. A theorem of Minkowski says, if C is strictly convex, then holds $k(C, L) \leq 2^{d+1} - 2$. Since $2^{d+1} - 2 < 2d^2$ for $d = 2, 3, 4$, a strictly convex smooth o-symmetric body can have an ultra-maximum lattice only if $d \geq 5$ – if at all. This explains why it is difficult to specify examples.

The proofs of Theorems 8 and 9 are more complicated than those of Theorems 6 and 7 yet, in essence, follow the same line.

Baire Categories

In the following Baire categories will be used several times. A topological space is *Baire* if any of its meager subsets has dense complement, where a set is *meager* or *of first Baire category,* if it is a countable union of nowhere dense sets. A version of the Baire category theorem says that each locally compact or metrically complete space is Baire. When speaking of *most* or of *typical* elements of a Baire space, we mean all elements, with a meager set of exceptions. The space of all o-symmetric convex bodies endowed with its natural topology is locally compact according to a version of Blaschke's selection theorem and, thus, Baire. See the author [36, 37] for information on Baire type results in convex geometry.

A result of Klee [50] and the author [32] says that most o-symmetric convex bodies are smooth and strictly convex.

Open Problems

Problem 3. Is it true that in all sufficiently high dimensions, for most o-symmetric convex bodies

(i) the maximum and the ultra-maximum lattices coincide,
(ii) the kissing number of each maximum or ultra-maximum lattice equals $2d^2$?

Problem 4. If there are convex bodies with maximum lattices which are not ultra-maximum, characterize the maximum lattices.

What is the situation in the special case of lattices which provide lattice packings of maximum density? A result of the author [33] says that for most o-symmetric convex bodies the kissing number of any lattice packing of maximum density is at most $2d^2$. If, in addition, the lattice is ultra-maximum, then, by Theorem 9, the kissing number is at least $2d^2$ and, thus, equals $2d^2$. An estimate of Swinnerton-Dyer [78] implies that a lattice which provides a packing of maximum density, has kissing number at least $d(d+1)$. For many years I thought that for most o-symmetric convex bodies the kissing number of lattice packings of maximum density is $d(d+1)$. Recently I have changed the opinion:

Problem 5. Show that in all sufficiently high dimensions and for most o-symmetric convex bodies C, the lattice which provides a lattice packing of C of maximum density, has the following properties:

 (i) L is unique up to dilatations.
 (ii) L is eutactic and perfect and, thus, ultra-maximum.
 (iii) L has kissing number $2d^2$.
 (iv) the packing $\{\frac{\lambda}{2}C + l : l \in L\}$, $\lambda = \lambda(C, L)$ is (perhaps?) connected.

Proof of Theorem 7

In order to show the reader the simple yet effective idea underlying the proofs of Theorems 6–9, we present the proof of Theorem 7. We begin with some remarks. Since $\delta(B^d, L)$ does not change if L is replaced by a multiple of it, we may assume that $\lambda(B^d, L) = 1$ and thus, $l = n = u$, $\|l\| = 1$ for $l \in M_L$. Trivially,

$$\lambda(B^d, L) = \min\left\{\|l\| : l \in M_L\right\} < \min\left\{\|l\| : l \in L\backslash(M_L \cup \{o\})\right\}.$$

Note that

$$\|l + Al\|^2 = \|l\|^2\left(1 + 2A \cdot n \otimes n + A^2 \cdot n \otimes n\right) = 1 + 2A \cdot l \otimes l + O\left(\|A\|^2\right)$$

as $A \to O$, $A \in \mathcal{T}$, $l \in M_L$,

$$\det(I + A) = 1 - \frac{1}{2}\|A\|^2 + O\left(\|A\|^2\right) \quad \text{as} \quad A \to O, \quad A \in \mathcal{T}$$

and, since L is discrete,

$$\lambda\left(B^d, (I + A)l\right) = \min\left\{\|l + Al\| : l \in M_L\right\}$$

$$< \min\left\{\|l + Al\| : l \in L\backslash M_L \cup \{o\}\right\}$$

$$\text{if} \quad A \in \mathcal{T} \quad \text{has sufficiently small norm.}$$

Thus we get,

$$\delta\big(B^d, (I+A)L\big) = \frac{\lambda\big(B^d, (I+A)L\big)^d V(B^d)}{2^d d\big((I+A)L\big)}$$

$$= \frac{V(B^d)}{2^d d(L)} \frac{\min\big\{\, \|l + Al\|^2 \,:\, l \in M_L \big\}^{\frac{d}{2}}}{\det(I+A)}\,.$$

$$= \frac{V(B^d)}{2^d d(L)} \min\big\{\, 1 + 2A \cdot n \otimes n + A^2 \cdot n \otimes n \,:\, l \in M_L \big\}^{\frac{d}{2}}$$

$$\times \big(1 - \tfrac{1}{2}\|A\|^2 + O\big(\|A\|^3\big)\big)^{-1}$$

$$= \frac{\lambda(B^d, L)^d V(B^d)}{2^d d(L)} \big(\min\{1 + dA \cdot l \otimes l \,:\, l \in M_L\} + O\big(\|A\|^2\big)\big)$$

$$= \delta(B^d, L)\big(1 + d\min\{A \cdot l \otimes l \,:\, l \in M_L\} + O\big(\|A\|^2\big)\big)$$

as $A \to O,\ A \in \mathcal{T}.$

(i) $\Leftrightarrow$ (iii):

L is ultra-maximum

$\Leftrightarrow \delta\big(B^d, (I+A)L\big) = \delta(B^d, L)\big(1 + \min\{dA \cdot l \otimes l \,:\, l \in M_L\} + O\big(\|A\|^2\big)\big)$

$\leq \delta(B^d, L)\big(1 - \mathrm{const}\|A\|\big)$ as $A \to O, A \in \mathcal{T}$

$\Leftrightarrow 1 + d\min\{A \cdot l \otimes l\} \leq 1 - \mathrm{const}\|A\|$ as $A \to O,\ A \in \mathcal{T}$

$\Leftrightarrow \min\{A \cdot l \otimes l \,:\, l \in M_L\} \leq -\mathrm{const}\|A\|$ for $A \in \mathcal{T}$

$\Leftrightarrow \min\big\{A \cdot (l \otimes l)^{\mathcal{T}} \,:\, l \in M_L\big\} < 0$ for all $A \in \mathcal{T}\backslash\{O\}$

$\Leftrightarrow O = I^{\mathcal{T}} \in \mathrm{relint}_{\mathcal{T}} \mathrm{conv}\big\{(l \otimes l)^{\mathcal{T}} \,:\, l \in M_L\big\}$

$\Leftrightarrow I \in \mathrm{pos}\{l \otimes l \,:\, l \in M_L\}$ since $I \cdot l \otimes l = l \cdot l > 0,$

$$\mathbb{E}^{\frac{1}{2}d(d+1)} = \mathrm{lin}\left\{l \otimes l \,:\, l \in M_L\right\}$$

$\Leftrightarrow L$ is eutactic and perfect.

(ii) $\Leftrightarrow$ (iii):

    L is maximum

$\Leftrightarrow \delta\big(B^d, (I+A)L\big) \leq \delta(B^d, L)$ for $A \to O$, $A \in \mathcal{T}$

$\Leftrightarrow \min\left\{1 + 2A \cdot l \otimes l + A^2 \cdot l \otimes l \,:\, l \in M_L\right\}$

$$\leq \left(1 - \frac{1}{2}\|A\|^2 + O\big(\|A\|^3\big)\right)^{\frac{2}{d}} \quad \text{as } A \to O,\ A \in \mathcal{T}$$

$\Leftrightarrow \min\left\{A \cdot l \otimes l + \dfrac{1}{2}A^2 \cdot l \otimes l \,:\, l \in M_L\right\} \leq 1 - \dfrac{1}{d}\|A\|^2 + O\big(\|A\|^3\big)$

    as $A \to O$, $A \in \mathcal{T}$

$\Rightarrow \min\{A \cdot l \otimes l \,:\, l \in M_L\} \leq -\dfrac{1}{d}\|A\|^2$ as $A \to O$, $A \in \mathcal{T}$

$\Rightarrow \min\{A \cdot l \otimes l \,:\, l \in M_L\} < 0$ for $A \in \mathcal{T}\backslash\{O\}$

    $\cdots$

$\Rightarrow L$ is eutactic and perfect

$\Rightarrow L$ is ultra-maximum. ∎

5. MAXIMUM PROPERTIES OF THE PRODUCT OF THE LATTICE PACKING DENSITY AND ITS POLAR

This section deals with refined maximum properties of the expressions

$$\delta(C, \cdot)\, \delta(C^*, \cdot^*)$$

in a neighborhood of a lattice L, where * indicates polarity. In particular, we consider the case when $C = B^d$ and thus $C^* = B^d$, which has been studied

before by Bergé and Martinet [12]. There are many results of a related type
in the geometry of numbers, see [46], Sect. 14, in convex geometry, see [37],
and in the asymptotic theory of normed spaces, see Gruber [38], and Sects. 8
and 9. Related results hold for

$$\frac{\delta(C, \cdot)}{\delta(C, L)} + \frac{\delta(C^*, \cdot^*)}{\delta(C^*, L^*)}.$$

Since the results for the weighted sum are very similar to those for the
product of the densities, only results for the latter will be presented.

Let C be an o-symmetric, smooth and strictly convex body and L a
lattice in $\mathbb{E}^d$. The *polar body* C^* and the *polar lattice* L^* are defined by

$$C^* = \{y \in \mathbb{E}^d : x \cdot y \leq 1 \text{ for } x \in C\}, \quad L^* = \{m \in \mathbb{E}^d : l \cdot m \in \mathbb{Z} \text{ for } l \in L\}.$$

Then $d(L)\, d(L^*) = 1$ and $(BL)^* = B^{-T}L^*$ for non-singular $B \in \mathbb{E}^{d^2}$.

Dual Maximum and Voronoĭ Type Properties

The lattice L is *dual semi-stationary, dual stationary, dual maximum*, or
dual ultra-maximum with respect to the product $\delta(B^d, \cdot)\, \delta(B^d, \cdot^*)$, if

$$\frac{\delta\big(B^d, (I+A)L\big)\, \delta\big(B^d, ((I+A)L)^*\big)}{\delta(B^d, L)\, \delta(B^d, L^*)} \left\{ \begin{array}{l} \leq 1 + o(\|A\|) \\ = 1 + o(\|A\|) \\ \leq 1 \\ \leq 1 - \text{const}\|A\| \end{array} \right\}$$

as $A \to O, \ A \in \mathcal{S}$.

The lattice L or its first layer M_L is *dual semi-eutactic, dual eutactic, dual
strongly eutactic* or *dual perfect* with respect to B^d, if

$$\sum_{l \in M_L} \lambda_l\, l \otimes l = \sum_{m \in M_{L^*}} \mu_m\, m \otimes m \neq O \text{ with suitable } \lambda_l,$$

$$\mu_m \left\{ \begin{array}{l} \geq 0 \\ > 0 \\ = \text{const} \end{array} \right\}, \quad \text{resp.}$$

$$\mathbb{E}^{d^2} = \text{lin}\left(\{l \otimes l : l \in M_L\} \cup \{m \otimes m : m \in M_{L^*}\}\right),$$

where const for λ_l may be different from const for μ_m.

Characterization of Dual Semi-stationary and Dual Ultra-Extreme Lattices

In analogy to Theorems 6 and 7, we have the following results:

Theorem 10. *The following properties (i) and (ii) of $\delta(B^d, \cdot)\, \delta(B^d, \cdot^*)$ and L are equivalent:*

(i) *L is dual semi-stationary.*
(ii) *L is dual semi-eutactic.*

Theorem 11. *The following properties of $\delta(B^d, \cdot)\, \delta(B^d, \cdot^*)$ and L are equivalent:*

(i) *L is dual ultra-maximum.*
(ii) *L is dual maximum.*
(iii) *L is dual perfect and dual eutactic.*

The equivalence of (ii) and (ii i) is due to Bergé and Martinet [12].

Extension to Smooth Convex Bodies

Theorems 10 and 11 continue to hold with C instead of B^d, omitting statement (ii) in Theorem 11.

Proof of Theorem 11

To show the additional arguments needed for the proofs of these results, we present the proof of Theorem 11.

First, some tools are put together. Since $A = A^T$ for $A \in \mathcal{T}$, we have

$$\left((I+A)L\right)^* = (I + A)^{-T}L^* = \left(I - A + A^2 - + \cdots\right)L^* \quad \text{for} \quad A \in \mathcal{T},\ \|A\| < 1.$$

Thus,

$$(l + Al)^2 = \|l\|^2\left(1 + 2A \cdot n \otimes n + (An)^2\right),$$

$$\left(m - Am + A^2m - + \cdots\right)^2$$

$$= \|m\|^2\left(1 - 2A \cdot p \otimes p + 3A^2 \cdot p \otimes p + O\left(\|A^3\|\right)\right),$$

where $n = l/\|l\|$, $p = m/\|m\|$. This, together with the definitions of λ and δ, yields the following equalities,

$$\delta\big(B^d, (I+A)L\big)$$

$$= \delta(B^d, L) \min\big\{(l + Al)^2 \;:\; l \in M_L\big\}^{\frac{d}{2}} \det (I+A)^{-1}$$

$$= \delta(B^d, L) \min\big\{1 + 2dA \cdot n \otimes n + O(\|A\|^2) \;:\; l \in M_L\big\}$$

$$\Big(1 - \frac{1}{2}\|A\|^2 + O(\|A\|^3)\Big)^{-1}$$

$$= \delta(B^d, L)\big(1 + d\min\{A \cdot n \otimes n \;:\; l \in M_L\} + O(\|A\|^2)\big)$$

as $A \to O$, $A \in \mathcal{T}$, where $O(\|A\|^2) \ge 0$,

$$\delta\big(B^d, ((I+A)L)^*\big)$$

$$= \delta(B^d, L^*) \min\big\{1 - dA \cdot p \otimes p + O(\|A\|^2) \;:\; m \in M_{L^*}\big\}$$

$$\big(\det (I+A)^{-T}\big)^{-1}$$

$$= \delta(B^d, L^*)\big(1 + d\min\{-A \cdot p \otimes p \;:\; m \in M_{L^*}\} + O(\|A\|^2)\big)$$

as $A \to O$, $A \in \mathcal{T}$, where $O(\|A\|^2) \ge 0$.

(i) $\Leftrightarrow$ (iii):

L is dual ultra-maximum

$\Leftrightarrow \delta\big(B^d, (I+A)L\big)\, \delta\big(B^d((I+A)L)^*\big)$

$\qquad \le \delta(B^d, L)\, \delta(B^d, L^*)\big(1 - \text{const}\|A\|\big)$

$\qquad$ as $A \to O$, $A \in \mathcal{T}$

$\Leftrightarrow \big(1 + d\min\{\ \} + O(\|A\|^2)\big)\big(1 - d\max\{\ \} + O(\|A\|^2)\big)$

$\qquad \le 1 - \text{const}\|A\|$

$\qquad$ as $A \to O$, $A \in \mathcal{T}$

$\Leftrightarrow \min\{A \cdot n \otimes n \;:\; l \in M_L\} - \max\{A \cdot p \otimes p \;:\; m \in M_{L^*}\}$

$\qquad \le -\text{const}\|A\|$

as $A \to O, \quad A \in \mathcal{T}$

$\Leftrightarrow \min \left\{ A \cdot (n \otimes n)^{\mathcal{T}} : l \in M_L \right\}$

$\quad < \max \left\{ A \cdot (p \otimes p)^{\mathcal{T}} : m \in M_{L^*} \right\}$ for $A \in \mathcal{T} \backslash \{O\}$

$\Leftrightarrow \operatorname{relint conv} \left\{ (n \otimes n)^{\mathcal{T}}, \, l \in M_L \right\}$

$\quad \cap \operatorname{relint conv} \left\{ (p \otimes p)^{\mathcal{T}} : m \in M_{L^*} \right\} \neq \emptyset$

$\Leftrightarrow \displaystyle\sum_{l \in M_L} \lambda_l \, n \otimes n = \sum_{m \in M_{L^*}} \mu_m \, p \otimes p \neq O$ with suitable $\lambda_l, \, \mu_m > O,$

$\quad \mathbb{E}^{\frac{1}{2}d(d+1)} = \operatorname{lin} \left(\{ n \otimes n : l \in M_L \} \cup \{ p \otimes p : m \in M_{L^*} \} \right)$

$\Leftrightarrow L$ is dual eutactic and dual perfect.

(ii) $\Leftrightarrow$ (iii): See [12]. ∎

6. Minimum Properties of Zeta Functions

Let L be a lattice in $\mathbb{E}^d$ with $d(L) = 1$. The *Epstein zeta function* of L then is defined by

$$\zeta(L, s) = \sum_{l \in L \backslash \{o\}} \frac{1}{\|l\|^s} \quad \text{for} \quad s > d.$$

It plays an important role in crystal physics, hydrodynamics, numerical integration and other areas. It has been investigated ever since its discovery by Epstein and its re-discovery by Sobolev in his work on numerical integration. For several applications and in the context of the geometry of numbers a major problem on the zeta function is to study for a fixed $s > d$, for all sufficiently large s, or for all $s > d$ the lattices L with $d(L) = 1$ for which $\zeta(\cdot, s)$ is (locally) minimum.

A *layer* of L consists of all vectors of $L \backslash \{o\}$ with the same norm. Order the layers by the norm of their vectors. The first layer then coincides with the set of minimum points of L with respect to B^d. Delone and Ryshkov [21] showed that a lattice L is minimum with respect to $\zeta(\cdot, s)$ for all sufficiently large s if and only if L is perfect and each layer is strongly eutactic or, in a different terminology, a spherical 2-design. If each layer of L is a spherical 4-design, then L minimizes $\zeta(\cdot, s)$ for each $s > d$, as shown by Coulangeon [19]. A different sufficient condition is due to Sarnak and Strömbergsson [68]. These authors show that many important lattices are minimum for each $s > d$, one example is the Leech lattice.

We characterize the lattices which, for given $s > d$ are stationary and quadratic minimum with respect to $\zeta(\cdot, s)$. This yields characterizations in other cases. Perhaps more important for applications are simple sufficient conditions. We state several such conditions, including one using automorphism groups. Finally, a relation to lattice packing of balls is mentioned. Most of these results can be extended to general zeta functions $\zeta_C(\cdot, s)$.

Minimum and Voronoĭ Type Properties, Spherical Designs and Automorphism Groups

Remark 3. Since $\zeta(\cdot, s)$ and $\zeta_C(\cdot, s)$ have the additional parameter s, it is not surprising that there are more properties needed than mere eutaxy, strong eutaxy, or perfection, to characterize the lattices L which are stationary, minimum, or quadratic minimum with respect to $\zeta(\cdot, s)$ or $\zeta_C(\cdot, s)$. The following stronger forms of eutaxy and perfection, together with automorphism groups, seem to be appropriate tools for such characterizations.

Let $s > d$. Then L is said to be *stationary, minimum,* or *quadratic minimum* with respect to $\zeta(\cdot, s)$, if

$$\frac{\zeta\left(\dfrac{I + A}{\det (I + A)^{\frac{1}{d}}} L, s\right)}{\zeta(L, s)} \quad \left\{ \begin{array}{l} = 1 + o(\|A\|) \\[2mm] \geq 1 \\[2mm] \geq 1 + \mathrm{const}\|A\|^2 \end{array} \right\} \quad \text{as } A \to O, \ A \in \mathcal{T}.$$

Let M be a finite, o-symmetric subset of S^{d-1} and put $\zeta = \zeta(L, s)$. The set M is a *spherical n-design* if the following identity holds for any polynomial $p : \mathbb{E}^d \to \mathbb{R}$ of degree at most n:

$$\int_{S^{d-1}} p(u) \, d\sigma(u) = \frac{1}{\#M} \sum_{l \in M} p(l).$$

Here σ is the usual rotation invariant area measure on S^{d-1}, normalized so that $\sigma(S^{d-1}) = 1$ and $\#$ stands for cardinal number. Venkov [79] showed that

the set M is a spherical n-design if and only if

$$\sum_{l \in M} (l \cdot x)^n = \text{const} \|x\|^n \quad \text{for} \quad x \in \mathbb{E}^d.$$

Let M be a layer of L. Then M is *strongly eutactic* or, after a suitable normalization, a *spherical 2-design*, if it satisfies one of the following equivalent conditions:

$$\sum_{l \in M} \frac{l \otimes l}{\|l\|^2} = \lambda I, \quad \text{or}$$

$$\sum_{l \in M} \frac{(l \cdot x)^2}{\|l\|^2} = \lambda \|x\|^2 \quad \text{for} \quad x \in \mathbb{E}^d, \quad \text{where} \quad \lambda = \frac{\#M}{d}, \quad \text{or}$$

$$\sum_{l \in M} A \cdot l \otimes l = 0 \quad \text{for} \quad A \in \mathcal{T}.$$

The lattice L is *strongly eutactic*, if its first layer is. It is *fully eutactic* with respect to $\zeta(\cdot, s)$, if one of the following equivalent conditions holds:

$$\sum_{l \in L \setminus \{o\}} \frac{l \otimes l}{\|l\|^{s+2}} = \frac{\zeta}{d} I, \quad \text{or} \quad \sum_{l \in L \setminus \{o\}} \frac{(l \cdot x)^2}{\|l\|^{s+2}} = \frac{\zeta}{d} \|x\|^2 \quad \text{for} \quad x \in \mathbb{E}^d, \quad \text{or}$$

$$\sum_{l \in L \setminus \{o\}} \frac{A \cdot l \otimes l}{\|l\|^{s+2}} = 0 \quad \text{for} \quad A \in \mathcal{T}.$$

Refined versions of these notions are the following: a layer M is *ultra-eutactic* or a *spherical 4-design*, if one of the following conditions holds:

$$\sum_{l \in M} \frac{(l \cdot x)^4}{\|l\|^4} = \mu \|x\|^4 \quad \text{for} \quad x \in \mathbb{E}^d, \quad \text{or, equivalently,}$$

$$\sum_{l \in M} \frac{(A \cdot l \otimes l)^2}{\|l\|^4} = \frac{2\mu}{3} \|A\|^2 + \frac{\mu}{3} (\text{tr} A)^2 \quad \text{for} \quad A \in \mathbb{E}^{\frac{1}{2} d(d+1)},$$

$$\text{where} \quad \mu = \frac{3\#M}{d(d+2)}.$$

The lattice L is *completely eutactic* with respect to $\zeta(\cdot, s)$, if the one of the following properties holds:

$$\sum_{l \in L \setminus \{o\}} \frac{(l \cdot x)^4}{\|l\|^{s+4}} = \nu \, \|x\|^4 \text{ for } x \in \mathbb{E}^d, \text{ or, equivalently,}$$

$$\sum_{l \in L \setminus \{o\}} \frac{(A \cdot l \otimes l)^2}{\|l\|^{s+4}} = \frac{2\nu}{3} \, \|A\|^2 + \frac{\nu}{3} (\operatorname{tr} A)^2 \text{ for } A \in \mathbb{E}^{\frac{1}{2}d(d+1)},$$

$$\text{where } \nu = \frac{3\zeta}{d(d+2)}.$$

Finally, the layer M is *perfect*, if

$$\mathbb{E}^{\frac{1}{2}d(d+1)} = \lin \{l \otimes l \; : \; l \in M\}.$$

If the first layer of L is perfect, then L is *perfect*. The *automorphism* or *symmetry group* $\mathcal{A} = \mathcal{A}(L)$ of L is the group of all orthogonal transformations of $\mathbb{E}^d$ which map L onto itself.

Characterization of Stationary and Quadratic Minimum Lattices

Note that in contrast to the situation for densities, for zeta functions a semi-stationary lattice is already stationary.

Theorem 12. *Let $s > d$. Then the following properties of $\zeta(\cdot, s)$ and L are equivalent:*
  (i) *L is stationary for s.*
  (ii) *L is fully eutactic for s.*

Corollary 8. *The following properties of $\zeta(\cdot, \cdot)$ and L are equivalent:*
  (i) *L is stationary for each $s > d$.*
  (ii) *Each layer of L is strongly eutactic.*

Theorem 13. *Let $s > d$. Then the following properties of $\zeta(\cdot, s)$ and L are equivalent:*
  (i) *L is quadratic minimum for s.*
  (ii) *L is fully eutactic for s and satisfies the inequality*

$$\sum_{l \in L \setminus \{o\}} \frac{(A \cdot l \otimes l)^2}{\|l\|^{s+4}} > \frac{2\zeta(L, s)}{d(s+2)} \|A\|^2 \text{ for } A \in \mathcal{T} \setminus \{O\}.$$

Corollary 9. *The following properties of $\zeta(\cdot,\cdot)$ and L are equivalent:*

(i) *L is quadratic minimum for each $s > d$.*

(ii) *Each layer of L is strongly eutactic and*

$$\sum_{l\in L\setminus\{o\}} \frac{(A\cdot l\otimes l)^2}{\|l\|^{s+4}} > \frac{2\zeta(L,s)}{d(s+2)}\|A\|^2 \text{ for each } s > d \text{ and } A\in\mathcal{T}\setminus\{O\}.$$

This yields, in particular, Coulangeon's criterion.

While Theorem 13 yields a characterization of the lattices, which are quadratic minimum for arbitrarily large s, in several cases the following sufficient conditions are more convenient to apply.

Corollary 10. *Each of the following two conditions is sufficient for L to be quadratic minimum with respect to $\zeta(\cdot, s)$ for all sufficiently large s:*

(i) *L is perfect and the automorphism group $\mathcal{A}(L)$ is transitive on the first layer of L.*

(ii) *L is perfect and each layer is strongly eutactic.*

Similarly, each of the following two conditions is sufficient for L to be quadratic minimum for each $s > d$:

(iii) *Each layer of L is ultra-eutactic.*

(iv) *L is completely eutactic for each $s > d$.*

Bravais Types of Lattices with Minimum Properties

The next result is a consequence of Theorem 4 and Corollaries 8 and 10:

Corollary 11. *In $\mathbb{E}^2$ and $\mathbb{E}^3$, it is precisely the following lattices of determinant 1 which are stationary, resp. quadratic minimum with respect to $\zeta(\cdot, s)$ for all $s > d$.*

$d = 2:$ *tp square*		$d = 3:$ *cP*	*cubic primitive*
	hp hexagonal	*cF*	*cubic face centered*
		cI	*cubic body centered,*

resp.

$d = 2:$ *hp hexagonal*	$d = 3:$ *cF*	*cubic face centered.*

For general d, there are, up to orthogonal transformations, only finitely many lattices of determinant 1 which are stationary, resp. quadratic minimum with respect to $\zeta(\cdot, s)$ for all $s > d$.

Zeta Functions and Ball Packing

A lattice which is quadratic minimum with respect to $\zeta(\cdot, s)$ for all sufficiently large s is perfect and each layer is strongly eutactic as can be shown by means of Theorem 13. Hence, by Theorem 7, the following remark holds:

Corollary 12. *Each lattice which is quadratic minimum with respect to $\zeta(\cdot, s)$ for all sufficiently large s, is ultra-maximum with respect to $\delta(B^d, \cdot)$.*

General Lattice Zeta Functions

Our next aim is to extend the above results to a more general type of zeta functions on lattices. Let C be a smooth, o-symmetric convex body, $\| \cdot \|_C$ the corresponding norm on $\mathbb{E}^d$ and L a lattice with $d(L) = 1$. The function ζ_C, defined by

$$\zeta_C(L, s) = \sum_{l \in L \setminus \{o\}} \frac{1}{\|l\|_C^s} \text{ for } s > d,$$

is called a *lattice zeta function* on $\mathbb{E}^d$.

Minimum and Voronoĭ Type Properties

Let $s > d$. The concepts of *stationary, minimum* or *quadratic minimum lattice* with respect to $\zeta_C(\cdot, s)$ are defined as earlier for B^d with $\mathcal{T}$ replaced by the subspace

$$\mathcal{S} = \left\{ A \in \mathbb{E}^{d^2} : \operatorname{tr} A = A \cdot I = 0 \right\}.$$

Similarly, the notions of *layer* and *eutactic, strongly eutactic, fully eutactic* and *perfect lattice* with respect to C or $\zeta_C(\cdot, s)$ are defined as before, but with C, $\| \cdot \|_C$, M_L, $u \otimes n$, ζ_C, $\mathcal{S}$, $\mathbb{E}^{d^2}$ instead of B^d, $\| \cdot \|$, M_L, $n \otimes n$, ζ, $\mathcal{T}$ and $\mathbb{E}^{\frac{1}{2}d(d+1)}$, respectively.

Characterization of Stationary and Quadratic Minimum Lattices

The extension of Theorem 12 is as follows:

Theorem 14. *Let $s > d$. Then the following properties of $\zeta_C(\cdot, s)$ and L are equivalent:*

  (i) *L is stationary for s,*
  (ii) *L is fully eutactic for s.*

Also Theorem 13 can be extended, but the corresponding necessary and sufficient condition for L to be quadratic minimum with respect to $\zeta_C(\cdot, s)$ for given $s > d$ is difficult to check. Thus we prefer to state the following result.

Theorem 15. *Let C be of class C^2. Then the following properties of $\zeta_C(\cdot, \cdot)$ and L are equivalent:*

  (i) *L is quadratic minimum for all sufficiently large s.*
  (ii) *L is perfect and each layer is strongly eutactic.*

Corollary 10, with suitable modifications, holds also for C instead of B^d.

Open Problems

The earlier characterizations and sufficient conditions guarantee in a series of cases that L is stationary, minimum, or quadratic minimum with respect to $\zeta(\cdot, s)$, or $\zeta_C(\cdot, s)$ for a given $s > d$, for all sufficiently large s, or for all $s > d$. The problem arises, to make this family of results complete. We state one particular problem.

Problem 6. If there are lattices, which are minimum (but not quadratic minimum) with respect to $\zeta(\cdot, s)$ or $\zeta_C(\cdot, s)$ for a given $s > d$, for all sufficiently large s, and for all $s > d$, characterize the minimum lattices.

The next problem is related to Problem 5.

Problem 7. Show that in all sufficiently high dimensions, for most *o*-symmetric convex bodies C there are lattices which are quadratic minimum with respect to $\zeta(\cdot, s)$ and $\zeta_C(\cdot, s)$ respectively, for a given s, for all sufficiently large s, or for all $s > d$.

A positive answer to this problem would settle also the question of the existence of convex bodies with eutactic and perfect lattices, see Theorem 15.

Problem 8. Is it true, that in all sufficiently high dimensions and for most o-symmetric convex bodies C, the lattice L with $d(L) = 1$, for which $\zeta_C(\cdot, s)$ attains its absolute minimum for a given s, for all sufficiently large s, or for all $s > d$, has the following properties:

(i) L is unique,
(ii) L is quadratic minimum?

Proof of Theorem 13

To show the idea of the proofs, we present the following proof of Theorem 13. The equality

$$\|l + Al\|^2 = \|l\|^2 (1 + 2A \cdot n \otimes n + A^2 \cdot n \otimes n)$$

$$\text{for} \quad A \in \mathbb{E}^{\frac{1}{2}d(d+1)}, \quad l \in \mathbb{E}^d \setminus \{o\}, \quad n = \frac{l}{\|l\|}$$

implies the formula

$$\frac{1}{\|l + Al\|^s} = \frac{1}{\|l\|^s} \left(1 + 2A \cdot n \otimes n + A^2 \cdot n \otimes n\right)^{-\frac{s}{2}}$$

$$= \frac{1}{\|l\|^s} \left(1 - sA \cdot n \otimes n - \frac{s}{2} A^2 \cdot n \otimes n\right.$$

$$\left. + \frac{s(s+2)}{2} (A \cdot n \otimes n)^2 + O(\|A\|^3)\right) \quad \text{as} \quad A \to O, \ A \in \mathcal{T},$$

which, in turn, yields the following identity, where the summation is over $l \in L \setminus \{o\}$ and ζ stands for $\zeta(L, s)$:

$$\zeta((I + A)L, s)$$

$$= \zeta - sA \cdot \sum \frac{l \otimes l}{\|l\|^{s+2}} - \frac{s}{2} A^2 \cdot \sum \frac{l \otimes l}{\|l\|^{s+2}} + \frac{s(s+2)}{2} \sum \frac{(A \cdot l \otimes l)^2}{\|l\|^{s+4}}$$

$$+ O(\|A\|^3)$$

$$= \zeta - sA \cdot \frac{\zeta}{d} I - \frac{s}{2} A^2 \cdot \frac{\zeta}{d} I + \frac{s(s+2)}{2} \sum \frac{(A \cdot l \otimes l)^2}{\|l\|^{s+4}} + O(\|A\|^3)$$

$$= \zeta - \frac{s\zeta}{2d} \|A\|^2 + \frac{s(s+2)}{2} \sum \frac{(A \cdot l \otimes l)^2}{\|l\|^{s+4}} + O(\|A\|^3)$$

as $A \to O$, $A \in \mathcal{T}$, if L is fully eutactic for s.

Note that

$$\zeta\left(\frac{I+A}{\det (I+A)^{\frac{1}{d}}} L, s\right) = \zeta\big((I+A)L, s\big) \det (I+A)^{\frac{s}{d}}.$$

(i) $\Leftrightarrow$ (ii): Since s is fixed, it is incorporated into const.

   L is quadratic minimum for s

$\Leftrightarrow$ L is stationary for s and

$$\zeta\left(\frac{I+A}{\det (I+A)^{\frac{1}{d}}} L, s\right) = \zeta\big((I+A)L, s\big) \det (I+A)^{\frac{s}{d}}$$

$$\geq \zeta(1 + \text{const}\|A\|^2)$$

$\Leftrightarrow$ L is fully eutactic with respect to $\zeta(\cdot, s)$ (by Theorem 12) and

$$\zeta - \frac{s\zeta}{2d} \|A\|^2 + \frac{s(s+2)}{2} \sum \frac{(A \cdot l \otimes l)^2}{\|l\|^{s+4}}$$

$$\geq \zeta(1 + \text{const}\|A\|^2) \left(1 - \frac{1}{2}\|A\|^2 + O(\|A\|^3)\right)^{-\frac{s}{d}} + O(\|A\|^3)$$

$\Leftrightarrow$ L is fully eutactic for s and

$$\zeta - \frac{s\zeta}{2d} \|A\|^2 + \frac{s(s+2)}{2} \sum \frac{(A \cdot l \otimes l)^2}{\|l\|^{s+4}}$$

$$\geq \zeta + \frac{s\zeta}{2d} \|A\|^2 + \zeta\text{const}\|A\|^2 + \zeta O(\|A\|^3)$$

$\Leftrightarrow$ L is fully eutactic for s and

$$\sum \frac{(A \cdot l \otimes l)^2}{\|l\|^{s+4}} > \frac{2\zeta}{d(s+2)}\|A\|^2 \text{ as } A \to O, \ A \in \mathcal{T}\backslash\{O\}.$$

In the last equivalence the implication $\Rightarrow$ is clear. To see the reverse implication $\Leftarrow$, note that the expression

$$\sum \frac{(A \cdot l \otimes l)^2}{\|l\|^{s+4}} - \frac{2\zeta}{d(s+2)} \|A\|^2$$

may be considered to be a quadratic form in the variable $A \in \mathcal{T}$. It is, obviously, positive definite and thus bounded below by $\text{const} \|A\|^2$ for a suitable constant. ∎

7. MINIMUM PROPERTIES OF THE PRODUCT OF ZETA FUNCTIONS AND THEIR POLARS

Let L be a lattice with $d(L) = 1$. This section deals with minimum properties of the quantity

$$\zeta(\cdot, s)\,\zeta(\cdot^*, s)$$

on the space of lattices of determinant 1. We characterize dual stationary and dual quadratic minimum lattices, both, for B^d and C. Similar results hold for

$$\zeta_C(\cdot, s) + \zeta_{C^*}(\cdot^*, s).$$

Minimum and Voronoï Type Properties

Let $s > d$. The lattice L is *dual stationary*, *dual minimum*, or *dual quadratic minimum* with respect to $\zeta(\cdot, s)\,\zeta(\cdot^*, s)$, if

$$\frac{\zeta\left(\dfrac{I+A}{\det\,(I+A)^{\frac{1}{d}}} L, s\right) \zeta\left(\left(\dfrac{I+A}{\det\,(I+A)^{\frac{1}{d}}} L\right)^*, s\right)}{\zeta(L, s)\,\zeta(L^*, s)} \begin{cases} = 1 + o(\|A\|) \\ \geq 1 \\ \geq 1 + \text{const}\,s^2 \|A\|^2 \end{cases}$$

$$\text{as} \quad A \to O, \quad A \in \mathcal{T}.$$

Let L_i, L_i^* be the layers of L, resp. L^*, $i = 1, 2, \ldots$. We use the abbreviations $\zeta = \zeta(L, s)$ and $\zeta^* = \zeta(L^*, s)$. Call L_i *dual strongly eutactic* with respect to C, C^*, if

$$\frac{1}{\#L_i} \sum_{l \in L_i} \frac{l \otimes l}{\|l\|^2} = \frac{1}{\#L_i^*} \sum_{m \in L_i^*} \frac{m \otimes m}{\|m\|^2}.$$

The lattice L is *dual fully eutactic* with respect to $\zeta\zeta^*$ for s, if

$$\frac{1}{\zeta} \sum_{l \in L \setminus \{o\}} \frac{l \otimes l}{\|l\|^{s+2}} = \frac{1}{\zeta^*} \sum_{m \in L^* \setminus \{o\}} \frac{m \otimes m}{\|m\|^{s+2}}, \quad \text{or, equivalently,}$$

$$\frac{1}{\zeta} \sum_{l \in L \setminus \{o\}} \frac{A \cdot l \otimes l}{\|l\|^{s+2}} = \frac{1}{\zeta^*} \sum_{m \in L^* \setminus \{o\}} \frac{A \cdot m \otimes m}{\|m\|^{s+2}} \quad \text{for} \quad A \in \mathcal{T}.$$

The layer L_i, is *dual ultra-eutactic* with respect to C, C^*, if

$$\frac{1}{\#L_i} \sum_{l \in L_i} \frac{(A \cdot l \otimes l)^2}{\|l\|^4} = \frac{1}{\#L_i^*} \sum_{m \in L_i^*} \frac{(A \cdot m \otimes m)^2}{\|m\|^4} \quad \text{for} \quad A \in \mathcal{T}.$$

Characterization of Dual Stationary and Dual Quadratic Minimum Lattices

From a series of results we select from each of our two extremality types a characterization result and a sufficient condition.

Theorem 16. *Let $s > d$. Then the following properties of $\zeta(\cdot, s)\,\zeta(\cdot^*, s)$ and L are equivalent:*

   (i) *L is dual stationary.*
   (ii) *L is dual fully eutactic.*

Corollary 13. *Each of the following conditions is sufficient for L to be dual stationary with respect to $\zeta(\cdot, s)\,\zeta(\cdot^*, s)$ for each $s > d$.*

   (i) *The first layer of L is perfect and $\mathcal{A}$ operates transitively on it.*
   (ii) *L is dual fully eutactic for any $s > d$.*

For quadratic minimality the result is rather lengthy.

Theorem 17. *Let $s > d$. Then the following properties of $\zeta(\cdot, s)\,\zeta(\cdot^*, s)$ and L are equivalent:*

(i) *L is dual quadratic minimum for s.*

(ii) *L is dual fully eutactic for s and satisfies the inequality,*

$$(s+2)\left(\frac{1}{\zeta}\sum_{l\in L\backslash\{o\}}\frac{(A\cdot l\otimes l)^2}{\|l\|^{s+4}} + \frac{1}{\zeta^*}\sum_{m\in L^*\backslash\{o\}}\frac{(A\cdot m\otimes n)^2}{\|m\|^{s+4}}\right)$$

$$> \frac{4}{\zeta}A^2\sum_{l\in L\backslash\{o\}}\frac{l\otimes l}{\|l\|^{s+2}} + \frac{2s}{\zeta}\left(A\cdot\sum_{l\in L\backslash\{o\}}\frac{l\otimes l}{\|l\|^{s+2}}\right)^2$$

as $A \to O, \; A \in \mathcal{T}\backslash\{O\}$.

Corollary 14. *The following condition is sufficient that L be dual quadratic minimum with respect to $\zeta(\cdot, s)\,\zeta(\cdot^*, s)$ for each $s > d$: Each layer of L is ultra-eutactic and dual ultra-eutactic.*

Zeta Functions and Ball Packing

Also in the duality case, there is a relation between products of zeta functions and densities of ball packings, see [45].

Extension to General Zeta Functions

Finally, we mention that a good many of the duality results for the Epstein zeta function can be extended to the more general lattice zeta functions ζ_C defined by means of a smooth and strictly convex o-symmetric convex body C.

8. JOHN TYPE RESULTS AND MINIMUM ELLIPSOIDAL SHELLS

This and the next section contain results of John type and minimum position results from the asymptotic theory of normed spaces. If not stated otherwise, the results are from the article [45] of the author. Let C be a convex body. Then there is an inscribed ellipsoid of maximum volume and a circumscribed ellipsoid of minimum volume. The uniqueness of both ellipsoids was proved by Danzer, Laugwitz and Lenz [20]. John [49] specified conditions which an inscribed ellipsoid of maximum volume must satisfy. That these conditions are sufficient was shown by Pełczyński [58] and Ball [5].

We state and prove a precise version of John's theorem, specify for typical convex bodies the number of contact points between (the boundaries of) the body and the unique inscribed ellipsoid of maximum volume. Analogous results are considered for minimal ellipsoidal shells. Minimal ellipsoidal shells are unique for typical, but not for all convex bodies.

John's Ellipsoid Theorem

In the case when C is o-symmetric, the result is as follows:

Theorem 18. *Let $B^d \subseteq C$. Then the following properties are equivalent:*

(i) *B^d is the unique ellipsoid of maximum volume contained in C.*
(ii) *There is a finite set $M = \{\pm u_1, \ldots, \pm u_k\}$ of common boundary points of B^d and C – such points are called* contact points *of B^d, C – such that*

$$I = \sum_{u \in M} \lambda_u \, u \otimes u \quad \text{with suitable } \lambda_u > 0 \quad \text{and} \quad k \leq \frac{d(d+1)}{2}.$$

This result, or versions of it, was proved and refined many times. We mention Bastero and Romance [9], Giannopoulos, Peressinaki and Tsolomitis [29], Gordon, Litvak, Meyer and Pajor [31] and the author and Schuster [47]. The later proof is taken from [47] and fits into the present context.

John's theorem and its dual counterpart, the characterization of the unique circumscribed ellipsoid of minimum volume, has generated a voluminous literature both in convex geometry and the asymptotic theory of normed spaces. It includes various versions, extensions and new proofs of these characterizations, and applications to normed spaces, in particular, the following one, where the *Banach-Mazur distance* between two norms $\| \cdot \|_C, \| \cdot \|_D$ on $\mathbb{E}^d$ with unit balls C, D, is defined by

$$\delta^{BM} \left(\| \cdot \|_C, \| \cdot \|_D \right) = \delta^{BM}(C, D)$$

$$= \inf \left\{ \lambda \geq 1 : C \subseteq AD \subseteq \lambda C, \ A \in \mathbb{E}^{d^2} \right\}.$$

Corollary 15. *Let* $\| \cdot \|_C$ *be an arbitrary norm and* $\| \cdot \|$ *the usual Euclidean norm on* $\mathbb{E}^d$. *Then*

$$\delta^{BM} \left(\| \cdot \|, \| \cdot \|_C \right) \leq \sqrt{d}.$$

This result and its proof based on John's theorem are well known. For a reason which will be explained later, it is a bit surprising that John's theorem yields this estimate, see Corollary 16.

The Contact Number of Typical Convex Bodies

Given a convex body, the question arises, how many contact points are there between the convex body and its volume maximizing inscribed, resp. its volume minimizing circumscribed ellipsoid. A result of the author [34] gives the following answer:

Theorem 19. *For most o-symmetric convex bodies* C *the unique o-symmetric inscribed ellipsoid of maximum volume and the unique o-symmetric circumscribed ellipsoid of minimum volume, both have precisely* $\frac{1}{2}d(d+1)$ *pairs* $\pm u$ *of contact points with* C.

For an alternative proof of this result see Rudelson [63].

Minimum Ellipsoidal Shells

A pair of solid o-symmetric ellipsoids $\langle E, \varrho E \rangle$ is called a *minimal ellipsoidal shell* of C, if $E \subseteq C \subseteq \varrho E$, where $\varrho \geq 1$ is minimal. It is easy to see that $\varrho = \delta^{BM} \left(\| \cdot \|, \| \cdot \|_C \right)$. Maurey [55] (unpublished) showed that a minimal ellipsoidal shell need not be unique, see Lindenstrauss and Milman [53] and Praetorius [60].

In analogy to John's theorem and its dual, we have the following results due to Gruber [38]:

Theorem 20. *Let* $B^d \subseteq C \subseteq \varrho B^d$. *Then the following properties are equivalent:*

  (i) $\langle B^d, \varrho B^d \rangle$ *is a (not necessarily unique) minimal ellipsoidal shell of* C.
  (ii) *There are contact points* $\pm u_1, \ldots, \pm u_k \in \operatorname{bd} B^d \cap \operatorname{bd} C$ *and* $\pm v_1, \ldots,$ $\pm v_l \in \operatorname{bd} C \cap \operatorname{bd} \varrho B^d$ *and reals* $\lambda_1, \ldots, \lambda_k, \mu_1, \ldots, \mu_l > 0$, *such that*
  (a) $2 \leq k, l$ *and* $k + l \leq \frac{1}{2}d(d+1) + 1$,

(b) $\displaystyle\sum_{i=1}^{k} \lambda_i \, u_i \otimes u_i = \sum_{j=1}^{l} \mu_j \, v_j \otimes v_j \neq O,$

(c) $\text{lin}\{u_1, \ldots, u_k\} = \text{lin}\{v_1, \ldots, v_l\}.$

While there are examples of convex bodies with more than one minimal ellipsoidal shell, this is a rare event, as the next result shows.

Theorem 21. *Most o-symmetric convex bodies C have a unique minimal ellipsoidal shell $\langle E, \varrho E \rangle$. The contact sets $\text{bd}\, E \cap \text{bd}\, C$ and $\text{bd}\, C \cap \text{bd}\, \varrho E$, each consist of at least 2 and at most $\frac{1}{2}d(d+1) - 1$, together of $\frac{1}{2}d(d+1) + 1$ pairs of points $\pm u$.*

Theorems 19 and 21 yield the following proposition:

Corollary 16. *For most o-symmetric convex bodies C neither the inscribed ellipsoid of maximum, nor the circumscribed ellipsoid of minimum volume, give rise to a minimum ellipsoidal shell.*

Remark 4. By Corollary 16, it is a happy, rather unexpected event, that John's theorem leads to a proof of Corollary 15. Being a characterization of minimal ellipsoidal shells, Theorem 20 should readily imply Corollary 15. This is, in fact, the case as the later proof of Corollary 15 shows.

Proofs of Theorem 18 and Corollary 15

Theorem 18: Let $h_C(v) = \max\{v \cdot x \, : \, x \in C\}$, $v \in \mathbb{E}^d$, be the *support function* of C. Then

$$C = \{x \in \mathbb{E}^d \, : \, v \cdot x \le h_C(v) \text{ for } v \in S^{d-1}\}.$$

The set

$$\mathcal{E} = \left\{A \in \mathbb{E}^{\frac{1}{2}d(d+1)} \, : \, AB^d \subseteq C\right\} \cap \mathcal{P}^d$$

$$= \bigcap_{\substack{u \in \text{bd}\, B^d \\ v \subseteq S^{d-1}}} \left\{A \in \mathbb{E}^{\frac{1}{2}d(d+1)} \, : \, Au \cdot v = A \cdot v \otimes u \le h_C(v)\right\} \cap \mathcal{P}^d$$

represents the set all o-symmetric ellipsoids contained in C. Since $\mathcal{E}$ is the intersection of a family of closed halfspaces and the open convex cone $\mathcal{P}^d$, the set $\mathcal{E}$ is a convex subset of $\mathcal{P}^d$, which is closed in $\mathcal{P}^d$. To the ellipsoid B^d corresponds the matrix $I \in \mathcal{E}$. Let $\mathcal{D}^d = \mathcal{D}^d(1) = \{A \in \mathcal{P}^d \, : \, \det A \ge 1\}$.

(i) $\Leftrightarrow$ (ii):

B^d is the unique o-symmetric ellipsoid of maximum volume in C

$\Leftrightarrow \mathcal{E} \cap \mathcal{D}^d = \{I\}$, i.e. the convex body $\mathcal{E}$ touches the smooth and strictly convex body $\mathcal{D}^d$ only at I

$\Leftrightarrow$ the interior normal vector I of $\mathcal{D}^d$ at I is contained in the (exterior) normal cone $\mathcal{N}$ of $\mathcal{E}$ at I. Note that $\mathcal{N}$ is generated by the exterior normal vectors of those of the defining halfspaces $\{A : Au \cdot v = A \cdot v \otimes u \leq h_C(v)\}$ of $\mathcal{E}$ which contain I as a boundary point. Thus $1 \geq u \cdot v = I \cdot v \otimes u = h_C(v) \geq 1$, and therefore $u \cdot v = h_C(v) = 1$, or $u = v$, $h_C(u) = 1$, or $u \otimes u = v \otimes u$, $u \in \operatorname{bd} B^d \cap \operatorname{bd} C = B^d \cap \operatorname{bd} C$. Thus $\mathcal{N} = \operatorname{pos}\{u \otimes u : u \in B^d \cap \operatorname{bd} C\}$

$\Leftrightarrow$ by Carathéodory's theorem for cones, we may choose a set of contact points $M = \{\pm u_1, \ldots, \pm u_k\} \subseteq B^d \cap \operatorname{bd} C$ such that

$$I = \sum_{u \in M} \lambda_u\, u \otimes u \quad \text{with suitable } \lambda_u > 0 \ \text{ and } \ k \leq \frac{1}{2}d(d+1). \qquad \blacksquare$$

Corollary 15: We may assume that $B^d \subseteq C \subseteq \varrho B^d$, where

$$\varrho = \delta^{BM}\left(\|\cdot\|, \|\cdot\|_C\right).$$

It is sufficient to show that $\varrho \leq \sqrt{d}$. Equating the traces of the two sides of the equality in Theorem 20(iib) implies that

$$\sum_i \lambda_i = \sum_i \lambda_i\, u_i \cdot u_i = \sum_j \mu_j\, v_j \cdot v_j = \varrho^2 \sum_j \mu_j.$$

Noting that $\frac{1}{\sqrt{d}}I$ has norm 1 and that $(u_i \cdot v_j)^2 \leq 1$, this yields that

$$\left\| \sum_i \lambda_i\, u_i \otimes u_i \right\| \geq \sum_i \lambda_i\, u_i \otimes u_i \cdot \frac{1}{\sqrt{d}}I = \frac{1}{\sqrt{d}} \sum_i \lambda_i\, u_i \cdot u_i = \frac{1}{\sqrt{d}} \sum_i \lambda_i,$$

$$\left(\sum_i \lambda_i\, u_i \otimes u_i \right)^2 = \sum_i \lambda_i\, u_i \otimes u_i \cdot \sum_j \mu_j\, v_j \otimes v_j = \sum_{i,j} \lambda_i \mu_j\, (u_i \cdot v_j)^2$$

$$\leq \sum_i \lambda_i \sum_j \mu_j = \frac{1}{\varrho^2}\left(\sum_i \lambda_i \right)^2,$$

$$\text{or} \quad \frac{1}{\varrho^2} \geq \frac{1}{d}, \quad \text{or} \quad \varrho \leq \sqrt{d}. \qquad \blacksquare$$

9. Minimum Position Problems

Related to John's theorem is the following question: Consider a real function F on the space of all convex bodies or on a suitable subspace of it, for example on the space of all o-symmetric convex bodies, and a group $\mathcal{G}$ of affinities. Assume that this subspace is invariant under the affinities of $\mathcal{G}$. Characterize for a given convex body C in this subspace those among its images under affinities from $\mathcal{G}$, for which F is minimum, the *minimum F-positions of C* with respect to the group $\mathcal{G}$. For numerous pertinent results and applications see Milman and Pajor [56], Giannopoulos and Milman [26, 27], Gordon, Litvak, Meyer and Pajor [31] and the author [38] and the references there.

In the following we state minimum position results of the author [38] which were proved using ideas in the sense of Voronoĭ, while the classical proofs rely on a variational argument, see Giannopoulos and Milman [26]. We characterize circumscribed ellipsoids of minimum surface area and minimum positions for polar moments, mean width and surface area. Let C be an o-symmetric convex body.

Circumscribed Ellipsoids of Minimum Surface Area

In the light of John's theorem and its dual, the following question arises naturally: Given a convex body C, characterize the inscribed and circumscribed ellipsoids of maximum, resp. minimum surface area. Are these unique? Moreover, what are the corresponding minimum positions with respect to the group of volume preserving linear transformations? Can the surface area be replaced by general quermassintegrals?

Theorem 22. *There is a unique ellipsoid of minimum surface area containing C.*

The original proof of the author [38] was rather complicated and made use of projection bodies and Alexandrov's projection theorem. Much easier is the recent proof by Schröcker [76].

Assign to a convex body C the minimum surface area $S_m(C)$ of a circumscribed ellipsoid.

Theorem 23. *Up to rotations, C has a unique minimum S_m-position with respect to the group of volume-preserving linear transformations and the following properties are equivalent:*

(i) *C is in minimum S_m-position and B^d is the circumscribed ellipsoid of minimum surface area.*

(ii) *There are contact points $\pm u_1, \ldots, \pm u_k \in \operatorname{bd} B^d \cap \operatorname{bd} C$ and $\lambda_1, \ldots,$ $\lambda_k > 0$ such that*

    (a) $d \le k \le \frac{1}{2}d(d+1)$,

    (b) $I = \sum \lambda_i\, u_i \otimes u_i$,

    (c) $\mathbb{E}^d = \operatorname{lin}\{u_1, \ldots, u_k\}$.

Comparing the dual counterpart of Theorem 18 together with some addenda (see the author and Schuster [47]) and Theorem 23 yields the next result.

Corollary 17. *Let $C \subseteq B^d$. Then the following properties are equivalent:*

(i) *B^d is the unique circumscribed ellipsoid of C of minimum volume.*

(ii) *C is in minimum S_m-position with respect to volume-preserving linear transformations and B^d is the unique circumscribed ellipsoid of C with minimum surface area.*

These results can be extended to general quermassintegrals. Then we see that the minimum positions for all quermassintegrals – except for the volume – coincide.

Polar f-Moments

Let $f : [0, +\infty) \to [0, \infty)$ be a non-decreasing function. Then

$$M(C, f) = \int_C f(\|x\|)\, dx$$

is the *polar f-moment* of C. If $f(t) = t^2$, then $M(C, t^2)$ is the *polar moment of inertia.*

Theorem 24. *Let f be convex and assume that $f(t) = 0$ only for $t = 0$. Then C has, up to rotations, a unique minimum polar f-moment position with respect to volume-preserving linear transformations and the following properties are equivalent:*

(i) C is in minimum polar f-moment position.

(ii) $I = \lambda \int_C \dfrac{f'(\|x\|)}{\|x\|} \, x \otimes x \, dx$ for suitable $\lambda > 0$.

The integral here is to be understood entry-wise. We now minimize the product $M(AC, t^2) \, M\big((AC)^*, t^2\big)$, where A ranges over all non-singular linear transformations.

Theorem 25. *Up to similarities which keep o fixed, C has a unique minimum $M(\,\cdot\,C, t^2) \, M\big((\cdot\,C)^*, t^2\big)$-position with respect to non-singular linear transformations and the following properties are equivalent:*

(i) C is in minimum $M(\cdot\,C, t^2) \, M\big((\cdot\,C)^*, t^2\big)$-position.

(ii) $\displaystyle\int_C x \otimes x \, dx = \lambda \int_{C^*} x \otimes x \, dx \in \mathcal{P}^d$ for a suitable $\lambda > 0$.

Mean Width and Surface Area

The *mean width* of a convex body C is defined by

$$W(C) = \frac{2}{S(B^{d-1})} \int_{S^{d-1}} h_C(u) \, d\sigma(u),$$

where $S(\cdot)$ and $\sigma(u)$ denote the usual surface area measure on S^{d-1}.

Theorem 26. *Up to rigid motions, C has a unique minimum mean width position with respect to volume preserving affinities and the following properties are equivalent:*

(i) C is in minimum mean width position.

(ii) $I = \lambda \displaystyle\int_{S^{d-1}} \big\{ \operatorname{grad} h_C(u) \otimes u + u \otimes \operatorname{grad} h_C(u) \big\} \, d\sigma(u)$ for a suitable $\lambda > 0$.

A first characterization of the minimum surface area position of C with respect to volume-preserving affinities is due to Giannopoulos and Papadimitrakis [28]. A different result can be described as follows: The *projection body* ΠC of C is the o-symmetric convex body with support function

$$h_{\Pi C}(u) = v(C \mid u^{\perp}),$$

where $C \mid u^\perp$ is the orthogonal projection of C onto the subspace $u^\perp$ orthogonal to u of codimension 1 and $v(\cdot)$ the volume in $d-1$ dimensions. Since by Cauchy's surface area formula the mean width of the projection body is, up to a multiplicative constant, the surface area of the original body, Theorem 26 implies the following result:

Corollary 18. *Up to rigid motions, C has a unique minimum surface area position with respect to volume-preserving affinities and the following properties are equivalent:*

(i) *C is in minimum surface area position.*

(ii) $I = \lambda \int_{S^{d-1}} \left\{ \operatorname{grad} h_{\Pi C}(u) \otimes u + u \otimes \operatorname{grad} h_{\Pi C}(u) \right\} d\sigma(u)$ *for a suitable* $\lambda > 0$.

There are similar results for $W(C)\,W(C^*)$ and $S(C)\,S(C^*)$.

Remark 5. The above characterizations of convex bodies in minimum position and similar results in the literature should permit one to prove all possible properties of the minimizing bodies. This seems to have been one of the objectives at the beginning of the development. So far, these expectations have not materialized, a minor exception being the proof of Corollary 15.

Acknowledgements. For his great help in the preparation of this article I am obliged to Tony Thompson.

REFERENCES

[1] Ash, A., On eutactic forms, *Canad. J. Math.*, **29** (1977), 1040–1054.

[2] Ash, A., On the existence of eutactic forms, *Bull. London Math. Soc.*, **12** (1980), 192–196.

[3] Bachoc, C., Designs, groups and lattices, *J. Théor. Nombres Bordeaux*, **17** (2005), 25–44.

[4] Bachoc, C. and Venkov, B., Modular forms, lattices and spherical designs. *Réseaux euclidiens, designs sphériques et formes modulaires*, 87–111, Monogr. Enseign. Math. **37**, Enseignement Math., Geneva, 2001.

[5] Ball, K. M., Ellipsoids of maximal volume in convex bodies, *Geom. Dedicata,* **41** (1992), 241–250.

[6] Bambah, R. P., On lattice coverings by spheres, *Proc. Nat. Inst. Sci. India,* **A 23** (954), 25–52.

[7] Barnes, E. S. and Dickson, T. J., Extreme coverings of *n*-space by spheres, *J. Austral. Math. Soc.,* **7** (1967), 115–127.

[8] Barvinok, A., *A course in convexity,* Amer. Math. Soc., Providence, RI, 2002.

[9] Bastero, J. and Romance, M., John's decomposition of the identity in the non-convex case, *Positivity,* **6** (2002), 1–16.

[10] Bavard, C., Systole et invariant d'Hermite, *J. Reine Angew. Math.,* **482** (1997), 93–120.

[11] Bavard, C., Théorie de Voronoï géométrique. Propriétés de finitude pour les familles de réseaux et analogues, *Bull. Soc. Math. France,* **133** (2005), 205–257.

[12] Bergé, A.-M. and Martinet, J., Sur un probléme de dualité lié aux sphéres en géométrie des nombres, *J. Number Theory,* **32** (1989), 14–42.

[13] Bergé, A.-M. and Martinet, J., Sur la classification des réseaux eutactiques, *J. London Math. Soc. (2),* **53** (1996), 417–432.

[14] Bertraneu, A. and Fichet, B., Étude de la frontière de l'ensemble des formes quadratiques positives, sur un espace vectoriel de dimension finie, *J. Math. Pures Appl. (9),* **61** (1982), 207–218.

[15] Böröczky, K., Jr. and Schneider, R., A characterization of the duality mapping for convex bodies, *Geom. Funct. Anal.,* **18** (2008), 657-667.

[16] Cassels, J. W. S., On a problem of Rankin about the Epstein zeta-function, *Proc. Glasgow Math. Assoc.,* **4** (1959), 73–80.

[17] Cohn, H. and Kumar, A., The densest lattice in twenty-four dimensions, *Electron. Res. Announc. Amer. Math. Soc.,* **10** (2004), 58–67 (electronic).

[18] Conway, J. H. and Sloane, N. J. A., *Sphere packings, lattices and groups,* 3rd ed. With additional contributions by E. Bannai, R. E. Borcherds, J. Leech, S. P. Norton, Λ. M. Odlyzko, R. A. Parker, L. Queen and B. B. Venkov, Grundlehren Math. Wiss. 290, Springer-Verlag, New York, 1999.

[19] Coulangeon, R., Spherical designs and zeta functions of lattices, *Int. Math. Res. Not. Art. ID,* **49620** (2006), 16pp.

[20] Danzer, L., Laugwitz, D. and Lenz, H., Über das Löwnersche Ellipsoid und sein Analogon unter den einem Eikörper einbeschriebenen Ellipsoiden, *Arch. Math.,* **8** (1957), 214–219.

[21] Delone, B. N. and Ryshkov, S. S., A contribution to the theory of the extrema of a multi-dimensional ζ-function, *Dokl. Akad. Nauk SSSR,* **173** (1967), 991–994, *Soviet Math. Dokl.,* **8** (1967), 499–503.

[22] Delone, B. N., Dolbilin, N. P., Ryshkov, S. S. and Shtogrin, M. I., A new construction of the theory of lattice coverings of an n-dimensional space by congruent balls, *Izv. Akad. Nauk SSSR Ser. Mat.,* **34** (1970), 289–298.

[23] Engel, P., Geometric crystallography, in: *Handbook of convex geometry,* **B** 989–1041, North-Holland, Amsterdam, 1993.

[24] Ennola, V., On a problem about the Epstein zeta-function, *Proc. Cambridge Philos. Soc.,* **60** (1964), 855–875.

[25] Erdős, P., Gruber, P. M. and Hammer, J., *Lattice points,* Longman Scientific, Harlow, Essex, 1989.

[26] Giannopoulos, A. A. and Milman, V. D., Extremal problems and isotropic positions of convex bodies, *Israel J. Math.,* **117** (2000), 29–60.

[27] Giannopoulos, A. A. and Milman, V. D., Euclidean structure in finite dimensional normed spaces, in: *Handbook of the geometry of Banach spaces,* **I** 707–779, North-Holland, Amsterdam, 2001.

[28] Giannopoulos, A. A. and Papadimitrakis, M., Isotropic surface area measures, *Mathematika,* **46** (1999), 1–13.

[29] Giannopoulos, A. A., Perissinaki, I. and Tsolomitis, A., John's theorem for an arbitrary pair of convex bodies, *Geom. Dedicata,* **84** (2001), 63–79.

[30] Goethals, J.-M. and Seidel, J. J., Spherical designs, in: *Relations between combinatorics and other parts of mathematics* (Proc. Sympos. Pure Math., Columbus 1978) 255–272, Proc. Sympos. Pure Math. XXXIV, Amer. Math. Soc., Providence RI, 1979.

[31] Gordon, Y., Litvak, A. E., Meyer, M. and Pajor, A., John's decomposition in the general case and applications, *J. Differential Geom.,* **68** (2004), 99–119.

[32] Gruber, P. M., Die meisten konvexen Körper sind glatt, aber nicht zu glatt, *Math. Ann.,* **228** (1977), 239–246.

[33] Gruber, P. M., Typical convex bodies have surprisingly few neighbours in densest lattice packings, *Studia Sci. Math. Hungar.,* **21** (1986), 163–173.

[34] Gruber, P. M., Minimal ellipsoids and their duals, *Rend. Circ. Mat. Palermo (2)* **37** (1988), 35–64.

[35] Gruber, P. M., The space of convex bodies, in: *Handbook of convex geometry* **A**, 301–318, North-Holland, Amsterdam, 1993.

[36] Gruber, P. M., Baire categories in convexity, in: *Handbook of convex geometry* **B**, 1327–1346, North-Holland, Amsterdam, 1993.

[37] Gruber, P. M., Convex and discrete geometry, *Grundlehren Math. Wiss.*, **336**, Springer, Berlin, Heidelberg, New York, 2007.

[38] Gruber, P. M., Application of an idea of Voronoĭ to John type problems, *Adv. in Math.*, **218** (2008), 299–351.

[39] Gruber, P. M., Geometry of the cone of positive quadratic forms, *Forum Math.*, **21** (2009), 147–166.

[40] Gruber, P. M., On the uniqueness of lattice packings and coverings of extreme density, *Adv. in Geom.*, **11** (2011), 691–710.

[41] Gruber, P. M., Voronoĭ type criteria for lattice coverings with balls, *Acta Arith.*, **149** (2011), 371–381.

[42] Gruber, P. M., John and Löwner ellipsoids, *Discrete Comput. Geom.*, **46** (2011), 776–788.

[43] Gruber, P. M., Lattice packing and covering of convex bodies, *Proc. Steklov Inst. Math.*, **275** (2011), 229–238.

[44] Gruber, P. M., Application of an idea of Voronoĭ to lattice packing, in preparation.

[45] Gruber, P. M., Application of an idea of Voronoĭ to lattice zeta functions, *Proc. Steklov Inst. Math.*, **276** (2012), to appear.

[46] Gruber, P. M. and Lekkerkerker, C. G., *Geometry of numbers,* 2nd ed., North–Holland, Amsterdam, 1987, Nauka, Moscow, 2008.

[47] Gruber, P. M. and Schuster, F. E., An arithmetic proof of John's ellipsoid theorem, *Arch. Math. (Basel)*, **85** (2005), 82–88.

[48] Grünbaum, B., *Convex polytopes,* 2nd ed., Prepared by V. Kaibel, V. Klee, G. M. Ziegler, Springer, New York, 2003.

[49] John, F., Extremum problems with inequalities as subsidiary conditions, in: *Studies and Essays,* Presented to R. Courant on his 60th Birthday, January 8, 1948, 187–204, Interscience, New York, 1948.

[50] Klee, V., Some new results on smoothness and rotundity in normed linear spaces, *Math. Ann.*, **139** (1959), 51–63.

[51] Klein, F., Die allgemeine lineare Transformation der Linienkoordinaten, *Math. Ann.*, **2** (1870), 366–370, Ges. Math. Abh. I, Springer, Berlin, 1921.

[52] Lim, S. C. and Teo, L. P., On the minima and convexity of Epstein zeta function, *J. Math. Phys.*, **49** (2008), 073513, 25pp.

[53] Lindenstrauss, J. and Milman, V. D., The local theory of normed spaces and its applications to convexity, in: *Handbook of convex geometry* **B**, 1154–1220, North-Holland, Amsterdam, 1993.

[54] Martinet, J., *Perfect lattices in Euclidean spaces,* Grundlehren Math. Wiss. **325**, Springer, Berlin, Heidelberg, New York, 2003.

[55] Mauray, Unpublished note.

[56] Milman, V. D. and Pajor, A., Isotropic position and inertia ellipsoids and zonoids of the unit ball of a normed *n*-dimensional space, in: *Geometric aspects of functional analysis* (1987–88) 64–104, Lecture Notes in Math., **1376**, Springer, Berlin, 1989.

[57] Montgomery, H. L., Minimal theta functions, *Glasgow Math. J.,* **30** (1988), 75–85.

[58] Pełczyński, A., Remarks on John's theorem on the ellipsoid of maximal volume inscribed into a convex symmetric body in $\mathbf{R}^n$, *Note Mat.,* **10** (1990), 395–410.

[59] Plücker, J., *Neue Geometrie des Raumes gegründet auf die Betrachtung der geraden Linie als Raumelement,* I. Abth. 1868 mit einem Vorwort von A. Clebsch., II. Abth. 1869, herausgegeben von F. Klein, Teubner, Leipzig, 1868.

[60] Praetorius, D., *Ellipsoide in der Theorie der Banachräume,* Master Thesis, U Kiel, 2000.

[61] Praetorius, D., Remarks and examples concerning distance ellipsoids, *Colloq. Math.,* **93** (2002), 41–53.

[62] Rankin, R. A., A minimum problem for the Epstein zeta-function, *Proc. Glasgow Math. Assoc.,* **1** (1953), 149–158.

[63] Rudelson, M., Contact points of convex bodies, *Israel J. Math.,* **101** (1997), 93–124.

[64] Ryshkov, S. S., On the question of the final ζ-optimality of lattices that yield the densest packing of *n*-dimensional balls, *Sibirsk. Mat. Zh.,* **14** (1973), 1065–1075, 1158.

[65] Ryshkov, S. S., Geometry of positive quadratic forms, in: *Proc. Int. Congr. of Math.* (Vancouver, 1974) **1**, 501–506, Canad. Math. Congress, Montreal, 1975.

[66] Ryshkov, S. S., On the theory of the cone of positivity and the theory of the perfect polyhedra $\Pi(n)$ and $n(m)$, *Chebyshevskii Sb.,* **3** (2002), 84–96.

[67] Ryshkov S. S. and Baranovskiĭ, E. P., Classical methods of the theory of lattice packings, *Uspekhi Mat. Nauk,* **34** (1979), 3–63, 236, *Russian Math. Surveys,* **34** (1979), 1–68.

[68] Sarnak, P. and Strömbergsson, A., Minima of Epstein's zeta function and heights of flat tori, *Invent. Math.,* **165** (2006), 115–151.

[69] Schmutz, P., Riemann surfaces with shortest geodesic of maximal length, *Geom. Funct. Anal.,* **3** (1993), 564–631.

[70] Schmutz Schaller (Schmutz), P., Systoles on Riemann surfaces, *Manuscripta Math.,* **85** (1994), 428–447.

[71] Schmutz Schaller, P., Geometry of Riemann surfaces based on closed geodesics, *Bull. Amer. Math. Soc. (N.S.)*, **35** (1998), 193–214.

[72] Schmutz Schaller, P., Perfect non-extremal Riemann surfaces, *Canad. Math. Bull.*, **43** (2000), 115–125.

[73] Schneider, R., *Convex bodies: the Brunn–Minkowski theory,* Cambridge Univ. Press, Cambridge, 1993.

[74] Schneider, R., The endomorphisms of the lattice of closed convex cones, *Beiträge Algebra Geom.*, **49** (2008), 541–547.

[75] Schürmann, A., *Computational geometry of positive definite quadratic forms,* Amer. Math. Soc., Providence, RI, 2009.

[76] Schröcker, H.-P., Uniqueness results for minimal enclosing ellipsoids, *Comput. Aided Geom. Design,* **25** (2008), 756–762.

[77] Schürmann, A. and F. Vallentin, F., Computational approaches to lattice packing and covering problems, *Discrete Comput. Geom.*, **35** (2006), 73–116.

[78] Swinnerton-Dyer, H. P. F., Extremal lattices of convex bodies, *Proc. Cambridge Philos. Soc.,* **49** (1953), 161–162.

[79] Venkov, B., Réseaux et designs sphériques, in: Réseaux euclidiens, designs sphériques et formes modulaires 10–86, *Monogr. Enseign. Math.*, **37**, Enseignement Math., Geneva, 2001.

[80] Voronoĭ (Voronoï, Woronoi), G. F., Nouvelles applications des paramètres continus à la théorie des formes quadratiques. Première mémoire: Sur quelques propriétés des formes quadratiques positives parfaites, *J. Reine Angew. Math.*, **133** (1908), 97–178, *Coll. Works,* **II**, 171–238.

[81] Voronoĭ, G. F., Nouvelles applications des paramètres continus à la théorie des formes quadratiques. Deuxième mémoire. Recherches sur les paralléloèdres primitifs I, II, *J. Reine Angew. Math.*, **134** (1908), 198–267 et **136** (1909), 67–181, *Coll. Works,* **II**, 239–368.

[82] Voronoĭ, G. F., Collected works **I–III**, Izdat. Akad. Nauk Ukrain. SSSR, Kiev, 1952.

[83] Wickelgren, K., Linear transformations preserving the Voronoi polyhedron, Manuscript, 2001.

[84] Zong, C., Sphere packings, Springer, New York, 1999.

Peter M. Gruber

Institut für Diskrete Mathematik und Geometrie
Technische Universität Wien
Wiedner Hauptstraße 8–10/1046
A–1040 Vienna, Austria

e-mail: peter.gruber@tuwien.ac.at

BOLYAI SOCIETY
MATHEMATICAL STUDIES, 24

Geometry –
Intuitive, Discrete, and Convex
pp. 159–186.

UNIFORM POLYHEDRALS

BRANKO GRÜNBAUM

This survey is meant to honor Laszlo Fejes Tóth, who for many years was one of few proponents of visual geometry

Half a century ago H. S. M. Coxeter, M. S. Longuet-Higgins and J. C. P. Miller published a very influential paper on "Uniform Polyhedra" [7]. These are finite polyhedra with regular polygons as faces, and with vertices in a single orbit under symmetries. *Uniform polyhedrals* are defined by the same conditions, but with *finite* replaced by *locally finite,* and the additional explicit requirement that there are no coinciding elements (vertices, edges or faces); this was self-understood in [7]. Coplanar faces, collinear edges, and partial overlaps are allowed for uniform polyhedrals, as they are for uniform polyhedra. It is somewhat surprising that no systematic study of infinite uniform polyhedrals has been undertaken so far. There are three distinct classes of such polyhedrals – rods, slabs, and sponges. The beginnings of their investigation form the core of this article, and many open problems become evident. Illustrations serve to shorten the explanations, but also to highlight the difficulty of presenting polyhedrals graphically. Applications of such polyhedrals and their relatives in fields such as architecture, biology, engineering, and others are discussed as well, as are the shortcomings of the mathematical reviewing journals in reporting these and related applications of geometry.

1. INTRODUCTION

The study of polyhedra more general than the convex ones received a very significant boost in 1954 by the appearance of the long paper "Uniform polyhedra" by H. S. M. Coxeter, M. S. Longuet-Higgins, and J. C. P. Miller [7]. A century of contributions to the topic was surveyed, as well as extended and systematized in a far-reaching manner. The authors' hope that their enumeration of the polyhedra in question was complete was vindicated independently by S. P. Sopov [34], J. Skilling [33], and I. Szepesváry [35]. Apart from the actual enumeration of uniform polyhedra, the most significant contribution of [7] was the explicit definition of the concept of "uniform polyhedron". It is well known that the general idea of "polyhedron" has been left rather murky ever since ancient times, and is still being discussed in many variants. However, the definition of the topic in [7] is simple, crisp and convenient. To quote:

> A *polyhedron* is a finite set of polygons such that every side of each belongs to just one other, with the restriction that no subset has the same property. [It is] *uniform* if its faces are regular while its vertices are equivalent under symmetries of the polyhedron.

Uniform polyhedrals, the objects discussed in the following pages, could be defined in exactly the same way except that instead of requiring that the set of polygons is finite, we require only that it is *locally finite,* that faces incident with a vertex form a single cycle, and no two vertices, or two edges, or two faces can coincide.

An independent and more generally applicable definition is:

A *polyhedral* is a locally finite and edge-sharing family of polygons, locally and globally strongly connected, with the property that no two vertices, or two edges, or two faces coincide. A *uniform polyhedral* has regular polygons as faces, and its vertices are equivalent under isometric symmetries of the polyhedral.

Since no other kinds are considered here, we shall simplify the language by frequently omitting "uniform" in the following discussion of uniform polyhedrals. It is understood that we are dealing exclusively with polyhedrals in the Euclidean 3-dimensional space.

We still need to clarify what are the regular polygons we admit for our polyhedrals. In contrast to various generalizations that have been considered (for example in [13], [15], [16], and other papers) and that could be used in the study of uniform polyhedrals, here we restrict attention to the classically accepted polygons denoted by $\{n/d\}$, for relatively prime n and d, with $1 \leq d < n/2$, where n is the number of edges (and vertices) and d the "density" (see, for example, [7]).

Since the finite uniform polyhedrals have been adequately described in the literature – there are three infinite families and 75 individual ones – most of our attention will go to the infinite polyhedrals. It is easy to see that it is convenient to distinguish three classes of infinite uniform polyhedrals: *rods* are infinite in one direction, *slabs* are infinite in two independent directions, and *sponges* are infinitely extended in three independent directions. Simple examples are shown in Fig. 1. The following sections will discuss each of these classes in considerable detail. But it seems worth mentioning that it appears strange that the finite polyhedrals, and the three classes of infinite polyhedrals, have not been considered together, even though they share many characteristics. One of the main aims of the following pages is the enumeration of the different kinds of polyhedrals – to the extent that this is known. Besides the *acoptic* (selfintersection-free) infinite polyhedrals, we shall pay particular attention to the non-acoptic ones; these seem not to have been considered in the literature at all.

The primary characteristics of all uniform polyhedrals is the local specification of the neighborhood of each (hence every) vertex by its *vertex star* – the circuit of polygons (faces) that are incident with the vertex. The vertex star can be recorded by the *vertex symbol,* such as (4.6.8); exponents are sometimes used to shorten the symbol. Additional examples are given in Fig. 1.

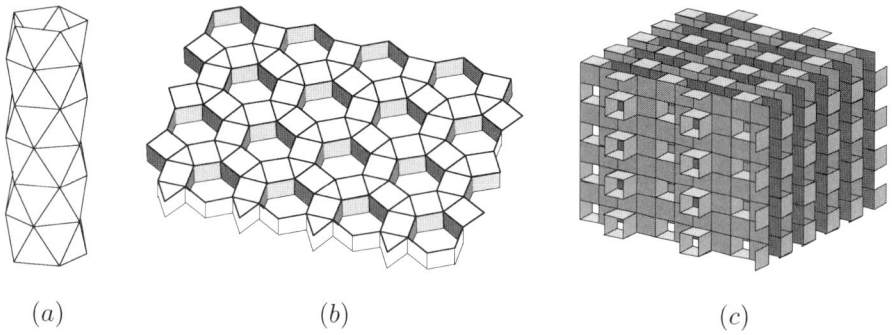

(a)   (b)   (c)

Fig. 1. (*a*) A rod with vertex symbol (3^6). (*b*) A slab with vertex symbol (3.4^4). (*c*) A sponge with vertex symbol (4^5)

Finite polyhedrals are uniquely determined by their vertex stars. Hence the vertex symbol, such as (3.4.5.4) or (6.6.5/2), is generally accepted as the identification symbol of the polyhedral.

For infinite polyhedrals this is not the case in general, and additional information needs to be specified for identification. This information requires two distinct components. First, the *shape* of the vertex star should be specified by listing the dihedral angles between the successive faces of the vertex star. In some cases this is determined by the other data – for example, for the unique polyhedral with vertex symbol (5.5.5.5.5), see Section 5. The other identifying component is the *adjacency symbol* of the polyhedral. This establishes the relation between adjacent vertex stars, that is, those that share an edge. By the definition of polyhedrals, the adjacency symbol is the same for all pairs of adjacent vertices, hence it is associated with the polyhedral itself. Together with the identification of the class (rod, slab, or sponge), these data determine the polyhedral uniquely. However, this way of describing individual polyhedrals is rather inconvenient and laborious, and except in the case of sponges, simpler possibilities exist. Hence we shall delay discussing the adjacency symbols till Section 6.

The three classes of infinite uniform polyhedrals are analogous to the infinite isohedral polyhedra the study of which was proposed in [18] thirty years ago, along with that of the infinite uniform polyhedrals in the sense of the present paper. However, it appears that this publication has not had any influence on later developments.

2. FINITE UNIFORM POLYHEDRALS

The best introduction to the finite polyhedrals is still the long paper [7] by Coxeter *et al.,* supplemented by the images of physical models in Wenninger's book [42]. The Internet has many excellent presentations, such as those of V. Bulatov [2], G. W. Hart [20], [21], Maeder [27], E. W. Weisstein [42], Wikipedia [43], [44], and many other pages.

We shall not enlarge upon the 75 particular finite uniform polyhedrals, but need to describe the three infinite families – the prisms and the antiprisms – since we shall have to refer to them in connection with the uniform rods.

Each of the prisms and antiprisms has as its two bases congruent regular polygons $\{n/d\}$, where $1 \leq d < n/2$ and d is relatively prime to n. The prisms have vertex symbols $(4.4.n/d)$, while the antiprisms have vertex symbols $(3.3.3.n/d)$. However, there are two varieties of antiprisms. The *ordinary* antiprism $(3.3.3.n/d)$ exists for all the n and d satisfying these conditions. The *crossed* antiprism $(3.3.3.n/d)$ exists only if, in addition, $n/3 < d < n/2$. The difference between the two varieties is that the triangles of the ordinary antiprisms do not cross the axis of rotational symmetry of the antiprisms, while the triangles of the crossed antiprisms do cross it. In [7] and many other publications the crossed antiprism with $\{n/d\}$ bases is denoted $\left(3.3.3.n/(n-d)\right)$. We shall adhere to this notation. We illustrate prisms and antiprisms in Fig. 2.

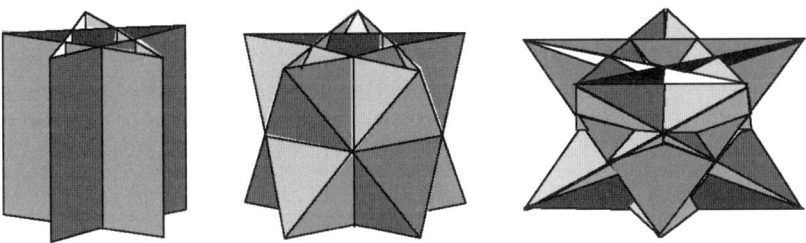

Fig. 2. The pentagram-based prism (4.4.5/2) and antiprisms (3.3.3.5/2) and (3.3.3.5/3)

3. UNIFORM RODS

There are three classes of uniform rods: stacked, ribboned, and helical. (However, see Note in Section 7.) Due to the assumed isogonality of each rod, all its vertices must be on a cylindrical surface. Hence uniform rods can have only squares or equilateral triangles as faces. The squares must have edges parallel and perpendicular to the axis of the prism. The first publication that illustrates all three kinds of (acoptic) rods is [41].

Each *stacked rod* is formed by an infinite stack of modules (modular units); each module consists of the faces that form the mantle of a prism or antiprism. The stacked rod is *unary* if all its modules are of the same kind (hence are congruent, see Fig. 3), and it is *binary* if its modules are of different kinds. In the latter case, for each pair of modules there are always two possibilities: Adjacent modules are either on opposite sides of the plane of the common bases, or on the same side – see Fig. 4. The same-side variant is possible without violating the non-coincidence restriction since the altitudes of different modules (with the same basis) are unequal.

Ribboned rods have vertical ribbons of either squares or triangles. *Unary* ribboned rods with squares are the same as stacked rods $(4.4.4.4)_n$. Hence the only unary ribboned rods that need to be considered are the ones with triangles. Some examples of acoptic (selfintersection-free) rods $R(3.3.3.3.3.3)_n$ of this kind are shown in Fig. 5(a). It should be noted that adjacent ribbons are translates of each other by a half-length of a side. Hence the number of ribbons must be even, regardless of whether the rod is acoptic or not. If the intended cross-section of the rod is $\{n/d\}$ with

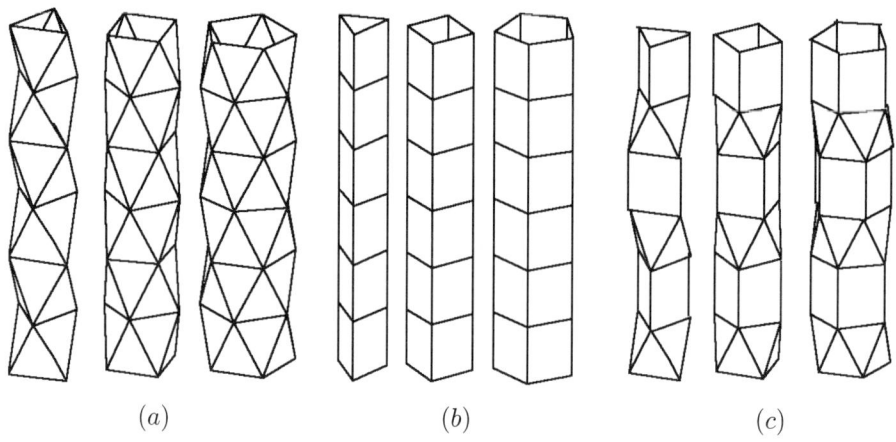

Fig. 3. Uniform acoptic stacked rods with $\{n\}$ bases, $n = 3, 4, 5$. (a) Unary triangle-faced rods $(3.3.3.3.3.3)_n$. (b) Unary square-faced rods $(4.4.4.4)_n$. (c) Binary rods $(3.3.3.4.4)_n$

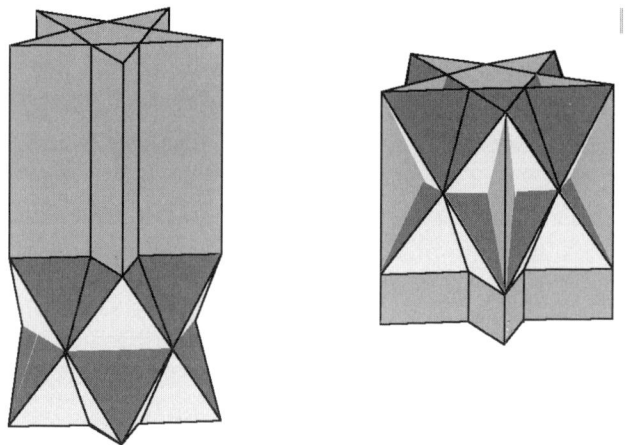

Fig. 4. Two possible pairs of modules with $\{5/2\}$ bases, leading to binary stacked rods that can be denoted $(3.3.3.4.4)_{5/2}$ and $(3.3.3. - 4. - 4)_{5/2}$, respectively. Notice that in the latter rod there are several simultaneous overlaps of squares. There are four other binary stacked rods with pentagrammatic bases: $(3.3.3.4.4)_{5/3}$ and $(3.3.3. - 4. - 4)_{5/3}$, $(3.3.3.3.3.3.)_{5/2,5/3}$ and $(3.3.3. - 3. - 3. - 3)_{5/2,5/3}$

n odd, the number of ribbons must be doubled. Then the ribbons coincide in pairs, but with a displacement of half an edge-length in the direction of the axis of the rod. Thus the evenness requirement is satisfied, as is the non-coincidence of faces. Hence the rod $R(3.3.3.3.3.3)_{n/d}$ with odd n has $2n$ ribbons. I have found no satisfactory way of presenting meaningful illustrations of such ribboned rods; attempts are made in the simplest cases in

Fig. 5(*b*). Equally hard to show are selfintersecting (multiply wound) ribboned rods $R(3.3.3.3.3.3)_{n/d}$ in which the cross-section is a regular polygon $\{n/d\}$, with $d > 1$ and with d and n relatively prime. Fig. 6 illustrates the unary ribboned rod $R(3.3.3.3.3.3)_{8/3}$.

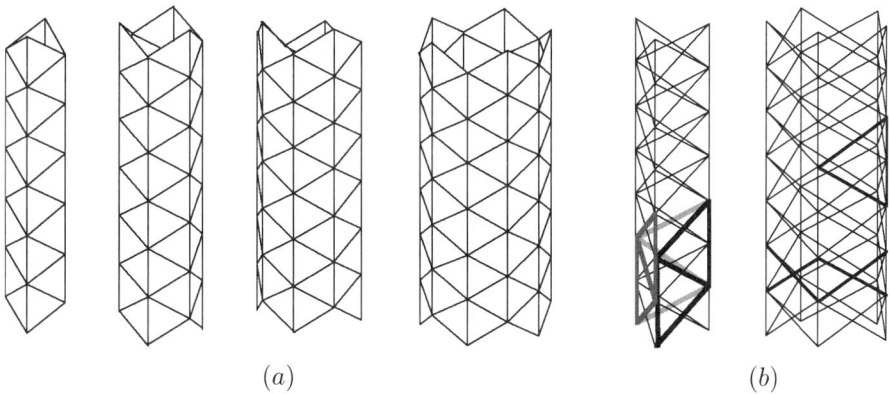

Fig. 5. (*a*) Acoptic unary ribboned rods $R(3.3.3.3.3.3)_n$, for $n = 4, 6, 8, 10$. (*b*) Unary ribboned rods $R(3.3.3.3.3.3)_3$ and $R(3.3.3.3.3.3)_5$

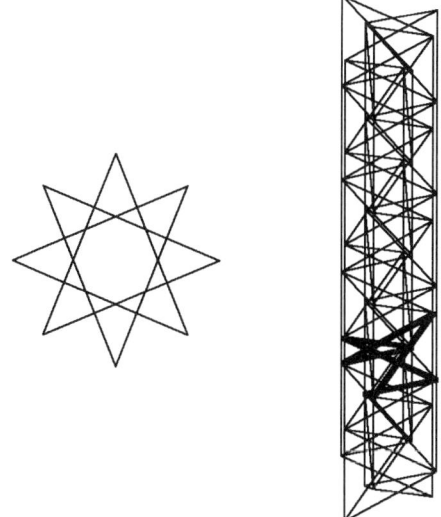

Fig. 6. The ribboned rod $R(3.3.3.3.3.3)_{8/3}$ shown alongside its cross-section

Binary ribboned rods $R(3.3.3.4.4)_{n/d}$ have n ribbons with squares and n ribbons with triangles, if n is even, and twice these numbers if n is odd. Examples of acoptic binary ribboned rods are shown in Fig. 7. For selfintersecting binary ribboned rods it is most appropriate to show their cross-section, see Fig. 8; each of these is an *isogonal* polygon, with sides alternating in length in ratio $\sqrt{3}/2$.

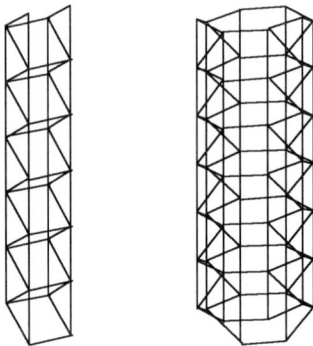

Fig. 7. Acoptic binary ribbons $R(3.3.3.4.4)_2$ and $R(3.3.3.4.4)_4$

Fig. 8. Isogonal polygons with ratio of sides $\sqrt{3}/2 = 0.866\ldots$. These are cross-sections of binary ribbons $R(3.3.3.4.4)_n$ with $n = 4/2, 8/3, 6/2, 10/2, 10/3$, respectively

Helical rods are possible only with triangles. The rods shown in Fig. 9 are the simplest acoptic ones. They are obtainable by selecting a suitable strip from the regular tiling by triangles, and wrapping it *once* (for acoptic rods) or *several* times around a cylinder. The process is explained in Fig. 10, but see the comments in Section 7. As far as I am aware, non-acoptic helical rods have not been mentioned in the literature; my technological limitations have prevented me from showing any here.

It is worth mentioning that an extension of the notation explained in Fig. 10 could also be used to designate ribboned rods $R(3.3.3.3.3.3)_n$ as $(n, n, 2n)$, and the stacked rods $(3.3.3.3.3.3)_n$ as $(0, n, n)$.

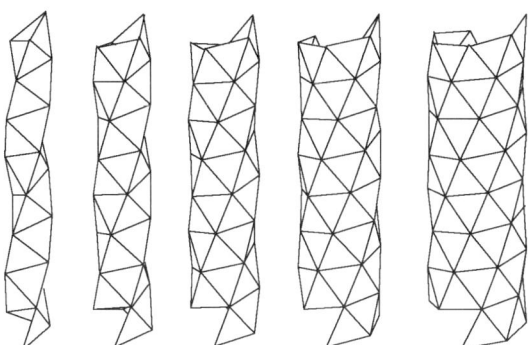

Fig. 9. Acoptic helical rods. In the notation explained in Fig. 10, they can be described by $(1, m, m + 1)$, for $m = 2, 3, 4, 5, 6$

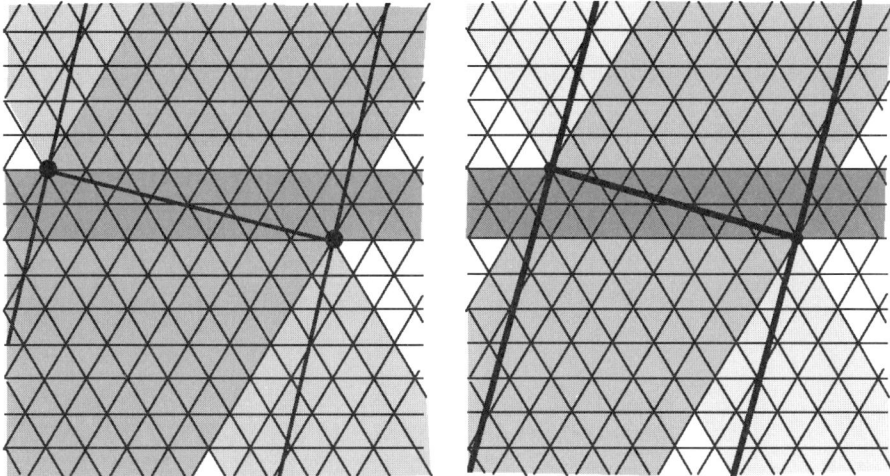

Fig. 10. The construction of the helical rods $(1, 4, 5)$ and $(2, 5, 7)$. In each case the strip between the parallel heavily drawn lines should be used to generate the rod by making the two heavy dots coincide. The perpendicular to these lines (which becomes an equatorial section of the rod) is intersected by 1, 4, and $5 = 1 + 4$, and by 2, 5, and $7 = 2 + 5$ helical strips of triangles, respectively; this explains the notation $(1, 4, 5)$ and $(2, 5, 7)$ for these helical rods. The general acoptic helical rod has symbol of the form $(m, n, m + n)$, with $1 < m < n$; the notation for non-acoptic helical rods is $(m, n, m + n)/k$ if the triangles of the rod form a rod of type $(m, n, m + n)$, that winds $k \geq 1$ times about the axis

4. UNIFORM SLABS

To describe uniform slabs it is convenient to distinguish four kinds: Tilings, corrugates, crinkles, and tunneled slabs.

Uniform tilings are slabs that are isogonal collections of regular polygons filling up a plane without overlaps. There are eleven acoptic ones – the traditional three *regular* tilings with vertex symbols (4.4.4.4), (6.6.6), (3.3.3.3.3.3), and eight *Archimedean* tilings with vertex symbols (3.3.3.4.4), (3.3.4.3.4), (4.8.8), (3.3.3.3.6), (4.6.12), (3.4.6.4), (3.12.12), (3.6.3.6) – as well as fourteen non-acoptic ones. Some of the latter were first investigated in the 19th century by A. Badoureau [3], and the full enumeration was carried out by J. C. P. Miller [28] in 1933. The first actual publication of the complete list (without illustrations) was by Coxeter *et al.* [7, Table 8]; the full enumeration with illustrations is accessible in [17]. The fourteen non-acoptic ones can be characterized by their vertex stars shown in [17], and the corresponding (modified) vertex symbols; the modification consists in a change of sign if two adjacent polygons are not oriented in the same way. The vertex symbols (see [17]) are:

$(3.3.3. - 4. - 4)$ illustrated in Fig. 11; $(3.3. - 4.3. - 4)$; $(4. - 8.8/3)$; $(8.8/3. - 8. - 8/3)$; $(-4.8/3.8/3)$; $(6. - 12.12/5)$; $(4. - 6.12/5)$; $(-3.12.6.12)$; $(4.12. - 4. - 12)$; $(3. - 4.6. - 4)$; $(3.12/5. - 6.12/5)$; $(4.12/5. - 4. - 12/5)$; $(12.12/5. - 12. - 12/5)$; $(-3.12/5.12/5)$.

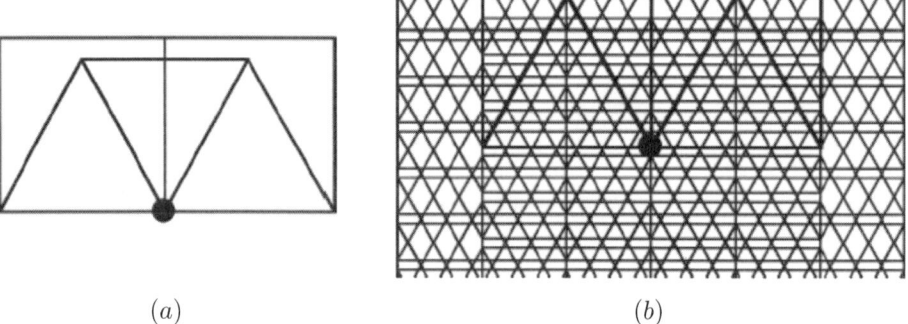

(a)   (b)

Fig. 11. The non-acoptic uniform tiling $(3.3.3. - 4. - 4)$. (a) The vertex star.
(b) A patch of the tiling, illustrating the multiple overlaps of the tiles

Corrugations or *corrugated tilings* are slabs obtained from appropriate uniform tilings (the ones with only triangles or squares as tiles) by bending them out of the plane along complete lines formed by edges of the tiling.

Hence only tilings (3.3.3.3.3.3), (4.4.4.4), (3.3.3.4.4), and (3.3.3. − 4. − 4) can be used. The amount of bending can be specified by an angle; in Fig. 12 we indicate the lines that are used for the bends by marking the *ridges* and *valleys*, and the cross-sections perpendicular to the lines of bends for the (4.4.4.4) tiling, which show the angles α of bend for each slab. The corrugated slabs arising from (3.3.3.3.3.3) are analogous. The corrugated slabs arising from (3.3.3.4.4) are the same as those arising from (3.3.3. − 4. − 4); there are three parametrized families, indicated in Fig. 13. Thus there are precisely seven parametrized families; in the acoptic case these were first specified in [41]. Hughes Jones [23] provides examples of the corrugated slabs (3.3.3.3.3.3) for a specific value of α.

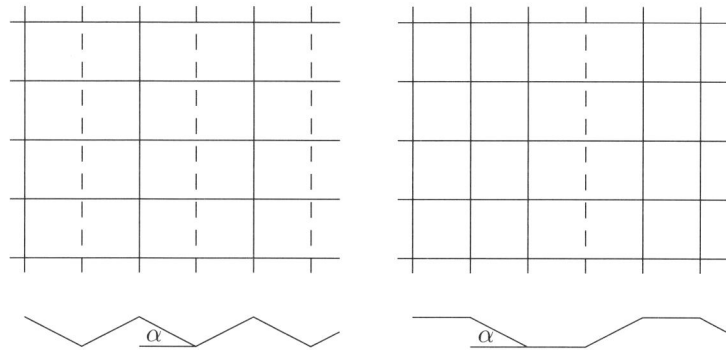

Fig. 12. The two families of corrugated slabs derived from the (4.4.4.4) tiling. (*a*) Here $0 < \alpha < \pi/2$. (*b*) Here $0 < \alpha < \pi$; note that for $\alpha \geq 2\pi/3$ the slab is selfintersecting

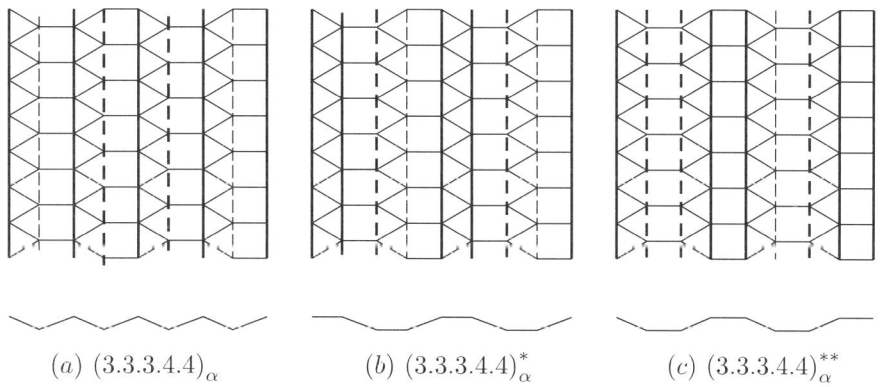

(*a*) $(3.3.3.4.4)_\alpha$ (*b*) $(3.3.3.4.4)_\alpha^*$ (*c*) $(3.3.3.4.4)_\alpha^{**}$

Fig. 13. The three corrugations that arise from (3.3.3.4.4) (and also from (3.3.3. − 4. − 4)). (*a*) The angle of bend satisfies $0 < \alpha < \pi/2$. (*b*) Here $0 < \alpha < \pi$; note that for $\alpha \geq 115.66°$ the slab is selfintersecting. (*c*) Here $0 < \alpha < \pi$ as well; note that for $\alpha \geq 125.26°$ the slab is selfintersecting

Crinkles or *crinkled tilings* are obtained in a way similar to corrugations, by appropriately bending certain tilings. The tilings which can be used are (3.3.3.3.3.3), (3.3.3.4.4), and (3.3.4.3.4). The four families of crinkles are parametrized by a real parameter. Details are shown in Fig. 14. The term "crinkle" seems to appear first in [23], where it is used to denote the crinkle $(3.3.3.3.3.3)^{\#}_{\alpha}$ for a specific value of α. The crinkle in Fig. 14(a) was devised by the present writer in the early 1990's; the other three crinkles were invented by William Webber shortly thereafter. They have not been published so far, and I am grateful to Professor Webber for allowing their inclusion here.

The last kind of uniform slabs are the *tunneled slabs*. A uniform tunneled slab consists of two copies of a tiling (usually a uniform tiling), placed in parallel planes, and from which an appropriate family of tiles is removed. The resulting holes in the two tilings are connected by suitable mantles of Archimedean prisms or antiprisms. (In fact, among the antiprisms only the mantle of the 3-antiprism – that is, the octahedron – is usable.) There are 18 acoptic tunneled slabs. All these appear in [41] – I am not aware of any other publication or internet page that presents them all. The non-acoptic ones are presented here for the first time.

There are two simple ways of symbolically presenting tunneled tilings. The generating symbol starts from the symbol of the tiling, by indicating in its vertex symbol the tile that is being omitted; this can be done, for example, by underlining the symbol of the omitted tile. As an illustration, the tunneled slab in Fig. 1(b) has the symbol (3.4.$\underline{6}$.4). The other way is by simply listing the vertex symbol of the tunneled slab. The former is more intuitive but there are two drawbacks. On the one hand, ($\underline{3}$.3.3.3.3.3) can denote two distinct tunneled slabs, with vertex symbols (4.4.3.3.3.3.3) and (3.3.3.3.3.3.3.3); similarly, ($\underline{3}$.6.3.6) corresponds to both (4.4.6.3.6) and (3.3.3.6.3.6). On the other hand, the two tunneled slabs that correspond to ($\underline{4}$.4.4.4), that are shown in Fig. 15, both have vertex symbol (4.4.4.4.4), hence cannot be distinguished by either unless one adds asterisks or some other ad hoc notation. Adjacency symbols (which we shall discus in the next section) distinguish between these two – the more symmetric slab has adjacency symbol $(ab^+c^+c^-b^-; ab^-c^-c^+b^+)$, the other slab has symbol $(a^+b^+c^+d^+e^+; a^-b^+c^-d^-e^-)$.

The complete list of 18 acoptic tunneled slabs is:

$$(3.3.\underline{4}.3.4) = (3.3.4.4.3.4), \quad (\underline{4}.8.8) = (4.4.8.8), \quad (4.\underline{8}.8) = (4.4.4.8),$$

$$(\underline{4}.4.4.4) = (4.4.4.4.4)^*, \quad (\underline{4}.4.4.4) = (4.4.4.4.4)^{**},$$

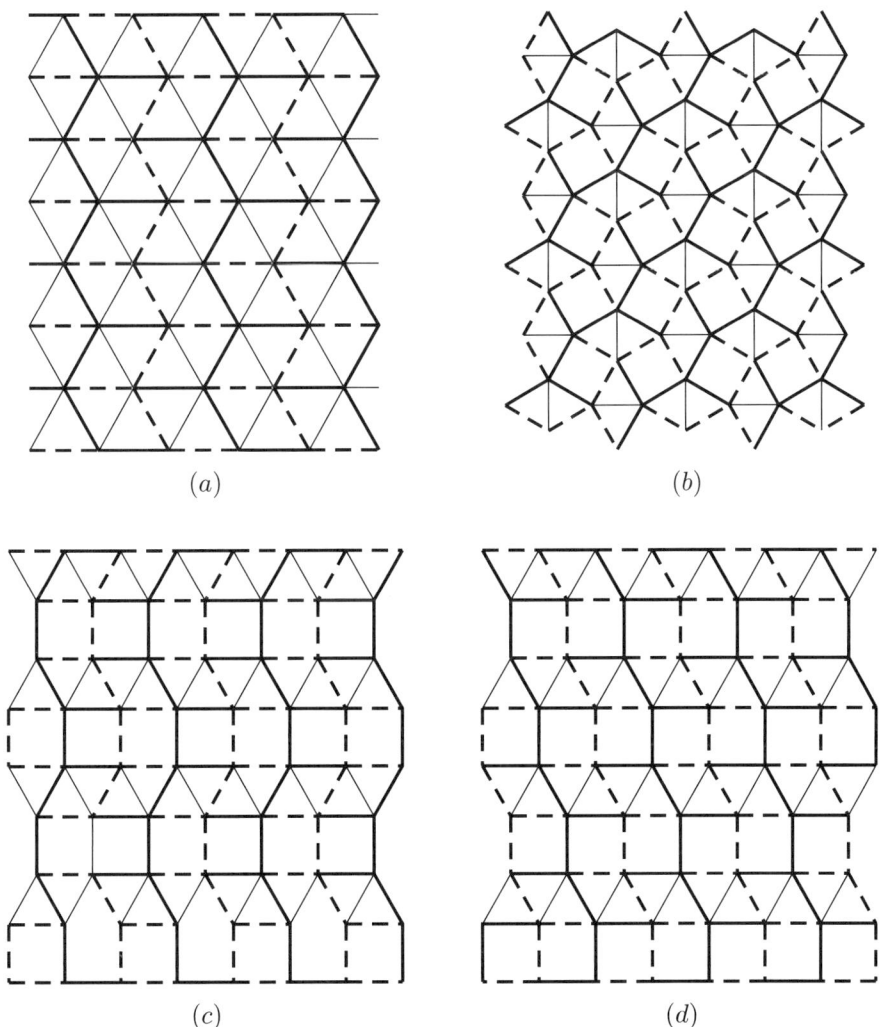

Fig. 14. The four parametrized families of crinkles. (a) $(3.3.3.3.3.3)^{\#}_{\alpha}$; (b) $(3.3.4.3.4)^{\#}_{\alpha}$; (c) $(3.3.3.4.4)^{\#}_{\alpha}$; (d) $(3.3.3.4.4)^{\#\#}_{\alpha}$, where α denotes the angle of the bend. For $\alpha \to 0$ each tends to the corresponding tiling. If $\alpha \rangle \pi$, the crinkle $(3.3.3.3.3.3)^{\#}_{\alpha}$ has a discontinuity: $(3.3.3.3.3.3)^{\#}_{\alpha}$ is a (2-dimensional) crinkle, while $(3.3.3.3.3.3)^{\#}_{\pi}$ is just a strip of triangles. If $\alpha \to \pi$, $(3.3.4.3.4)^{\#}_{\alpha} \to (3.3.3. - 4. - 4)$, which is the tiling illustrated in Fig. 11. The crinkles in (c) and (d) have in fact three different angles of bend; α denotes the bend along edges between triangles. For $\alpha \to \pi$, each of these crinkles tends to a non-acoptic structure that is not a uniform polyhedral

$$(3.3.3.3.\underline{6}) = (3.3.3.3.4.4), \quad (3.3.\underline{3}.3.6) = (3.3.4.4.3.6),$$

$$(\underline{4}.6.12) = (4.4.6.12), \quad (4.\underline{6}.12) = (4.4.4.12), \quad (4.6.\underline{12}) = (4.4.4.6),$$

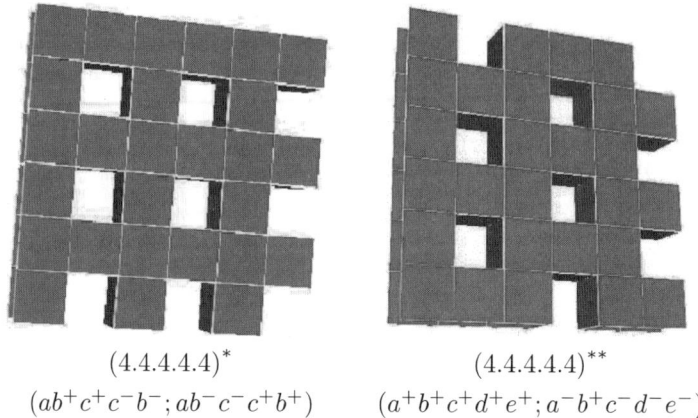

$$(4.4.4.4)^*$$
$$(ab^+c^+c^-b^-; ab^-c^-c^+b^+)$$

$$(4.4.4.4)^{**}$$
$$(a^+b^+c^+d^+e^+; a^-b^+c^-d^-e^-)$$

Fig. 15. Two distinct tunneled slabs that cannot be distinguished by their vertex symbols without the asterisks. These slabs, as well as the illustrations in Section 5, were created by Steven Gillispie

$$(\underline{3}.4.6.4) = (4.4.4.4.6), \quad (3.4.\underline{6}.4) = (3.4.4.4.4), \quad (\underline{3}.12.12) = (4.4.12.12),$$

$$(\underline{6}.6.6) = (4.4.6.6), \quad (3.3.3.3.3.\underline{3}) = (3.3.3.3.3.4.4),$$

$$(3.3.3.3.3.\underline{3}) = (3.3.3.3.3.3.3.3), \quad (3.6.\underline{3}.6) = (3.6.4.4.6),$$

$$(3.6.\underline{3}.6) = (3.3.3.6.3.6).$$

Known selfintersecting tunneled slabs are:

$$(3.3. - 4.3.\underline{-4}) = (3.3. - 4.3. - 4. - 4), \quad (\underline{4}. - 8.8/3) = (4.4. - 8.8/3),$$

$$(4.\underline{-8}.8/3) = (4. - 4. - 4.8/3), \quad (4. - 8.\underline{8/3}) = (4.4.4. - 8),$$

$$(\underline{8}.8/3. - 8. - 8/3) = (4.4.8/3. - 8. - 8/3),$$

$$(8.\underline{8/3}. - 8. - 8/3) = (8.4.4. - 8. - 8/3),$$

$$(\underline{-4}.8/3.8/3) = (-4. - 4.8/3.8/3), \quad (-4.\underline{8/3}.8/3) = (-4.4.4.8/3),$$

$$(\underline{6}. - 12.12/5) = (4.4. - 12.12/5), \quad (6.\underline{-12}.12/5) = (6. - 4. - 4.12/5),$$

$$(6. - 12.\underline{12/5}) = (6. - 12.4.4), \quad (\underline{4}. - 6.12/5) = (4.4. - 6.12/5),$$

$$(4.\underline{-6}.12/5) = (4. - 4. - 4.12/5), \quad (4. - 6.\underline{12/5}) = (4.4.4. - 6),$$

$$(\underline{-3}.12.6.12) = (-4. - 4.12.6.12), \quad (-3.12.\underline{6}.12) = (-3.12.4.4.12),$$

$$(\underline{3}. - 4.6. - 4) = (4.4. - 4.6. - 4), \quad (3. - 4.\underline{6}. - 4) = (3. - 4.4.4. - 4),$$

$$(\underline{3}.12/5. - 6.12/5) = (4.4.12/5. - 6.12/5),$$
$$(3.12/5.\underline{-6}.12/5) = (3.12/5. - 4. - 4.12/5),$$
$$(\underline{-3}.12/5.12/5) = (-4. - 4.12/5.12/5).$$

It is an unconfirmed conjecture that this list of 21 selfintersecting tunneled slabs is complete.

One additional observation about the tunneled slabs is worth mentioning. Several of them are not rigid – they admit of continuous deformations in the 3-dimensional space. However, in contrast to the corrugations and crinkles, that stay uniform through the deformation, the deformed tunneled slabs are not uniform; they are not even *monogonal* (that is, the vertex stars of the deformed slabs are not all congruent, regardless of symmetries).

5. UNIFORM SPONGES

Uniform sponges come in a wide variety of types; they are only poorly known, and no overall classification has been proposed at this time. Two kinds that we are able to characterize in a reasonable way are the *isotropic* sponges and the *layered* sponges; see Fig. 16. An example of a uniform sponge that is of neither of these kinds is shown in Fig. 17. The present account of uniform sponges is based mainly on [10], and on unpublished joint research with Steven Gillispie.

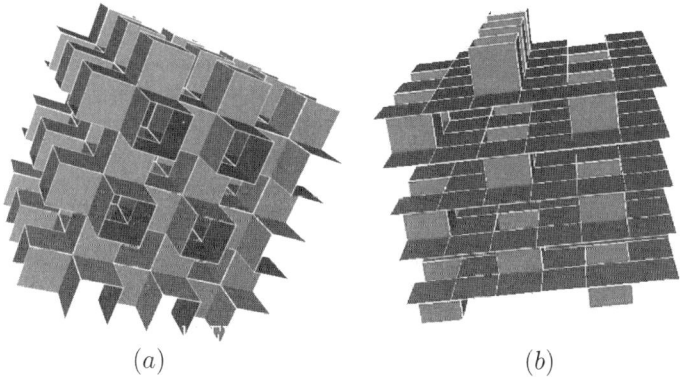

(a)   (b)

Fig. 16. (a) An isotropic sponge (4.4.4.4.4.4) with incidence symbol $(a\ a^\wedge aa^\wedge aa^\wedge; a)$. This is one of the three regular Coxeter–Petrie sponges; all are isotropic. (b) A layered sponge (4.4.4.4.4) with incidence symbol $(a^+b^+c^+d^+e^+; a^-b^-c^-d^{\wedge+}e^-)$

Fig. 17. A uniform sponge $(4.4.4.4.4) = (a^+b^+c^+d^+e^+; a^-b^+c^-d^+e^-)$ from [10], where it is denoted $N1$. Like several other sponges from [10], this is neither isotropic nor layered. It was first described in [14]

An *isotropic sponge* is periodic in three independent directions, that are equivalent under symmetries of the sponge.

A *layered sponge* consists of a series of copies of a family of polygons in parallel planes, connected by "tunnels" or "barriers". The family of polygons is usually a tiling of the plane from which some tiles have been removed, so that the "tunnels" or "barriers" can be attached. The "tunnels" and "barriers" are formed either by squares (as in the tunneled example below), or else by triangles, just as in the case of the tunneled slabs. For each plane, some of these tunnels go above and some below the plane itself. Each "barrier" is a strip, straight or zigzag, and either perpendicular to the planes or inclined at a suitable angle; again, some go up and some go down from the plane.

Here is a brief survey of the history of uniform sponges.

The first mention of such objects is in Coxeter's paper [5], where he describes specific sponges $(4.4.4.4.4.4)$, $(6.6.6.6)$, and $(6.6.6.6.6.6)$. These are *regular* in the sense that the symmetries act transitively on the flags of each, where a *flag* is a triple consisting of a vertex, an edge, and a face, all mutually incident; in that context the appropriate Schläfli symbols are $\{4,6\}$, $\{6,4\}$, and $\{6,6\}$. Moreover, Coxeter proved that there are no other regular sponges.

Several individual sponges were described in the 1950's and later; however, in most cases there was no attempt at finding any general results or points of view. The same can be said for the various appearances of sponges on the Internet. More detailed references may be found in [10]. Of greater interest is the work of Gott [12], who described several sponges, including the three regular ones. The most interesting of Gott's sponges is a (5.5.5.5.5) uniform sponge shown in Fig. 18. He also illustrates the acoptic tunneled slab we denoted by $(3.3.3.3.3.\underline{3}) = (3.3.3.3.3.3.3.3.3)$.

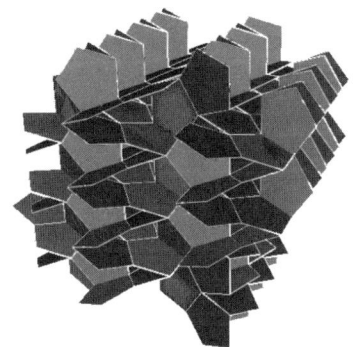

Fig. 18. The (5.5.5.5.5) uniform sponge discovered by Gott [12]

The largest collection of uniform sponges is [41], with photos of over eighty cardboard models of different sponges; the precise number depends on the definition of "different sponges", which will occupy us soon. This work introduced concepts close to isotropic and layered sponges, albeit with different terminology.

The next noteworthy collection of uniform sponges is that of Hughes Jones [23], which gives details on 26 special acoptic sponges (3^k) with triangular faces and with $7 \leq k \leq 12$. The faces of Hughes Jones' sponges are required to be among the faces of the tiling of 3-space by tetrahedra and octahedra. Of the slabs, he lists the two corrugations and one crinkle; he mentions that he knows 11 other uniform sponges with triangular faces, that are not derived from the tiling by tetrahedra and octahedra.

Goodman-Strauss and Sullivan [11] characterized the (4.4.4.4.4.4) sponges that have faces among the squares of the cubic lattice. They proved that there are precisely six distinct acoptic sponges of this kind. The approach taken in [11] excludes selfintersecting sponges.

The most recent addition to this list is the paper [10]. It presents an enumeration of the (4.4.4.4.4) acoptic uniform sponges with faces among the squares of the cubic lattice. There is a total of 15 different types; it should be noted that two of the types are slabs (see Fig. 15 above), which in the present paper are not considered to be sponges. However, the main importance of [10] is the presentation of a computational model that can be used in the study of arbitrary uniform sponges. The next several pages will be devoted to an explanation of this program.

As in many other contexts – for example, the enumeration of isohedral tilings or other symmetrical objects – the first step is the replacement of the geometric sponges by combinatorial objects which yield "candidates" for realization by geometric sponges; these can then be drawn and inspected, or investigated by other means. Several illustrations of this idea are detailed in [18].

In the case of sponges, we start by considering the vertex star of a family of sponges we wish to investigate, and label its edges incident with the central vertex. The procedure is illustrated by the example of the (4.4.4.4.4) sponges with faces in the cubic lattice; this is taken from [10], where additional details (omitted here for brevity) can be found. As indicated in Fig. 19, this particular vertex star needs two different labelings; each is shown with a canonical "vertex star label". Either no symmetry among the edges is invoked, or else the symmetry is encoded in the labeling. If an edge x^+ of a vertex star is mapped onto another by a reflection, the reflected edge needs to be marked x^-. An edge x^+ mapped onto itself by a reflection is relabeled as x. This leads to the vertex star labels shown in Fig. 19. (In cases where a vertex symbol is associated with several vertex stars, each

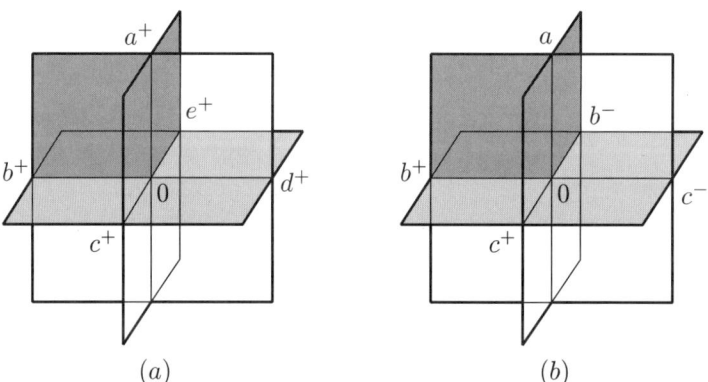

(a) (b)

Fig. 19. The two labeled vertex stars (4^5) possible in the cubic lattice. (a) Asymmetric vertex star label $a^+b^+c^+d^+e^+$. (b) Symmetric vertex star label $a^+b^+c^+c^-b^-$, using the mirror that contains the Oa edge and bisects the angle c^+Oc^-. Other symmetries that a vertex star may possess lead to other vertex star labels

vertex star has to be treated separately, for each of its (inequivalent) vertex star labels. For example, the symbol (4.4.4.4.4) in the cubic lattice is associated with the two vertex stars in Fig. 20, each of which admits several vertex star labels.) Among other possible maps of an edge onto itself we mention the *flip*, which exchanges the *sides* of the vertex star. This is indicated, for an edge labeled x, by $x^\wedge$, and similar notation is introduced for other possibilities. For example, the right part of Fig. 20 can have vertex star label $aa\widehat{\ }aa\widehat{\ }aa^\wedge$. Since we are not trying to reproduce [10] here, the reader is again urged to consult that paper for details and specifics.

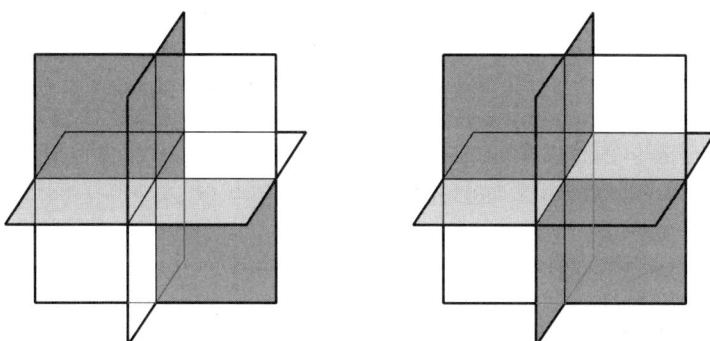

Fig. 20. The two vertex stars associated in the cubic lattice with the vertex symbol (4^6). Each admits several vertex star labels

The vertex star label constitutes the first part of the *incidence symbol*. The second part is the *adjacency symbol*. For each edge at a chosen vertex, the adjacency symbol specifies which symmetry carries its vertex star to the vertex star of the vertex at the other end of the edge. Since the labels given to the edges of one vertex star are, by the assumed isogonality, automatically transferred to all vertices, we only need to specify for each symbol of the vertex star label of the starting vertex which is the symbol of that edge in the vertex star label of the other end of the same edge. The reader is invited to verify the adjacency symbols we attached to the slabs and sponges in Figures 15, 16, and 17.

Once an incidence symbol has been chosen, the program in [10] attempts to build a valid polyhedral by combining multiple copies of the vertex star according to the incidence symbol. The various steps described in [10] apply in great generality.

6. THE "TYPE" OF SPONGES

Among uniform polyhedrals, the sponges are the least explored and under-stood. One critical question, which arises in enumerations of objects of any kind, is how to decide when two of the objects are "the same", or "of the same type", for the purposes of the enumeration. As long as we assume ahead of time that the sponges have faces among a preassigned set (such as the cubic lattice in [11] and [10], or the faces of the tiling of 3-space by tetrahedra and octahedra in [23]), there is no difficulty: Sponges that look different are assigned to different types. However, if one does not insist that the vertex stars be taken from a finite, discrete set of possibilities, this sit-uation changes. For example, the two sponges in Fig. 21 certainly appear different, but since they have the same adjacency symbol it is not obvious by what criteria they should be used to decide whether they are of the same type or different types. Clearly, no enumeration of possible types of a cer-tain kind of sponges – such as the $\left(4^5\right)$ sponges – can be attempted before this question is settled in a meaningful and practicable way.

This concern is not something abstract; it has occurred in actual at-tempts at enumeration, such as [41]. Besides two $\left(4^5\right)$ slabs, there are nine $\left(4^5\right)$ sponges shown in [41]. Two of these – which the authors de-note by $4^5(4)$ and $4^5(6)$ – are analogous to the two sponges in our Fig. 21, but with the tunnels in adjacent layers shifted by one step. Thus, although considered to be two different sponges, they are in fact two representatives of a continuum of sponges with the same adjacency symbol $(a^+b^+c^+d^+e^+;$ $a^-b^-c^-d^{\wedge-}e^-)$ and with the same symmetry group; it would appear rea-sonable to consider all these to be of the same "type". It would seem that a satisfactory definition of two sponges being of the same "type" would involve the equality of their adjacency symbols, and the continuous deformability of their vertex stars into each other, subject to certain reasonable and well-defined restrictions. However, the exact nature of these restrictions has not been determined so far. Moreover, there is still the question whether the intuitive feeling that the sponge in Fig. 21(a) is somehow different from the other sponges of the same "type" can be (or should be) rationalized and codified in some way.

As a consequence of these difficulties, it is at this time impossible to state for any vertex symbol $(a.b. \; \ldots \; .g)$ how many different "types" of sponges it admits; the only known exception is the sponge $\left(5^5\right)$ of Gott [12] that has the single realization as an unlabeled sponge, shown in Fig. 18. Even the more basic question whether there are *any* sponges with a given vertex *symbol* $(a.b. \; \ldots \; .g)$ does not have any general solution at this time. The algorithm in [10] at best only produces all candidate sponges for a given vertex *star*.

Only acoptic uniform sponges were considered in [14]. In this context, several conjectures made there are still open, and it is appropriate to mention them now.

- No vertex star of an acoptic sponge has more than 12 faces.
- If the vertex star of an acoptic sponge is incident with more than eight faces, then all faces are triangles.
- No acoptic sponge has only faces with seven or more sides.

There is no information on which to base guesses concerning the analogues of these conjectures if the sponges are not assumed to be acoptic.

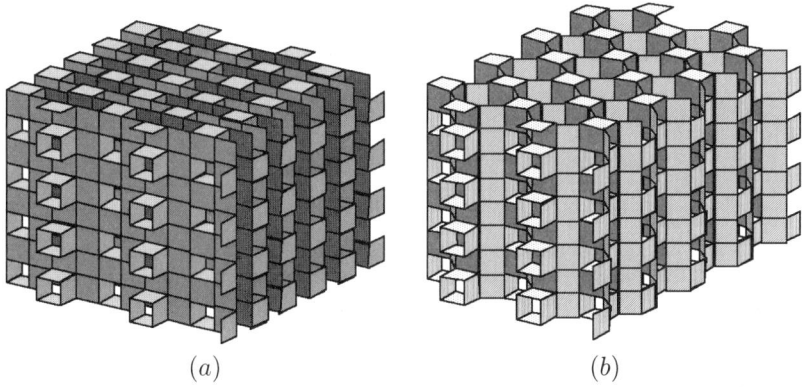

(a) (b)

Fig. 21. The sponge in (a) is another view of the sponge in Fig. 16(b). It can be "stretched" (or "flexed") in a continuous way to yield the one in (b). Both have the same adjacency symbol $(a^+b^+c^+d^+e^+; a^-b^-c^-d^{\wedge+}e^-)$, hence should be considered as being of the same "type". The sponge (a) appears in [41] as $4^5(8)$, but the sponge (b) is missing from that collection

7. NOTES AND COMMENTS

As noted earlier, an extension of the symbol $(m, n, m+n)$ for helical rods to the case $m = 0$ covers the stacked rods $(3.3.3.3.3.3)_n$. However, the very different properties of stacked rods as compared with helical rods make it reasonable to treat them separately. Analogously, the ribboned rods $R(3.3.3.3.3.3)_n$ can be interpreted as helical rods $(n, n, 2n)$; again, this is more of a sidelight than a contribution to understanding.

The study of helical rods developed from applications in biology. A detailed account was published by van Iterson [40] in 1907, although some aspects were investigated even earlier. The work [40] concentrates on symmetrically arranged points situated on spirals that wind on cylinders. In particular, van Iterson studies (among more general kinds) point sets that can be used as centers of congruent balls, each of which touches six other balls. It is obvious that connecting the centers of such families of balls yields uniform helical rods. On the other hand, there are connections to the distribution of leaves or florets on various plants (phyllotaxis), to flagella of various bacteria, to subunits of certain microtubules, and various other biological entities; for details see Thompson [37] or Erickson [8]. Many other writers (mostly non-mathematicians) dealt with this topic; it is amusing that Alan Turing also studied this topic, see [39, pp. 141–144], and, in particular, [38].

The helical rods have a rich history. In particular, the smallest one – $(1, 2, 3)$ in the notation of Section 3 – was repeatedly discovered by various workers; the name "tetrahelix" for it is reported to have been coined by R. Buckminster Fuller in [9]. As noticed by Hurley [24] and Coxeter [6], the tetrahelix can be obtained as a stack of regular tetrahedra, with adjacent tetrahedra related by reflection in the plane of the common face. Boerdijk [4] obtained it earlier, as a stack of regular tetrahedra related by screw-motion, as well as in the guise of a family of balls. He was not the first, as van Iterson [40] found the helical structures and the sphere packings in 1907. It is interesting that van Iterson credits (without explicit references) the tetrahelix of spheres (but not the other helices he investigates) to Federico Delpino; I was not able to determine which publication of Delpino's lead to the attribution.

Almost all uniform rods can be interpreted as resulting from suitable strips of uniform tilings of the plane. For the unary stacked and ribboned rods the square tiling and the tiling by triangles can be used, while binary ribboned rods and binary acoptic stacked rods require the (3.3.3.4.4) tiling.

A fine but non-obvious point needs to be mentioned; it is stressed (among others) by Erickson [8]. The often invoked picture (see Fig. 10 above) of the construction of uniform helical rods intimates that a suitable strip of the regular tiling (3.3.3.3.3.3) is bent into a cylinder. By this is meant that the vertices are placed on a cylinder, and the triangles are kept planar – so that the resulting rod *is not convex*. In particular, the equal length of the edges of the rod is *not compatible* with the assumption that each vertex is equally distant *in the geodesic distance on the cylinder* from all the adjacent vertices. Naturally, the vertices on each spiral are equidistantly disposed. Additional considerations of these matters can be found in Lord [25], where also various related constructions that lead to symmetric (but not uniform) rods are described. Lord considers, in particular, isohedral structures with polygonal regions (such as hexagons) on cylinders, stressing their occurrence in biology and structural chemistry. Related information appears in [26].

Note (Added January 12, 2010). After the present paper was submitted, Dr. William T. Webber informed me that he has found an additional type of uniform rods with vertex symbol (3.3.3.3.3.3). In analogy to the situation concerning uniform slabs, these objects deserve to be called *crinkled rods*. Webber is preparing a paper with details of the construction of these rods.

8. RELATED POLYHEDRA-LIKE STRUCTURES

In this section we shall discuss various modifications of "uniform polyhedrals"; hence the word "polyhedral" and related expressions will be used without the implication of uniformity. The results mentioned here show that interesting geometry may result by considering polyhedrals more general than the uniform ones. It is my hope that the present exposition may lead to such investigations.

Helical triangle-faced polyhedrals (with not necessarily equilateral triangles as faces) have been studied by engineers in several contexts. Guest and Pelegrino [19] study the cases in which a given triangle leads to helical isohedral rods that can be collapsed to planar sets of triangles, with relatively small stresses (or deformations) during the transition from one configuration to another.

Raskin's Ph.D. thesis [32] deals with deployable structures, that is, rods and slabs that can change their dimensions. These questions are addressed from an engineering point of view, but many of the examples are close to the topic of the present paper. For example, Figure 1.3 of [32] can be interpreted (in a manner not intended by Raskin) as a unary stacked rod $\left(3^6\right)_8$; it is attributed to I. Hegedus [22]; Figure 1.6 could inspire the stacked rod $\left(3^6\right)_3$, but a better example is given by Miura [30]. Figure 3.39 in [32] can be interpreted as the ribboned rod $R(3.3.3.3.3.3)_3$, with triangles that are isosceles but not equilateral; it is attributed to You and Pellegrino [45]. Many other examples are given in [32], together with a wealth of references.

Barker and Guest [1] describe several types of isogonal and isohedral rods with triangular faces, and investigate bucking patterns that present some features of this kind.

In a different engineering application, Tarnai [36] describes the buckling of cylindrical shells as giving rise to isogonal stacked rods of type $(3^6)_n$, but with isosceles triangles; this is based on Miura [29]. He also presents cylindrical rods obtainable from the uniform tiling (3.3.4.3.4); these are not isogonal (there are two or more orbits of vertices) but the faces are planar polygons. Miura (in [29]) also describes analogues of isohedral acoptic stacked rods, with faces that are isosceles trapezes.

A very interesting deformable isohedral slab is described by Miura [30], see Fig. 22; it is meant to be employed in space structures such as arrays of solar cells. Suitable changes of the angle α of bend along the ridges and valleys lead to a complete collapse. It may be noted that if the parallelograms are rhombi and $\tau = 60°$, then the rhombi can be interpreted as pairs of equilateral triangles and the crinkle $(3.3.3.3.3.3)_\alpha^\#$ shown in Fig. 14(a) is obtained. On the other hand, if the parallelograms are squares, and only the heavily drawn zigzags are taken as ridges and valleys (they are straight lines in this situation), then the corrugation $(4.4.4.4)_\alpha$ shown in Fig. 12(a) results. However, Miura's claim that the parallelogram tiling "will fold into a point" is clearly wrong; a similar error occurs in [29]. The precise properties of the folding transformation still need to be investigated, in terms of the dependence on the angle τ and on the ratio of the parallelogram sides. A partial analysis is given by Piekarski [31], who also considers variants that lead to curved structures.

From the facts mentioned above it is impossible to escape the conclusion that engineers, architects, biologists and others could have benefited from closer contact with mathematicians; even more evident is the fact that mathematicians could have been inspired in important ways by acquaintance with the works and problems in these applied disciplines. It should be borne in mind that the list of results mentioned here is very haphazard, since there is no reasonable way for a mathematician to access all of the relevant literature. The shocking failure in this matter of the reviewing journals (Math Reviews and Zentralblatt) is illustrated by the fact that almost none of the references we give – including the books [41] and [26], that clearly have mathematical importance and relevance – are even mentioned in either of these journals. Among the editorial policies of the refereeing journals there seems to be a deeply ingrained aversion to anything that has to do with the geometry of polyhedral objects in 3-space; this stand leads to a great loss to both pure and applied mathematics, and to culture in general.

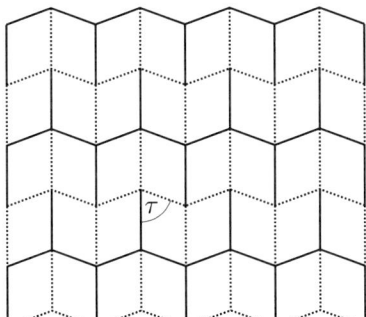

Fig. 22. An isohedral tiling by parallelograms, described by Miura [30]. It can be folded to yield a crinkled slab; heavy solid edges indicate ridges, dashed ones valleys

REFERENCES

[1] Barker, R. J. P. and Guest, S. D., Inflatable triangulated cylinders, in: *IUTAM-IASS Symposium on Deployable Structures: Theory and Applications,* S. Pellegrino and S. D. Guest, eds. Kluver, Dordrecht 2000, pp. 17–26.

[2] Bulatov, V., Uniform polyhedra.
http://bulatov.org/polyhedra/uniform/index.html

[3] Badoureau, A., Mémoire sur les figures isoscèles, *J. École Polytechn,* **30** (1881), 47–172.

[4] Boerdijk, A. H., Some remarks concerning close-packing of equal spheres, *Philips Research Reports,* **7** (1952), 303–313.

[5] Coxeter, H. S. M., Regular skew polyhedra in three and four dimensions, and their topological analogues, *Proc. London Math. Soc.* (2), **43** (1937), 33 62. Improved reprint in: *Twelve Geometric Essays,* Southern Illinois University Press, Carbondale II, 1968. Reissued by Dover, 1999.

[6] Coxeter, H. S. M., The simplicial helix and the equation $\tan nt = n \tan t$, *Canad. Math. Bull.,* **28** (1985), 385–393.

[7] Coxeter, H. S. M., Longuet-Higgins, M. S. and Miller, J. C. P., Uniform polyhedra, *Philos. Trans. Roy. Soc. London* (A), **246** (1953/54), 401–450.

[8] Erickson, R. O., Tubular packings of spheres in biological fine structure, *Science,* **181** (1973), No. 4101, pp. 705–716.

[9] Fuller, R. B., *Synergetics: Explorations in the Geometry of Thinking*, Macmillan, New York 1975, 1982.

[10] Gillispie, S. D. and Grünbaum, B., The {4, 5} isogonal sponges on the cubic lattice, *Electronic J. Combinatorics,* **16** (2009), R22.

[11] Goodman-Strauss, C. and Sullivan, J. M., Cubic polyhedra, *Discrete Geometry: in Honor of W. Kuperberg's 60th Birthday,* A. Bezdek, ed. Dekker, New York 2003, pp. 305–330.

[12] Gott, J. R., III, Pseudopolyhedrons, *The American Mathematical Monthly,* **74** (1967), pp. 497–504.

[13] Grünbaum, B., Regular polyhedra – old and new, *Aequationes Math,* **16** (1977), 1–20.

[14] Grünbaum, B., Infinite uniform polyhedra, *Geombinatorics,* **2** (1993), 53–60.

[15] Grünbaum, B., Metamorphoses of polygons, in: *The Lighter Side of Mathematics, Proc. Eugène Strens Memorial Conference,* R. K. Guy and R. E. Woodrow, eds. Math. Assoc. of America, Washington, D.C.. 1994 Pp. 35–48.

[16] Grünbaum, B., "New" uniform polyhedra, in: *Discrete Geometry: In Honor of W. Kuperberg's 60th Birthday,* Monographs and Textbooks in Pure and Applied Mathematics, vol. 253. Marcel Dekker, New York, 2003. pp. 331–350.

[17] Grünbaum, B., Miller, J. C. P. and Shephard, G. C., Uniform tilings with hollow tiles, *"The Geometric Vein – The Coxeter Festschrift",* C. Davis, B. Grünbaum and F. A. Sherk, eds. Springer-Verlag, New York – Heidelberg – Berlin 1982, pp. 17–64.

[18] Grünbaum, B. and Shephard, G. C., Incidence symbols and their applications, in: *Relations Between Combinatorics and Other Parts of Mathematics,* D. K. Ray-Chaudhuri, ed. Proc. Symposia Pure Math. vol. 34, pp. 199–244. Amer. Math. Soc., Providence, R.I., 1979.

[19] Guest, S. D. and Pellegrino, S., The folding of triangulated cylinders. Parts I, II, *J. Applied Mechanics,* **61** (1994), 773–777 and 778–783.

[20] Hart, G. W., Uniform polyhedra.
http://www.georgehart.com/virtual-polyhedra/uniform-info.html

[21] Hart, G. W., Archimedean polyhedra.
http://www.georgehart.com/virtual-polyhedra/archimedean-info.html

[22] Hegedus, I., Branching of equilibrium paths in a deployable column, *Internat. J. of Space Structures,* **8** (1993), 119–125.

[23] Hughes Jones, R., Enumerating uniform polyhedral surfaces with triangular faces, *Discrete Math.,* **138** (1995), 281–292.

[24] Hurley, A. C., Some helical structures generated by reflexions, *Austral. J. Physics,* **38** (1985), 299–310.

[25] Lord, E. A., Helical structures: The geometry of protein helices and nanotubes, *Structural Chemistry,* vol. **13**, nos. 3/4 (2002), 305–314.

[26] Lord, E. A., Mackay, A. L. and Ranganathan, S., *New Geometries for New Materials,* Cambridge Univ. Press, 2006.

[27] Maeder, R., All uniform polyhedra.
http://www.mathconsult.ch/showroom/unipoly/list.html

[28] Miller, J. C. P., *On Stellar Constitution, on Statistical Geophysics, and on Uniform Polyhedra (Part 3: Regular and Archimedean Polyhedra),* Ph.D. Thesis 1933. (Copy deposited in Cambridge University Library).

[29] Miura, K., Proposition of pseudo-cylindrical concave polyhedral shells, *IASS Symposium on folded plates and prismatic structures,* Vol. 1, Vienna, 1970.

[30] Miura, K., Concepts of deployable space structures, *International J. of Space Structures.,* **8** (1993), 3–16.

[31] Piekarski, M., Constructional solutions for two-way-fold-deployable space trusses, *IUTAM-IASS Symposium on Deployable Structures: Theory and Applications,* S. Pellegrino and S. D. Guest, eds. Kluver, Dordrecht, 2000, pp. 301–310.

[32] Raskin, I., *Stiffness and stability of deployable pantographic columns,* PhD thesis in Civil Engineering, University of Waterloo, 1998.

[33] Skilling, J., The complete set of uniform polyhedra, *Philos. Trans. Roy. Soc. London* (A), **278** (1975), 111–135.

[34] Sopov, S. P., Proof of the completeness of the enumeration of uniform polyhedra [in Russian], *Ukrain. Geom. Sbornik,* **8** (1970), 139–156.

[35] Szepesvári, I., On the number of uniform polyhedra. I, II. [in Hungarian], *Mat. Lapok,* **29** (1977/81), 273–328.

[36] Tarnai, T., Folding of uniform plane tessellations, *Origami Science and Art. Proc. of the Second Int. Meeting of Origami Science and Scientific Origami, Otsu, Japan, 1994* (ed.: K. Miura), Seian Univ. of Art and Design, 1997, 83–91.

[37] Thompson, D'A. W., *On Growth and Form,* Cambridge Univ. Press, 1948.

[38] Turing, A. M., The chemical basis of morphogenesis, *Philos. Trans. Roy. Soc. London, Ser. B,* **237** (1952), 37–72. Reprinted, together with previously unpublished manuscripts, in *Morphogenesis,* Collected Works of A. M. Turing, P. T. Saunders, ed. North-Holland, Amsterdam, 1992.

[39] Turing, S., *Alan M. Turing,* W. Heffer & Sons, Cambridge, 1959.

[40] van Iterson, G., Jun., *Mathematische und mikroskopisch-anatomische Studien über Blattstellungen, nebst Betrachtungen über den Schalenbau der Miliolinen,* Gustav Fischer, Jena, 1907.

[41] Wachman, A., Burt, M. and Kleinmann, M., *Infinite Polyhedra,* Technion, Haifa 1974; reprinted 2005.

[42] Weisstein, E. W., "Uniform Polyhedron." From *MathWorld* – A Wolfram Web Resource. http://mathworld.wolfram.com/UniformPolyhedron.html

[43] Wikipedia, List of uniform polyhedra.
http://en.wikipedia.org/wiki/List_of_uniform_polyhedra

[44] Wikipedia, Expanded list of uniform tilings.
http://en.wikipedia.org/wiki/Uniform_tiling#Expanded_lists_of_uniform_tilings

[45] You, Z. and Pelegrino, S., Cable-stiffened pantographic deployable structures. I. Triangular mast, *AIAA Journal,* vol. **34**, no. 4 (1996), 813–820.

Branko Grünbaum

Department of Mathematics
University of Washington
Box 354350
Seattle, WA 98195-4350
USA

e-mail: grunbaum@math.washington.edu

BOLYAI SOCIETY
MATHEMATICAL STUDIES, 24

Geometry –
Intuitive, Discrete, and Convex
pp. 187–203.

GEOMETRIC TRANSVERSAL THEORY: $T(3)$-FAMILIES IN THE PLANE

ANDREAS F. HOLMSEN

A line that intersects every member of a finite family F of convex sets in the plane is called a line transversal to F. In this paper we will survey the main results and open problems concerning $T(3)$-families: Finite families of convex sets in the plane in which every subfamily of size 3 admit a line transversal.

1. INTRODUCTION

Let F be a finite family of compact convex sets in the Euclidean plane. In fact we will restrict ourselves to the case when the members of F are *convex bodies,* i.e. compact convex sets with non-empty interiors, to avoid dealing with certain (uninteresting) degenerate situations. A *line transversal* (or just *transversal*) to F is a straight line that intersects every member of F. For a positive integer $k \geq 3$ we call F a $T(k)$-*family* if every subfamily of size at most k has a transversal. In this paper we will be concerned with some fundamental results concerning $T(k)$-families, and in particular the case when $k = 3$.

It is a well-known fact that there is no *Helly-type* theorem for line transversals to general families of convex bodies in the plane. This means that for every integer $k \geq 3$ one can find a $T(k)$-family that does not have a line transversal. In order to obtain Helly-type theorems one must make additional assumptions on the shape or the relative positions of the bodies. One such direction which has been very fruitful and produced many results is the case when F is family of (disjoint) translates of a convex body. The recent development of this line of research has been surveyed by the present author in [17].

The basic problem that we will discuss in this paper is the following: *For every positive integer $k \geq 3$, does there exist a real number $\alpha(k) \in (0,1)$ such that for any $T(k)$-family in the plane there is a line that meets at least $\alpha(k) \, |F|$ members of F?* Or in other words, does every $T(k)$-family F admit a *partial transversal* whose size is at least some fixed fraction of the size of F?

Such a function does indeed exist as was shown by Katchalski and Liu [22]. Moreover, they show that $\lim_{k\to\infty} \alpha(k) = 1$, which is an interesting phenomenon in view of the non-existence of a Helly-type theorem. Actually, the Katchalski-Liu result only deals with the case when $k \geq 4$. The existence of $\alpha(3)$ follows as a consequence of a result by Kramer [23], and his method was later improved by Eckhoff [9]. A different proof of the existence of $\alpha(3)$ is given by Matoušek [24] which uses an argument due to Alon and Kalai [1], but no attempt is made to optimize the value (and in fact it establishes a more general result).

In 1978 Katchalski conjectured that $\alpha(3)$ can be taken close to $\frac{2}{3}$. (See [12] for more detailed references). To make this more precise we define $\alpha(n,k) \in (0,1)$ to be the greatest fraction such that every $T(3)$-family of size n has a partial transversal of size $\alpha(n,k)n$. Let us then define

$$\alpha(k) = \lim_{n \to \infty} \alpha(n,k)$$

Katchalski's conjecture then states that $\alpha(3) = \frac{2}{3}$.

Recently Eckhoff [12] revisited the study of $T(k)$-families and showed that $\alpha(k) \geq 1 - \frac{\sqrt{2}}{\sqrt{k-1}}$. Furthermore, Eckhoff gives a construction which shows that Katchalski's conjecture is best possible, if true. Finally, Eckhoff conjectures that $\alpha(k) = \frac{k-1}{k}$ which is an extension of Katchalski's conjecture.

This paper will discuss some approaches towards improving the bound on $\alpha(k)$, and in particular the value of $\alpha(3)$. In Sections 2–4 we will use various results (some new and some old) to obtain incrementally better lower bounds of $\alpha(3)$. The best known bounds on $\alpha(3)$ are due to the present author [18] where it is shown that $\frac{1}{3} \leq \alpha(3) \leq \frac{1}{2}$, thus disproving Katchalski's (and Eckhoff's) conjecture. This approach will be reviewed in Section 5. Even though the method in [18] gives the current best lower bound for $\alpha(3)$, we feel it worthwhile to report on the other approaches.

In Section 2 we introduce a (new) condition on a family of convex bodies which we call tightness: A triple of convex bodies is called *tight* if the union of the convex hulls of the pairs is convex. A special case of this is when the triple has a line transversal. We show that if F is a family where every triple is tight then there is a partial transversal of size $\frac{1}{8}|F|$. Although this approach could at best give a fraction of $\frac{2}{5}$ we believe that it could yield some new insight to the problem. (See Figure 2 and Lemma 3 below).

In Section 3 we review the recent *colorful version of Hadwiger's transversal theorem* due to Arocha, Bracho, and Montejano [4]. We will show that their theorem implies that $\alpha(3) \geq \frac{1}{5}$.

In Section 4 we will review *Eckhoff's partition theorems* for $T(3)$- and $T(4)$-families. In general Eckhoff's theorems concern partitioning a $T(k)$-family into as few parts as possible such that each part has a transversal. Such theorems are often called *Gallai-type theorems*. An immediate consequence of these results is that $\alpha(3) \geq \frac{1}{4}$ and $\alpha(4) \geq \frac{1}{2}$.

In Section 5 we give a short outline of the recent results due Eckhoff [12] and the present author [18], and in Section 6 we conclude with discussions of further generalizations and extensions.

The topic dealt with in this paper is just a small part of the broader subject *geometric transversal theory*. There are several excellent surveys on the subject, and the interested reader should consult [10, 14, 16, 28] for more information and references.

2. TIGHT TRIPLES

It is a basic fact that a triple of convex bodies in the plane have a transversal if and only if there is a partition of the triple into two non-empty parts $\{A\}$ and $\{B, C\}$, say, such that $A \cap \text{conv}(B \cup C) \neq \emptyset$. In this section we will show that this property can be weakened and still we are guaranteed a large partial transversal.

Let A, B, and C be convex bodies in the Euclidean plane. We will say that the triple A, B, C is *tight* if the following holds:

$$\text{conv}(A \cup B) \cap \text{conv}(A \cup C) \cap \text{conv}(B \cup C) \neq \emptyset.$$

For instance, if two of the bodies, A and B, say, have non empty intersection, then A, B, C is tight. More generally, if A, B, C have a transversal, then A,

B, C is tight, but it should be clear that these conditions are not necessary to guarantee tightness. (See Figure 1).

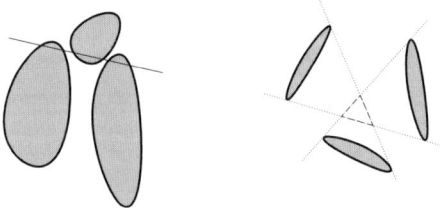

Fig. 1. Examples of tight triples

It can easily be shown that the following are equivalent:

1. *A, B, C is a tight triple.*

2. *The union* $\operatorname{conv}(A \cup B) \cup \operatorname{conv}(A \cup C) \cup \operatorname{conv}(B \cup C)$ *is convex.*

In this section we establish the following result.

Theorem 1. *Let F be a finite family of convex bodies in the Euclidean plane. If every triple of F is tight, then F has a partial transversal of size at least $\frac{1}{8}|F|$.*

The fraction $\frac{1}{8}$ in Theorem 1 could probably be improved. It is quite possible that $\frac{2}{5}$ is the correct fraction, and it is simple to see this would be best possible. (See Figure 2).

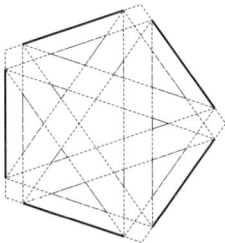

Fig. 2. Every triple among the five segments is tight, but any line intersects at most two of the segments. Each segment can be duplicated m times to obtain arbitrarily large families

Since a triple that has a transversal is also tight, Theorem 1 implies the following.

Corollary 2. $\alpha(3) \geq \frac{1}{8}$.

The main lemma. Here we present the main lemma that will be used to show Theorem 1. First we need to introduce some notions. Let $F = \{A_1, \ldots, A_{2k}\}$ be a family of convex bodies. By a *matching* of F we mean, as usual, a partition of F given by means of a splitting of $\{1, \ldots, 2k\}$ into k disjoint pairs. Each element of the matching is thus a pair of convex bodies $\{A_i, A_j\}$ and we will refer to $\mathrm{conv}(A_i \cup A_j)$ as a *component* of the matching.

Lemma 3. *For $k > 0$, let $F = \{A_1, \ldots, A_{2k}\}$ be a family of convex bodies in the plane, and suppose every triple of F is tight. There exists a matching of F such that the components of the matching have a point in common.*

In what follows we will speak of separation of convex bodies. It is a basic theorem of convexity that two disjoint convex bodies in the plane can be separated by a line, and it will be necessary to distinguish *weak separation* from *strict separation:* Two convex bodies are weakly (strictly) separated by a line l if they are contained in opposite *closed* (*open*) halfplanes bounded by l.

Proof of Lemma 3. First consider the degenerate case when the members of F are points: It is then clear that the condition of tightness implies that the points are all collinear, thus we have $2k$ points on the line, and the statement is trivial. For the general case of convex bodies, the crucial observation is the following.

Claim 4. Let $\{A_0, A_1, B_0, B_1\}$ be a family of convex bodies in the plane such that every triple is tight. Let D be a disk of positive radius which is tangent to $\mathrm{conv}(A_0 \cup A_1)$ and $\mathrm{conv}(B_0 \cup B_1)$ at distinct points, and denote by l_A and l_B the unique (weak) separating lines of D and $\mathrm{conv}(A_0 \cup A_1)$ and $\mathrm{conv}(B_0 \cup B_1)$, respectively. Then one of the following cases must occur:

(i) There is a matching of the members of $\{A_0, A_1, B_0, B_1\}$ such that the components intersect the interior of D.

(ii) Some member of $\{A_0, A_1, B_0, B_1\}$ is separated (weakly) from D by both l_A and l_B.

Proof of Claim 4. We first consider the case when l_A and l_B are non-parallel. We will assume that *(ii)* does not hold and exhibit the desired matching: First suppose one of the bodies, A_0, say, is tangent to D. Since B_0 and B_1 both contain points that are *strictly* separated from A_0 by l_A, it follows that $\mathrm{conv}(A_0 \cup B_i)$ intersects the interior of D for $i = 0$ and 1. The

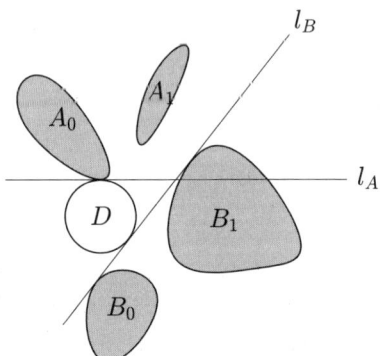

Fig. 3. A_0 is tangent to D

same argument applies (with the roles reversed) if one of the B_i is tangent to D, so we may assume that $B_i \cap D = \emptyset$ for $i = 0$ and 1. (See Figure 3).

It is now easily seen that if neither $\mathrm{conv}(A_1 \cup B_0)$ nor $\mathrm{conv}(A_1 \cup B_1)$ intersect the interior of D, then the triple $\{A_1, B_0, B_1\}$ cannot be tight: The point of tangency between l_B and D is (1) in the interior of $\mathrm{conv}(A_1 \cup B_0 \cup B_1)$, (2) on the boundary of $\mathrm{conv}(B_0 \cup B_1)$, and (3) disjoint from $\mathrm{conv}(A_1 \cup B_0)$ and $\mathrm{conv}(A_1 \cup B_1)$.

It remains to consider the case where no member of $\{A_0, A_1, B_0, B_1\}$ is tangent to D. Orient the lines such that D is contained in the third quadrant. Then the first quadrant does not contain any of the bodies of $\{A_0, A_1, B_0, B_1\}$, and we can assume (after relabeling if necessary) that l_A meets the bodies in order $A_0 D A_1$, and l_B meets the bodies in order $B_0 D B_1$. (See Figure 4). It is clear that $\{A_0, B_1\}$, $\{A_1, B_0\}$ is the desired matching.

The case when l_A and l_B are parallel follows by the exact same argument as above. In fact, we don't have to assume that *(ii)* does not occur, since it is clearly impossible! This concludes the proof of the claim. ∎

For a given matching of F, let $r \geq 0$ be the minimal radius of a disk that intersects each component of the matching. If $r = 0$, then we are done, so suppose $r > 0$. We will show that we can find a new matching with associated radius $r' < r$.

Let D be a disk of minimal positive radius that intersects every component of a given matching. Since D is of minimal radius, some components of the matching must be tangent to D, and let $P = \{p_1, \ldots, p_n\}$ denote these points of tangency. If P can be strictly separated from the center of D, then we could nudge D slightly towards the $\{p_i\}$ such that every p_i

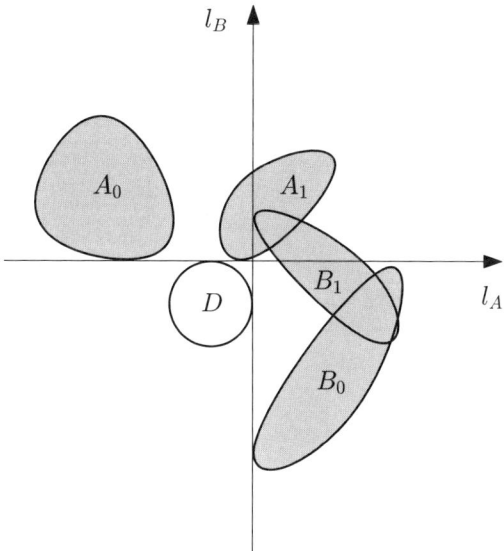

Fig. 4. No member of $\{A_0, A_1, B_0, B_1\}$ is tangent to D

is contained in the interior of D, which would still intersect the interior of all the remaining components, but this would imply that D was not minimal. It therefore follows by Carathéodory's theorem that the center of D is contained in $\mathrm{conv}(Q)$ for some $Q \subset P$ with $|Q| \leq 3$. If $|Q| = 2$ then it follows by Claim 4 that we can rearrange the matching of the four bodies involved, obtaining a matching with fewer tangency points. If $|Q| = 3$ then it also follows from Claim 4 that some two of the three components involved can be rearranged and again reducing the number of tangency points: If no pair of components involved can be rearranged, they all fall into case *(ii)* of Claim 4, which results in a triple which is not tight. In this way we can keep removing tangency points until the remaining ones can be strictly separated from the center of D. The resulting matching has an associated radius strictly smaller than the radius of D. This completes the proof of the Lemma 3. ∎

Proof of Theorem 1. First assume there is an even number of members in F (here we need disregard at most one member). By Lemma 3 there is a matching of F and a point p contained in every component. For each member of A of F let L_A denote the set of lines through p that intersect A. If $p \notin A$ the union of lines in L_A, denoted by $\{L_A\}$, is a double cone with apex at p and we refer to the half which contains A as the *positive* part of $\{L_A\}$

and the other half as the *negative* part. Let $M \subset F$ denote the subfamily whose cones $\{L_A\}_{A \in M}$ are minimal with respect to inclusion. Using, for instance, Proposition 19 of [16], we may assume that there are pairs X, Y in M for which $L_X \cap L_Y = \emptyset$. Among such pairs, choose one, A, B for which the angular distance between the positive parts of L_A and L_B is maximal. Let l_1 and l_2 be the boundary lines of L_A, and l_3 and l_4 the boundary lines of L_B. (See Figure 5). We will show that there is a subfamily $G \subset F$, with $|G| > \frac{1}{2}|F|$, such that every member of G meets at least one of the l_i, $i = 1, \ldots, 4$, which will complete the proof.

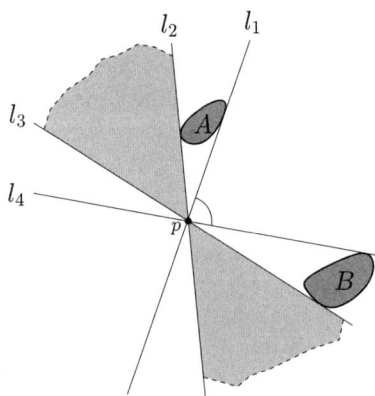

Fig. 5. $\{A, B\} \subset M \subset F$ are chosen so the angle β is maximal. This implies that the open double cone bounded by l_2 and l_3 (shaded region) cannot contain any members of F

Consider the open double cone bounded by the positive part of L_A (L_B) and the negative part of L_B (L_A): It cannot contain any member of F as this would contradict the maximal choice of angle between A and B. (See Figure 5). The open cone bounded by the negative parts of L_A and L_B cannot contain any member of F, as this would, together with A and B, violate the condition that every triple is tight. Thus every member of F which does not meet any l_i must be contained in the open cone bounded by the positive parts of L_A and L_B, but since we have a matching of F for which every component contains p, this open cone contains at most $\lfloor \frac{|F|}{2} \rfloor - 2$ members of F. ∎

3. THE COLORFUL HADWIGER THEOREM

One of the classical results of geometric transversal theory is Hadwiger's theorem from 1957 [15]: *A finite family of pairwise disjoint convex bodies has a line transversal if and only if the members of the family admit a linear ordering such that each three members of the family are intersected in the given order by a suitable line.*

We can think of Hadwiger's theorem to be dealing with a special class of $T(3)$-families. It is interesting to note that Hadwiger's theorem does not guarantee that there is a line that meets the members of the family in the given order. If one asks for a line that meets the members in the prescribed order, one must require that every *four* members can be intersected in the given order by a suitable line, as was noted by Tverberg [26] and Wenger [27].

Hadwiger's theorem has had a great influence on geometric transversal theory. In a sense it was a forerunner to the notion of a *geometric permutation* introduced by Katchalski, Lewis and Zaks [20] in 1985, almost 30 years later. Geometric permutations have been studied extensively as a tool in geometric transversal theory (see [17] for many references), and they also give rise to many questions that are interesting in their own right.

The ordering condition in Hadwiger's theorem can be reformulated as follows: If the members of the family are ordered $A_1, \ldots, A_n$, then for every $1 \le i < j < k \le n$ we have $\mathrm{conv}(A_i \cup A_k) \cap A_j \ne \emptyset$. It is easy to see that when the members of the family are pairwise disjoint this condition is equivalent to the ordering condition in Hadwiger's Theorem. The advantage of the latter formulation is that it still makes sense when we drop the condition of pairwise disjointness. This raises the question if there is a generalization of Hadwiger's transversal theorem for general families of convex bodies in the plane. This is indeed the case as was shown by Wenger [27] in 1990.

Theorem 5 (Wenger). *Let $F = \{A_1, \ldots, A_n\}$ be a family of convex bodies in the plane such that for every $1 \le i < j < k \le n$ we have $\mathrm{conv}(A_i \cup A_k) \cap A_j \ne \emptyset$. Then F has a transversal.*

The proof of Theorem 5 uses the fact that each disjoint pair $\{A_i, A_j\} \subset F$ $(i < j)$ gives rise to a pair of antipodal arcs of *separation directions* on the circle. Depending on whether the direction separates with A_i on the left and A_j on the right (or vice versa) the direction is labeled positive (or negative). If there is no transversal to the family then the separation arcs must cover the circle, and it follows that the circle is covered by two open sets (the union of the positive arcs and the union of the negative arcs). By the Borsuk–Ulam theorem one of these open sets contains an antipodal pair, and a simple case analysis of this situation yields a contradiction. Wenger's proof also generalizes to the setting of a topological affine plane as was noted in [7], which means that the straightness of the lines is not a necessary assumption.

In 2008, Arocha, Bracho and Montejano [4] discovered a generalization of Theorem 5 in the spirit of Bárány's generalization of Carathéodory's theorem [5]. They call this the *Colorful Hadwiger Theorem.*

Theorem 6 (Arocha–Bracho–Montejano). *Let* $F = F_1 \cup F_2 \cup F_3 = \{A_1, \ldots, A_n\}$ *be a family of convex bodies in the plane. If for every* $1 \leq i < j < k \leq n$ *where the* A_i, A_j, A_k *belong to distinct parts* (F_p's) *we have* $\mathrm{conv}(A_i \cup A_k) \cap A_j \neq \emptyset$, *then one of the* F_p *has a line transversal.*

Notice that Theorem 6 does not require the parts F_p to be disjoint: A convex body A_j could belong to more than one of the $F_p's$. In particular, if we let $F_1 = F_2 = F_3$ then we obtain Theorem 5. The proof of Theorem 6 resembles Wenger's proof but requires an additional combinatorial lemma, and it is easily seen that it also extends to the more general setting of a topological affine plane. We conclude with an application of Theorem 6 to general $T(3)$-families.

Corollary 7. $\alpha(3) \geq \frac{1}{5}$.

Proof. Let F be a finite family of compact convex bodies. Using a standard compactness argument we can assume that the bodies are $|F|$ different convex polygons, no two having vertices in common, and no three vertices being collinear. Then we can choose any direction as vertical and find two distinct vertical lines l and m having the following property:

If a, b, c denotes the number of bodies in F in the three open strips defined by l and m, then $\max\{a, b, c\} - \min\{a, b, c\} \leq 1$.

Assume that R_2 is the strip that is bounded by l and m. (See Figure 6 below). Let $G \subset F$ consist of the members of F that are contained in one of the regions R_i. If $|F \setminus G| \geq \frac{2}{5}|F|$, then we are done since either l or m intersects at least $\frac{1}{5}|F|$ members of F. We may therefore assume that $|G| \geq \frac{3}{5}|F|$. Partition $G = G_1 \cup G_2 \cup G_3$ such that the members of G_i are contained in region R_i. Label arbitrarily the members of G_1 by $\{1, \ldots, |G_1|\}$, the members of G_2 by $\{|G_1| + 1, \ldots, |G_1| + |G_2|\}$, and the members of G_3 by $\{|G_1| + |G_2| + 1, \ldots, |G_1| + |G_2| + |G_3|\}$. By construction,

G will satisfy the conditions of Theorem 6, therefore one of the G_i has a transversal. ∎

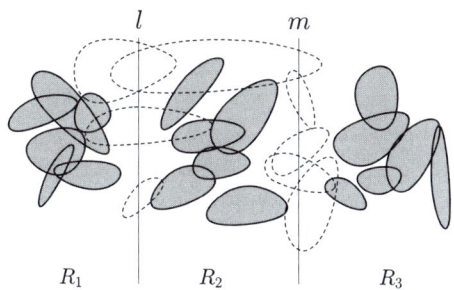

Fig. 6. Partitioning of G in the proof of Corollary 7

4. ECKHOFF'S PARTITION THEOREMS FOR $T(3)$- AND $T(4)$-FAMILIES

Consider the following *Gallai-type* question: *For a given integer $k \geq 3$ does there exist a finite positive integer $p(k)$ such that every $T(k)$-family F admits a partition into at most $p(k)$ parts such that each part has a transversal?*

An affirmative answer to this question would immediately imply $\alpha(k) \geq \frac{1}{p(k)}$. Here are two results due to Eckhoff [8, 9] which answer this question in two particular cases.

Theorem 8 (Eckhoff). *A $T(4)$-family can be partitioned into two parts such that each part has a transversal.*

Theorem 9 (Eckhoff). *A $T(3)$-family can be partitioned into four parts such that each part has transversal.*

Eckhoff's proof of Theorem 8 [8] relies on a lemma which states that if a family F of convex bodies satisfies the property that any pair of members in F can be intersected by a vertical or horizontal line then there exists a vertical line v and a horizontal line h such that any member of F meets v or h. The fact that a $T(4)$-family has the stated property was already shown by Hadwiger and Debrunner in [16], where they deduced that $p(4) \leq 4$.

Theorem 9 is much more complicated to establish and is based on ideas due to Kramer [23]. In the proof [9] Eckhoff starts by singling out four special lines and then makes several deductions concerning the relative positions of the sets with respect to these lines. What follows is an intricate series of rotations and translations of these special lines to obtain a position where each member of the family is met by at least one of these (new) lines.

Eckhoff has pointed out that it is strongly believed that *three* lines should be sufficient, but so far no proof is known. See Section 4 of [9] for a detailed discussion of this problem. In [8] Eckhoff gives an example that shows that three lines may be necessary, and therefore this approach could ultimately give a lower bound $\alpha(3) \geq \frac{1}{3}$. It would be of considerable interest to find a new (simpler) proof of Eckhoff's partition theorem for $T(3)$-families and/or a proof or counter-example to Eckhoff's (Gallai-type) conjecture. It should also be noted that the fact that there is a finite *piercing number* follows from the more general result by Alon and Kalai [1]. As pointed out earlier, Eckhoff's partition theorems give the following.

Corollary 10. $\alpha(3) \geq \frac{1}{4}$ *and* $\alpha(4) \geq \frac{1}{2}$.

5. IMPROVED BOUNDS FOR $T(k)$-FAMILIES

In [12] Eckhoff improves on the lower bound for $\alpha(k)$. In particular it is shown that $\alpha(k) \geq 1 - \frac{\sqrt{2}}{\sqrt{k-1}}$ when $k \geq 4$. The methods employed involve the *Fractional Helly Theorem* due to Katchalski and Liu [21] and an analogous result concerning intersections of arcs on the circle. The key to the argument is a transformation which maps a pair of convex bodies to an arc on the circle. This transformation has previously been used for transversal problems by Hadwiger and Debrunner (see for instance Proposition 26 of [16] and Eckhoff [8]).

Eckhoff [12] also constructs a family of disks which shows that $\alpha(3) \leq \frac{2}{3}$, which means that Katchalski's conjecture would be best possible if true. A slightly weaker version of Eckhoff's construction can be described as follows.

Let C denote the unit radius circle centered at the origin. It is easily verified that any strip of width $\frac{3}{2}$ can cover at most $\frac{2}{3}$ of the total length of C. If we place m points uniformly on C, then, as m tends to infinity, the greatest proportion of points which can be covered by a strip of width $\frac{3}{2}$ will tend to $\frac{2}{3}$. If we now replace each of the points on C by a disk of radius

$\frac{3}{4}$ centered at the respective point, it is easily seen that a line l intersects a disk if and only if the strip of width $\frac{3}{2}$ symmetric about l contains the center of the disk. Furthermore it is easily verified that the resulting family of disks is a $T(3)$-family. This proves that $\alpha(3) \leq \frac{2}{3}$.

Recently, the present author [18] has improved on Eckhoff's results. It is shown that

$$\alpha(k) \geq \left(\frac{2}{k(k-1)} \right)^{1/(k-2)} .$$

For $k = 3$ this gives $\alpha(3) \geq \frac{1}{3}$, which is currently the best known. Also, for $k \geq 6$ this bound is sharper then the one from Eckhoff [12]. It is interesting to note that the best known lower bounds for $\alpha(4)$ and $\alpha(5)$ still follow from Theorem 8. Also the upper bound for $\alpha(k)$ is improved in [18]. Constructions are given that give the following upper bounds. (The reader should consult [18] for more details).

$$\alpha(k) \leq \frac{k-2}{k-1}.$$

6. REMARKS

The results of Section 5 disprove Katchalski's conjecture on the value of $\alpha(3)$ and Eckhoff's conjecture regarding $\alpha(k)$. However it is still possible that Eckhoff's conjecture describes the correct asymptotic behavior of the function $\alpha(k)$. It is tempting to conjecture that $\alpha(k) = \frac{k-2}{k-1}$.

It should also be mentioned that the planar results discussed in this paper have higher-dimensional analogues (i.e. hyperplane transversals), something that follows from the previously mentioned results due to Alon and Kalai [1]. In fact their results imply the existence of higher-dimensional analogues of the Gallai-type results discussed in Section 4, and even much more general (p, q)-type theorems, however much less is known about optimal bounds. For more information the reader should consult Eckhoff's surveys [10, 11], as well as the paper by Alon et al. [2] which shows the importance of the Fractional Helly Theorem of Katchalski and Liu [21] with respect to these problems.

For the main result of Section 2, the crucial observation is Lemma 3. This Tverberg-type Lemma seems interesting in its own right, and there should be several ways it could be generalized. We first describe a higher-dimensional version due to R. Karasev [19].

Let us define for a given family F of convex sets in $\mathbb{R}^d$ the *Carathéodory number*, c_F, to be the least integer such that for any subset $G \subset F$ of size at least c_F we have the following.

$$\mathrm{conv}\left(\bigcup_{X \in G} X \right) = \bigcup_{H \subset G, |H| = c_F} \left(\mathrm{conv}\left(\bigcup_{X \in H} X \right) \right)$$

Note that Carathéodory's theorem implies that $c_F \leq d + 1$.

Theorem 11 (Karasev). *Let F be a family of convex sets (in $\mathbb{R}^d$) with Carathéodory number c, and suppose $|F| = ck + 1$. Then F can be partitioned into k non-empty parts $F = F_1 \cup \cdots \cup F_k$ such that $F_1 \cap \cdots \cap F_k \neq \emptyset$.*

It is not known whether the '+1' is necessary, and we conjecture that Theorem 11 also holds when $|F| = c_F k$.

It seems possible that there is a further generalization of Lemma 3, but we were unable to establish this. Here we state the planar version: For $k \geq 3$ call a k-tuple $\{A_1, \ldots, A_k\}$ of convex bodies in the plane *tight* if the following holds:

$$\bigcup_{i<j} \mathrm{conv}(A_i \cup A_j) \text{ is convex.}$$

By Caratheodory's Theorem it follows that if every k-tuple of a family of convex bodies is tight then every $(k+1)$-tuple is also tight. It seems plausible that the conclusion of Lemma 3 could hold (that there is a matching in which the components have a common point) under the weaker condition that every k-tuple is tight, for $k > 3$.

Problem 12. Does there exist an increasing function $f(n)$, such that every family $F = \{A_1, \ldots, A_n\}$ ($n = 2k$) for which every $f(n)$-tuple is tight admits a partition into disjoint pairs $F = B_1 \cup \cdots \cup B_k$ such that $\mathrm{conv}(B_1) \cap \cdots \cap \mathrm{conv}(B_k) \neq \emptyset$?

The first higher-dimensional analogue of Hadwiger's transversal theorem was discovered by Goodman and Pollack [13]. Subsequently Pollack and Wenger found a short proof based on the Borsuk–Ulam Theorem [25], and the most general version is due to Anderson and Wenger [3]. The key is to replace the linear ordering to a higher-dimensional analogue, namely an oriented matroid.

Theorem 13 (Anderson–Wenger). *A family F of compact convex sets in $\mathbb{R}^d$ has a hyperplane transversal if and only if for some k, $0 \le k < d$, there is and acyclic oriented matroid of rank $k+1$ such that every $k+2$ of the sets have a k-transversal meeting them consistently with that oriented matroid.*

An oriented k-flat (k-dimensional affine subspace) meets a family of convex sets consistently with a given acyclic oriented matroid, M, if one can choose a point from the intersection of each set and the k-flat such that the orientation of each $(k+1)$-tuple of points matches the orientation of the corresponding $(k+1)$-tuples of M.

It seems to be a challenging problem of geometric transversal theory to find a higher-dimensional version of the Colored Hadwiger Theorem. Only some very special cases in $\mathbb{R}^3$ have been obtained [4].

Note added in proof. Theorem 11 has appeared in [6].

REFERENCES

[1] N. Alon and G. Kalai, Bounding the piercing number, *Discrete Comput Geom.*, **13** (1995), 245–256.

[2] N. Alon, G. Kalai, J. Matoušek and R. Meshulam, Transversal numbers for hypergraphs arising in geometry, *Adv in Appl Math.*, **29** (2002), 79–101.

[3] L. Anderson and R. Wenger, Oriented matroids and hyperplane transversals, *Adv. Math.*, **119** (1996), 117–125.

[4] J. Arocha, J. Bracho and L. Montejano, A colorful theorem on transversal lines to plane convex sets, *Combinatorica*, **28** (2008), 379–384.

[5] I. Bárány, A generalization of Carathéodory's theorem, *Discrete Math.*, **40** (1982), 141–152.

[6] I. Bárány and R. Karasev, Notes about the Caratheodory number, *Discrete Comput. Geom.*, **48** (2012), 783–792.

[7] S. Basu, E. Goodman, A. Holmsen and R. Pollack, The Hadwiger transversal theorem for pseudolines, *MSRI Publications Volume*, **52** (2005), 87–97.

[8] J. Eckhoff, Transversalenprobleme in der Ebene, *Arch. Math.*, **24** (1973), 195–202.

[9] J. Eckhoff, A Gallai-type transversal problem in the plane, *Discrete Comput. Geom.*,
     9 (1993), 203–214.

[10] J. Eckhoff, Helly, Radon, and Carathéodory type theorems, *Handbook of convex
     geometry*, Vol. A, 389 –448, North-Holland Amsterdam, 1993.

[11] J. Eckhoff, A survey of the Hadwiger–Debrunner (p, q)-problem, *Discrete and
     Computational Geometry – The Goodman–Pollack Festschrift,* 347–377, Algorithms
     Combin. 25, Springer – Berlin, 2003.

[12] J. Eckhoff, Common transversals in the plane: The fractional perspective, *Europ.
     J. Combinatorics,* **29** (2008), 1872–1880.

[13] J. E. Goodman and R. Pollack, Hadwiger's transversal theorem in higher dimensions,
     J. Amer. Math. Soc., **1** (1998), 301–309.

[14] J. E. Goodman, R. Pollack and R. Wenger, Geometric transversal theory, *New trends
     in discrete and computational geometry,* 163–198, Algorithms Combin., 10, Springer-
     Berlin, 1993.

[15] H. Hadwiger, Über Eibereiche gemeinsamer Treffgeraden, *Portugal Math.,* **16** (1957),
     23–29.

[16] H. Hadwiger, H. Debrunner and V. Klee, *Combinatorial Geometry in the Plane,*
     Holt, Rinehart & Winston, New York, 1964.

[17] A. F. Holmsen, Recent Progress on line transversals in the plane, *Contemp Math.,*
     453 (2008), 283–297.

[18] A. F. Holmsen, New results for $T(3)$-families in the plane, *Mathematika,* **56** (2010),
     86–92.

[19] R. Karasev, Personal communication.

[20] M. Katchalski, T. Lewis and J. Zaks, Geometric permutations for convex sets,
     Discrete Math., **54** (1985), 271–284.

[21] M .Katchalski and A. Liu, A problem of geometry in $\mathbb{R}^n$, *Proc. Amer. Math. Soc.,*
     75 (1979), 284–288.

[22] M. Katchalski and A. Liu, Symmetric twins and common transversals, *Pacific
     J. Math.,* **86** (1980), 513–515.

[23] D. Kramer, Transversalenprobleme vom Hellyschen und Gallaischen Typ. Disserta-
     tion, Universität Dortmund, 1974.

[24] J. Matoušek, *Lectures on Discrete Geometry,* Graduate Text in Mathematics 212,
     Springer.

[25] R. Pollack and R. Wenger, Necessary and sufficient conditions for hyperplane
     transversals, *Combinatorica,* **10** (1990), 307–311.

[26] H. Tverberg, On geometric permutations and the Katchalski-Lewis conjecture on partial transversals for translates, *DIMACS Series in Discrete Mathematics and Theoretical Computer Science*, **6** (1991), 351–361.

[27] R. Wenger, A generalization of Hadwiger's transversal theorem to intersecting sets, *Discrete Comput. Geom.*, **5** (1990), 383–388.

[28] R. Wenger, Helly-type theorems and geometric transversals, *Handbook of discrete and computational geometry*, Second Edition, 73–96, CRC Press Ser Discrete Math Appl, 1997.

Andreas F. Holmsen
KAIST
Daejeon, South Korea

BOLYAI SOCIETY
MATHEMATICAL STUDIES, 24

Geometry –
Intuitive, Discrete, and Convex
pp. 205–218.

TRANSVERSALS, TOPOLOGY AND COLORFUL GEOMETRIC RESULTS

L. MONTEJANO*

1. INTRODUCTION

Suppose we have two convex sets A and B in euclidean d-space $\mathbb{R}^d$. Assume the only information we have about A and B comes from the space of their transversal lines. Can we determine whether A and B have a point in common? For example, suppose the space of their transversal lines has an essential curve; that is, suppose there is a line that moves continuously in $\mathbb{R}^d$, always remaining transversal to A and B, and comes back to itself with the opposite orientation. If this is so, then A must intersect B, otherwise there would be a hyperplane H separating A from B; but it turns out that our moving line becomes parallel to H at some point on its trip, which is a contradiction to the fact that the moving line remains transversal to the two sets. If we have three convex sets A, B and C, for example, in $\mathbb{R}^3$, then our essential curve does not give us sufficient topological information. In this case, to detect whether $A \cap B \cap C \neq \phi$, we need a 2-dimensional cycle. So, for example, if we can continuously choose a transversal line parallel to every direction, then there must be a point in $A \cap B \cap C$, otherwise if not, the same is true for $\pi(A) \cap \pi(B) \cap \pi(C)$, for a suitable orthogonal projection $\pi : \mathbb{R}^3 \to H$ where H is a plane through the origin (see [4, Lemma 3.1]). Hence clearly there is no transversal line orthogonal to H.

Suppose now we have three convex sets A, B and C in euclidean 3 space $\mathbb{R}^3$. Assume the only information we have about A, B and C comes from the space of their transversal planes. Can we determine whether A,

*Supported by CONACYT, 41340.

B and C have a transversal line? For example, suppose the space of their transversal planes has an essential curve; that is, suppose there is a plane that moves continuously in $\mathbb{R}^3$, always remaining transversal to A, B and C, and comes back to itself with the opposite orientation. If this is so, then there must be a transversal line to $\{A, B, C\}$. This time the proof is slightly more complicated from the topological point of view. Let me present it here.

Suppose there is no transversal line to $\{A, B, C\}$. Denote by T_2 the space of transversal planes to $\{A, B, C\}$. Consider the continuous map $\psi : A \times B \times C \to T_2$ given by $\psi(a, b, c)$, the unique plane containing $\{a, b, c\}$. The continuous map ψ is well defined precisely because there is no transversal line to $\{A, B, C\}$. Furthermore, if $H \in T_2$, then $\psi^{-1}(H) = (A \cap H) \times (B \cap H) \times (C \cap H)$, which is contractible by the convexity of the two sets. The fact that the fibers of ψ are contractible implies that ψ is a homotopy equivalence. This implies that T_2 is contractible, contradicting the hypothesis that there is an essential curve in T_2.

We claim that for a sufficiently small family of convex sets, the topology of its transversals provide enough information to derive geometric information. To be more precise, let us state the following definition.

Let F be a family of compact, convex sets. We say that $\mathcal{F}$ has a *topological ρ-transversal of index* (m, k), $\rho < m$, $0 < k \leq d - m$, if there are, homologically, as many transversal m-planes to $\mathcal{F}$ as m-planes through a fixed ρ-plane in $\mathbb{R}^{m+k}$. Clearly, if $\mathcal{F}$ has a ρ-transversal plane, then $\mathcal{F}$ has a topological ρ-transversal of index (m, k), for $\rho < m$ and $k \leq d - m$. The converse is not true. It is easy to give examples of families with a topological ρ-transversal but without a ρ-transversal plane. We conjecture that for a family $\mathcal{F}$ of $k + \rho + 1$ compact, convex sets in euclidean d-space $\mathbb{R}^d$, there is a ρ-transversal plane if and only if there is a topological ρ-transversal of index (m, k). A good reference for the algebraic topology needed in this paper is [8], and [7] for the geometric transversal theory.

The purpose of this paper is to use the structure of the topology of the space of transversals to obtain geometric results in the spirit of the colourful theorems of Lovász and Bárány.

2. THE STRUCTURE OF THE SPACE OF TRANSVERSALS

The purpose of this section is to state several results about the structure of the topology of the space of transversals to a family of convex sets. For the proofs see [1], [4], [5] and [9].

Let $\mathcal{F}$ be a family of compact, convex sets in $\mathbb{R}^d$. By $M(d, m)$ we denote the space of m-planes in $\mathbb{R}^d$. It can be considered as an open subset of $G(d + 1, m + 1)$ and retractible to the classic Grassmanian space, $G(d, m)$, of m-dimensional linear subspaces of $\mathbb{R}^d$. For $0 < m < d$, we denote by $\mathcal{T}_m(\mathcal{F})$ the subspace of $M(d, m) \subset G(d+1, m+1)$ consisting of all m-planes transversal to $\mathcal{F}$.

We say that $\mathcal{F}$ has a *topological ρ-transversal of index* (m, k), $\rho < m$, $0 < k \leq d - m$, if there are homologically as many transversal m-planes to $\mathcal{F}$ as m-planes through a fixed ρ-plane in R^{m+k}.

More precisely, for $\rho < m$, $0 < k \leq d - m$, the family $\mathcal{F}$ has a topological ρ-transversal of index (m, k) if

$$i^*([0, \ldots, 0, k, \ldots, k]) \in H^{(m-\rho)k}\big(\mathcal{T}_m(\mathcal{F}), \mathbb{Z}_2\big) \quad \text{is not zero,}$$

where $i^* : H^{(m-\rho)k}\big(G(d + 1, m + 1), \mathbb{Z}_2\big) \to H^{(m-\rho)k}\big(\mathcal{T}_m(\mathcal{F}), \mathbb{Z}_2\big)$ is the cohomology homomorphism induced by the inclusion $\mathcal{T}_m(\mathcal{F}) \subset M(d, m) \subset G(d + 1, m + 1)$, and

$$[0, \ldots, 0, k, \ldots, k] \in H^{(m-\rho)k}\big(G(d + 1, m + 1), \mathbb{Z}_2\big)$$

is the Schubert-cocycle, in which the last symbol starts with $\rho + 1$ zeros (see [6] for the definition of Schubert cocycle).

Clearly, if $\mathcal{F}$ has a ρ-transversal plane, then $\mathcal{F}$ has a topological ρ-transversal of index (m, k), for $\rho < m$ and $k \leq d - m$. The converse is not true. It is easy to give examples of families with a topological ρ-transversal but without a ρ-transversal plane. We conjecture that for a family $\mathcal{F}$ of $k + \rho + 1$ compact, convex sets in euclidean d-space $\mathbb{R}^d$, there is a ρ-transversal plane if and only if there is a topological ρ-transversal of index (m, k).

The proof of the following theorem follows the ideas of the proof, given in the introduction, that the space of transversal planes to three convex sets, without transversal lines, in 3-space is contractible. See the proof of Theorem 3.1 in [1].

Theorem 2.1. *Let $0 \leq \rho < m \leq d-1$. Let $\mathcal{F} = \{A_0, \ldots, A_{\rho+1}\}$ be a family of convex sets in $\mathbb{R}^d$ and let $\alpha_i \in A_i$, $i = 0, \ldots, \rho+1$. Suppose there is no ρ-plane transversal to $\mathcal{F}$. Then the inclusion*

$$\mathcal{T}_m(\{\alpha_0, \ldots, \alpha_{\rho+1}\}) \subset \mathcal{T}_m(\{A_0, \ldots, A_{\rho+1}\})$$

is a homotopy equivalence.

In particular, $T_m(F)$ has the homotopy type of $G(d - \rho - 1, m - \rho - 1)$.

As a corollary, we have the following theorem which proves our main conjecture when $k = 1$. This theorem will allow us to transform topological information into geometric information.

Theorem 2.2. *Let $0 \leq \rho < m$, and let $\mathcal{F}$ be a family of $\rho + 2$ compact convex sets in $\mathbb{R}^d$. Then there is a ρ-plane transversal to $\mathcal{F}$ if and only if there is a topological ρ-transversal plane of index $(m, 1)$.*

That is, there is a ρ-plane transversal to $\mathcal{F}$ if

$$[0, \ldots, 0, 1, \ldots, 1] \quad \text{is not zero in} \quad \mathcal{T}_m(\mathcal{F}),$$

where $[0, \ldots, 0, 1, \ldots, 1] \in H^{(m-\rho)}\big(G(m + 1, d + 1), \mathbb{Z}_2\big)$ is the $(m - \rho)$-Stiefel–Whitney characteristic class, in which the last symbol starts with $\rho + 1$ zeros.

All results in this paper can be stated in a more general setting, but to simplify the topological technicalities and to clarify the ideas, we will prove and state them only for dimensions 3 and 4. So, let us summarize in the following proposition the topology we will need in the next section.

Proposition 2.1. *Let F be a family of convex sets in $\mathbb{R}^4$. Let $D(F) \subset \mathbb{RP}^3$ be the set of directions in $\mathbb{R}^4$ orthogonal to a transversal hyperplane of F and let $d(F) \subset G(4, 2)$ be the set of directions in $\mathbb{R}^4$ orthogonal to a transversal plane of F. Then*

    a) *if the homomorphism induced by the inclusion $H^1\big(\mathbb{RP}^3, \mathbb{Z}_2\big) \to H^1\big(D(F), \mathbb{Z}_2\big)$ is not zero, there is a transversal plane to every quadruple of the convex sets of F,*

    b) *if the homomorphism induced by the inclusion $H^2\big(\mathbb{RP}^3, \mathbb{Z}_2\big) \to H^2\big(D(F), \mathbb{Z}_2\big)$ is not zero, there is a transversal line to every triple of the convex sets of F,*

c) *if the homomorphism induced by the inclusion $H^1(G(4,2),\mathbb{Z}_2) = \mathbb{Z}_2 \to H^1(d(F),\mathbb{Z}_2)$ is not zero, there is a transversal line to every triple of convex sets of F.*

Proof. Note that the classical retraction $M(4,3) \to \mathbb{R}P^3$ is a homotopy equivalence. Furthermore its restriction $T_3(F) \to D(F)$ is a homotopy equivalence because the fibers are contractible. So, if the homomorphism induced by the inclusion $H^1(\mathbb{R}P^3,\mathbb{Z}_2) \to H^1(D(F),\mathbb{Z}_2)$ is not zero, then the generator of $G(5,4)$ is not zero in $T_3(F)$ and hence by Theorem 2.2 there is a plane transversal to every quadruple of convex sets of F. The proofs of b) and c) are essentially the same. ∎

3. THE COLORFUL GEOMETRIC RESULTS

The purpose of this section is to use the topological results developed in the previous section to obtain geometric results in the spirit of the colorful theorems of Lovász and Bárány [3].

We state the colorful Helly Theorem.

Theorem 3.1. *Let F be a family of convex sets in $\mathbb{R}^d$ painted with $d+1$ colors. Suppose that every heterochromatic $d+1$-tuple of F is intersecting. Then there is a color with the property that the family of all convex sets of this color is intersecting.*

In particular, if we have a collection of red and blue intervals in the line and every red interval intersects every blue internal, then either there is a point in the intersection of all red intervals or there is a point in the intersection of all blue intervals.

The colorful Helly Theorem has the following geometric interpretation: any linear embedding of a combinatorial d-cube in $\mathbb{R}^d$ has, in every direction, a transversal line to two opposite faces.

Let us consider the configuration of lines in the plane that consists of nine points and six lines, in which the first three red lines, ℓ_1, ℓ_2, ℓ_3 are parallel and the next three blue lines L_1, L_2, L_3 are parallel and orthogonal to the red ones. So every line has exactly three points, and the intersection of a red and a blue line consists exactly of one point. Let us denote by G^3 the 2-dimensional simplicial complex describing this configuration, in which we have three red triangles corresponding to the red lines and three blue triangles corresponding to the blue lines.

Theorem 3.2. *In any linear embedding of G^3 in euclidean 3-space $\mathbb{R}^3$, there is either a transversal line to the red triangles or a transversal line to the blue triangles.*

Proof. The ingredients of the proof are: i) the fact that if $\mathbb{R}P^2$, the projective plane, is the union of two closed ANR sets R and B, then either R contains an essential cycle or B contains an essential cycle, and ii) the colorful Helly Theorem in the line.

Let $\mathcal{R} \subset \mathbb{R}P^2$ be the collection of directions orthogonal to transversal planes to the red triangles and let $\mathcal{B} \subset \mathbb{R}P^2$ be the collection of directions orthogonal to transversal planes to the blue triangles. First note that $\mathcal{R} \cup \mathcal{B} = \mathbb{R}P^2$, because if L is any line through the origin, we may project the three red triangles and the three blue triangles orthogonally onto L. Thus we have three red intervals and three blue intervals in L with the property that every red interval intersects a blue interval, but this means that either there is a point in the intersection of all red intervals or there is a point in the intersection of all blue intervals. Therefore there is, orthogonally to L, either a plane transversal to the three red triangles or a plane transversal to the three blue triangles. Since $\mathcal{R} \cup \mathcal{B} = \mathbb{R}P^2$, either $\mathcal{R}$ contains an essential cycle or $\mathcal{B}$ contains an essential cycle. This immediately implies that there is an essential cycle of planes (see the introduction) transversal to the red triangles or an essential cycle of planes transversal to the blue triangles. Thus there is either a transversal line to the red triangles or a transversal line to the blue triangles. ∎

We have essentially proved that if we have three red convex sets and three blue convex sets in 3-space and every red set intersects every blue set, then there is either a transversal line to the red sets or a transversal line to the blue sets. Now we want to prove a similar theorem but this time using more than two colors. For this purpose we need the following proposition, which essentially claims that $G(4,2)$ cannot be covered by three null homotopic sets.

Proposition 3.1. *Let $G(4,2) = A_1 \cup A_2 \cup A_3$ be a closed cover of the 4-dimensional Grassmanian space $G(4,2)$ of planes through the origin in $\mathbb{R}^4$. For some $i \in \{1,2,3\}$, the homomorphism induced by the inclusion $H^*\big(G(4,2), \mathbb{Z}_2\big) \to H^*\big(A_i, \mathbb{Z}_2\big)$ is not zero.*

Proof. The strategy is to prove first that there are $\gamma_i \in H^*\big(G(4,2), \mathbb{Z}_2\big)$, $i = 0, 1, 2$ such that $\gamma_0 * \gamma_1 * \gamma_2 \neq 0$. Recall (see [6] that the product structure in $H^*\big(G(4,2), \mathbb{Z}_2\big)$ can be totally described by the following formula:

$$[\lambda_1, \lambda_2][0, 1] = \sum [\xi_1, \xi_2],$$

where the summation extends over all combinations ξ_1, ξ_2 such that

i) $0 \leq \xi_1 \leq \xi_2 \leq 2$,

ii) $\lambda_1 \leq \xi_1 \leq \lambda_2$, $\lambda_2 \leq \xi_2 \leq 2$, and

iii) $\xi_1 + \xi_2 = \lambda_1 + \lambda_2 + 1$.

Let $\gamma_0 = [1, 1]$ and $\gamma_1 = [0, 1]$. Then $\gamma_0 * \gamma_1 = [1, 2]$ and then $\gamma_0 * \gamma_1 * \gamma_1 = [2, 2] \neq 0$.

Suppose that the homomorphism induced by the inclusion

$$H^*\big(G(4,2), \mathbb{Z}_2\big) \to H^*\big(A_i, \mathbb{Z}_2\big)$$

is zero, for $i \in \{1, 2, 3\}$. Hence by exactness

$$H^*\big(G(4,2), A_i; \mathbb{Z}_2\big) \to H^*\big(G(4,2), \mathbb{Z}_2\big)$$

is an epimorphism. We can pull γ_0, γ_1, γ_2 back to $H^*\big(G(4,2), A_i; \mathbb{Z}_2\big)$ and hence pull the product $\gamma_0 * \gamma_1 * \gamma_1$ back to $H^*\big(G(4,2), A_1 \cup A_2 \cup A_3; \mathbb{Z}_2\big) = 0$, which is a contradiction. ∎

We are ready for the following theorem:

Theorem 3.3. *Suppose we have three red convex sets, three blue convex sets and three green convex sets in $\mathbb{R}^4$ and every heterochromatic triple is intersecting. Then there is one color that has a line transversal to all convex sets of this color.*

Proof. The proof is essentially that of the previous theorem but using the colorful Helly Theorem in the plane and Proposition 3.3. Let $\mathcal{R} \subset G(4,2)$ be the set of directions in $G(4,2)$ orthogonal to transversal planes to the red convex sets, let $\mathcal{B} \subset G(4,2)$ be the set of directions orthogonal to transversal planes to the blue convex sets, and finally let $\mathcal{G} \subset G(4,2)$ be the set of directions orthogonal to transversal planes to the green convex sets. First note that $\mathcal{R} \cup \mathcal{B} \cup \mathcal{G} = G(4,2)$, because if H is any plane through the origin, we may project our nine convex sets orthogonally onto L. So, by the colorful Helly Theorem 3.2 in the plane, there is a color, say red, such that there is a point in common to the projection of all convex sets of that color. Therefore there is, orthogonally to H, a plane transversal to the three red sets. Since $\mathcal{R} \cup \mathcal{B} \cup \mathcal{G} = G(4, 2)$, by Proposition 3.3 one of these closed sets, say $\mathcal{B}$ without loss of generality, has the property that the homomorphism induced by the inclusion $H^*\big(G(4, 2), \mathbb{Z}_2\big) \to H^1(\mathcal{B}, \mathbb{Z}_2)$ is not zero, but if this is so $H^1\big(G(4, 2), \mathbb{Z}_2\big) \to H^1(\mathcal{B}, \mathbb{Z}_2)$ is not zero. By Proposition 2.3c), this implies that there is a transversal line to the three blue convex sets, as required. ■

Now we will use a variant of the colorful Helly Theorem.

Proposition 3.2. *Let F be a family of red, blue and green intervals in $\mathbb{R}^1$. Suppose that for every heterochromatic triple, one of the intervals intersects the other two. Then there is a color such that there is a point in common to all intervals of this color.*

Proof. If every pair of red and every blue intervals intersects, then by the colorful Helly Theorem in the line, either there is point common to all red intervals or there is a point common to all blue intervals. If not, there is a red interval $I_R \in F$ and a blue interval $I_B \in F$, such that $I_R \cap I_B = \phi$. Therefore every green interval of F intersects both I_R and I_B, which implies that there is point in common to all green intervals, as required. ■

This variant of the colorful Helly Theorem and the fact that $\mathbb{R}P^3$ can not be covered by three null homotopic closed sets together give rise to the following theorem:

Theorem 3.4. *Suppose we have four red convex sets, four blue convex sets and four green convex sets in $\mathbb{R}^4$ and for every heterochromatic triple one of the sets intersects the other two. Then there is a color such that there is a plane transversal to all convex sets of this color.*

Proof. Let $\mathcal{R} \subset \mathbb{R}P^3$ be the collection of directions orthogonal to transversal hyperplanes to the red convex sets, let $\mathcal{B} \subset \mathbb{R}P^3$ be the collection of directions orthogonal to transversal hyperplanes to the blue convex sets, and let $\mathcal{G} \subset \mathbb{R}P^3$ be the collection of directions orthogonal to transversal hyperplanes to the green convex sets. Note that $\mathcal{R} \cup \mathcal{B} \cup \mathcal{G} = \mathbb{R}P^3$, because if L is any line through the origin, we may project our twelve convex sets orthogonally onto L, obtaining four red intervals, four blue intervals and four green intervals in L with the property that for every heterochromatic triple, one of the intervals intersects the other two. Then by Proposition 3.5, there is a color such that there is a point common to all intervals of this color. Therefore there is, orthogonally to L, a hyperplane transversal to the three convex sets of this color. Since $\mathcal{R} \cup \mathcal{B} \cup \mathcal{G} = \mathbb{RP}^3$, the Lusternik Schnirelmann category $\mathbb{RP}^3$ implies that one of these closed sets, say $\mathcal{B}$ without loss of generality, has the property that the homomorphism induced by the inclusion $H^1(\mathbb{RP}^3, \mathbb{Z}_2) \to H^1(\mathcal{B}, \mathbb{Z}_2)$ is not zero. By Proposition 2.3a), this implies that there is a transversal plane to the blue convex sets as required. ∎

It is well known that the projective plane is not the union of two null homotopic closed sets, but can be the union of three null homotopic closed sets. As a consequence, the following topological proposition, whose proof is an interesting application of the Mayer–Vietoris exact sequence in homology, will allow us to obtain two interesting results.

Proposition 3.3. *Let $A \cup B \cup C = \mathbb{RP}^2$ be a closed, null homotopic cover of projective 2-space. Then $A \cap B \cap C$ is non-empty. Moreover, $A \cap B \cap C$ has at least four non-empty components.*

Theorem 3.5. *Suppose we have three red convex sets, three blue convex sets and three green convex sets in $\mathbb{R}^3$ and every heterochromatic triple is intersecting. Then either there is a color such that there is a line parallel to the xy-plane transversal to all convex sets of this color, or else there are three parallel transversal hyperplanes, one for the red sets, one for the blue sets and one for the green sets.*

Proof. As before, let $\mathcal{R} \subset \mathbb{RP}^2$ be the collection of directions orthogonal to transversal hyperplanes to the red convex sets, let $\mathcal{B} \subset \mathbb{RP}^3$ be the collection of directions orthogonal to transversal hyperplanes to the blue convex sets and let $\mathcal{G} \subset \mathbb{RP}^3$ be the collection of directions orthogonal to transversal hyperplanes to the green convex sets. Note that $\mathcal{R} \cup \mathcal{B} \cup \mathcal{G} = \mathbb{RP}^2$, because if L is any line through the origin, we may project our nine convex sets orthogonally onto L, obtaining three red intervals, three blue intervals and three green intervals in L with the property that every heterochromatic triple is intersecting. Then, by the colorful Helly Theorem 3.1 in the plane, there is a color such that there is a point in common to all intervals of this color. Therefore there is a hyperplane orthogonal to L transversal to the three convex sets of this color. Hence $\mathcal{R} \cup \mathcal{B} \cup \mathcal{G} = \mathbb{RP}^2$ is a closed cover.

Suppose that for any of the three colors, there is no line parallel to the xy-plane transversal to all convex sets of this color. Hence $\mathcal{R} \cup \mathcal{B} \cup \mathcal{G} = \mathbb{RP}^2$ is a closed, null homotopic cover of projective 2 space. By Proposition 3.7, there is at least one line L through the origin whose direction lies in $\mathcal{R} \cap \mathcal{B} \cap \mathcal{G}$. Then there is a transversal hyperplane to the red sets orthogonal to L, a transversal hyperplane to the blue sets and a transversal hyperplane to the green sets. ∎

Theorem 3.6. *Suppose we have three red convex sets, three blue convex sets and three green convex sets i $\mathbb{R}^3$ and for every heterochromatic triple one of the sets intersects the other two. Then either there is a color such that there is a line transversal to the all convex sets of this color or else there is a color, say green, and two parallel planes H_1 and H_2 such that H_1 is a transversal plane to all green and red convex sets and H_2 is a transversal plane to all green and blue convex sets.*

Proof. Define $\mathcal{R}$, $\mathcal{B}$ and $\mathcal{G}$ as in the proof of the previous theorem. Note that $\mathcal{R} \cup \mathcal{B} \cup \mathcal{G} = \mathbb{R}P^2$, because if L is any line through the origin, we may project our nine convex sets orthogonally onto L, obtaining three red intervals, three blue intervals and three green intervals in L with the property that for every heterochromatic triple, one of the intervals intersects the other two. Then by Proposition 3.5, there is a color such that there is a point common to all intervals of this color. Therefore there is a hyperplane orthogonal to L transversal to the three convex sets of this color. Hence $\mathcal{R} \cup \mathcal{B} \cup \mathcal{G} = \mathbb{RP}^2$. Suppose that for any of the three colors, there is no line transversal to all convex sets of this color. Hence $\mathcal{R} \cup \mathcal{B} \cup \mathcal{G} = \mathbb{RP}^2$ is a closed, null homotopic cover of projective 2-space. By Proposition 3.7, there is at least one line L through the origin whose direction lies in $\mathcal{R} \cap \mathcal{B} \cap \mathcal{G}$.

Let us project our nine convex sets orthogonally onto L, obtaining three red intervals, three blue intervals and three green intervals in L. Note that since the direction of L lies in $\mathcal{R} \cap \mathcal{B} \cap \mathcal{G}$, every pair of intervals of the same color intersect. If every pair of differently-colored intervals intersects, then the collection of our nine intervals intersects pairwise and hence by the Helly Theorem in the line, there is a point $x_0 \in L$ common to all nine intervals. Then if $H_1 = H_2$ is the plane orthogonal to L through x_0, we are done. If not, let I_1 and I_2 be the two intervals with different color that are farthest apart. Suppose without loss of generality that $I_1 = [a_1, b_1]$ is red, $I_2 = [a_2, b_2]$ is blue and $a_1 \le b_1 < a_2 \le b_2$. By the hypothesis, every green interval contains both b_1 and a_2. Furthermore, every red interval I_R contains b_1, otherwise the distance from I_R to I_2 would be greater than the distance from I_1 to I_2. Similarly, every blue interval I_B contains b_1. Consequently the plane H_1 orthogonal to L through b_1 and the plane H_2 orthogonal to L through a_2 satisfy our requirements. ∎

Proposition 3.4. *Let $A_1 \cup A_2 \cup A_3 = \mathbb{RP}^2$ be a closed cover and suppose that $A_1 \cap A_2 \cap A_3 = A_1 \cap A_2 = A_2 \cap A_3 = A_1 \cap A_3$. Then either $A_1 \cap A_2 \cap A_3$ is not null homotopic or there is $i \in \{1, 2, 3\}$ such that $A_i - (A_1 \cap A_2 \cap A_3)$ is not null homotopic.*

Proof. If $A_1 \cap A_2 \cap A_3$ is null homotopic, then by duality there is an essential curve α of $\mathbb{RP}^2$ contained in $\mathbb{RP}^2 - (A_1 \cap A_2 \cap A_3)$. This essential curve must lie in some connected component of $\mathbb{RP}^2 - (A_1 \cap A_2 \cap A_3)$. Therefore since $A_1 \cap A_2 \cap A_3 = A_1 \cap A_2 = A_2 \cap A_3 = A_1 \cap A_3$, there must be $i \in \{1, 2, 3\}$ such that α is contained in $A_i - (A_1 \cap A_2 \cap A_3)$. ∎

For the following theorems, we need a definition. Let F be a family of red, blue and green convex sets in $\mathbb{R}^3$. A transversal plane (resp. line) is a bicolor transversal plane (resp. line) if it cuts all convex sets of two different colors.

Theorem 3.7. *Let F be a family of red, blue and green convex sets in $\mathbb{R}^3$. Suppose every pair of convex sets of F with different color intersects and suppose that every bicolor transversal plane through the origin is a transversal plane to all convex sets of F. Then there is a transversal line to all convex sets of F.*

Proof. Let us begin by analyzing the situation in the line. Suppose we have a family of red, blue and green intervals in the line with the property that every pair of intervals of different color intersect. Then by Helly's Theorem in the line, either all intervals have a point in common or there is a pair of intervals of the same color that do not intersect. In the latter case, all the intervals of the other two colors have a point in common.

As always, let $\mathcal{R}_{RB} \subset \mathbb{RP}^2$ be the collection of directions orthogonal to transversal planes to the red and the blue convex sets of F. Similarly, we have $\mathcal{R}_{RG} \subset \mathbb{RP}^2$ and $\mathcal{R}_{GB} \subset \mathbb{RP}^2$ for the other two combinations of colors. Our first argument proves that $\mathcal{R}_{RB} \cup \mathcal{R}_{RG} \cup \mathcal{R}_{GB} = \mathbb{RP}^2$ is a closed cover. Now note that $\mathcal{R}_1 \cap \mathcal{R}_2 \cap \mathcal{R}_3 = \mathcal{R}_1 \cap \mathcal{R}_2 - \mathcal{R}_1 \cap \mathcal{R}_3 - \mathcal{R}_2 \cap \mathcal{R}_3$. Furthermore, our hypothesis implies that for $i \in \{1, 2, 3\}$, $\mathcal{R}_i - (\mathcal{R}_1 \cap \mathcal{R}_2 \cap \mathcal{R}_3)$ is null homotopic. Therefore by Proposition 3.10, and since the directions in which there are transversal planes to all convex sets of F coincide with

$\mathcal{R}_1 \cap \mathcal{R}_2 \cap \mathcal{R}_3$, there is an essential curve of transversal planes to all convex sets of F and consequently there is a transversal line to all convex sets of F.
∎

Theorem 3.8. *Let F be a family of red, blue and green convex sets in $\mathbb{R}^3$. Suppose every non-monochromatic triple is intersecting and every bicolor transversal line parallel to the xy-axis is a transversal line to all convex sets of F. Then there is, parallel to every plane of $\mathbb{R}^3$, a transversal line to all convex sets of F.*

Proof. Let us begin by analyzing the situation in the plane. Suppose we have a family of red, blue and green convex sets in the plane with the property that every non-monochromatic triple is intersecting. Then the family F is pairwise intersecting, and furthermore, by Helly's Theorem in the plane, either all convex sets have a point in common or there are three convex sets of the same color that do not intersect but which are pairwise intersecting. If this is so, then by Lemma 1 ($k = \lambda = 2$) of [10], all convex sets of the other two colors have a point in common.

As always, let $\mathcal{R}_1 \subset \mathbb{R}P^2$ be the collection of directions parallel to transversal lines to the red and the blue convex sets of F. Similarly, we have $\mathcal{R}_2 \subset \mathbb{R}P^2$ and $\mathcal{R}_3 \subset \mathbb{R}P^2$ for the other two combinations of colors. Our first argument proves that $\mathcal{R}_1 \cup \mathcal{R}_2 \cup \mathcal{R}_3 = \mathbb{R}P^2$ is a closed cover. Now note that $\mathcal{R}_1 \cap \mathcal{R}_2 \cap \mathcal{R}_3 = \mathcal{R}_1 \cap \mathcal{R}_2 = \mathcal{R}_1 \cap \mathcal{R}_3 = \mathcal{R}_2 \cap \mathcal{R}_3$. Furthermore, our hypothesis implies that for $i \in \{1, 2, 3\}$, $\mathcal{R}_i - (\mathcal{R}_1 \cap \mathcal{R}_2 \cap \mathcal{R}_3)$ is null homotopic. Therefore by Proposition 3.10, and since the directions in which there are transversal lines to all convex sets of F coincide with $\mathcal{R}_1 \cap \mathcal{R}_2 \cap \mathcal{R}_3$, there is an essential curve of transversal lines to all convex sets of F and consequently given a plane $H \subset \mathbb{R}^3$, one of these transversal lines must be parallel to H. ∎

A *system* Ω of *λ-planes* in $\mathbb{R}^d$ is a continuous selection of a unique λ-plane in every direction of $\mathbb{R}^d$. In [2], it is proved that $\lambda + 1$ systems of λ-planes in $\mathbb{R}^d$ coincide in some direction. We use this fact to prove the following theorem.

Theorem 3.9. *Let F be a family of red, blue, white and green convex sets in $\mathbb{R}^3$. Suppose that every non-heterochromatic triple is intersecting. Then there is a transversal line to all convex sets.*

Proof. By Helly's Theorem in the plane, there is, parallel to every direction, a transversal line to the red and blue convex sets. The same is true for white and green. So we have two different systems of lines. Consequently by Theorem 2 of [2], they must coincide in some direction. ■

A similar argument proves that a family of convex sets in $\mathbb{R}^4$ painted with six colors and with the property that every non-heterochromatic triple is intersecting has a transversal plane to all convex sets.

References

[1] J. Arocha, J. Bracho, L. Montejano, D. Oliveros and R. Strausz, *Separoids; their categories and a Hadwiger-type theorem for transversals.* Journal of Discrete and Computational Geometry, Vol. 27, No. 3 (2002), 377–385.

[2] J. Arocha, J. Bracho, L. Montejano and J. Ramirez-Alfonsin, *Transversals to the convex hull of all k-sets of discrete subsets of $\mathbb{R}^n$* (2009) preprint.

[3] I. Bárány and S. Onn, *Carathéodory's Theorem, colorful and applicable.* In Intuitive Geometry (Budapest, 1995) volume 6 of Bolyai Soc. Math. Stud., pp. 11–21. János Bolyai Math. Soc., Budapest, 1997.

[4] J. Bracho and L. Montejano, *Helly type theorems on the homology of the space of transversals.* Journal of Discrete and Computational Geometry, Vol. 27, No. 3 (2002), 387–393.

[5] J. Bracho, L. Montejano and D. Oliveros, *The topology of the space of transversals through the space of configurations.* Topology and Its Applications Vol. 120, No. 1–2 (2002), 92–103.

[6] S. S. Chern, *On the multiplication in the characteristic ring of a sphere bundle,* Annals of Math., 49, (1948), 362–372.

[7] J. E. Goodman, R. Pollack, and R. Wenger, *Geometric transversal theory.* In J. Pach, ed., *New Trends in Discrete and Computational Geometry,* vol. 10 of *Algorithms Combin.,* pp. 163–198. Springer-Verlag, Berlin, 1995.

[8] J. W. Milnor and J. D. Stasheff, *Characteristic Classes.* Annals of Mathematical Studies No. 76, Princeton University Press, N.J., 1974.

[9] L. Montejano and R. Karasev, *Topological Transversals to a Family of Convex Sets.* To appear in *Discrete and Computational Geometry,* 2010.

[10] L. Montejano and P. Soberon, *Piercing numbers for balanced and unbalanced families.* To appear in *Discrete and Computational Geometry,* 2010.

L. Montejano

Instituto de Matcmáticas,
Unidad Juriquilla,
Qro. Universidad Nacional Autnoma de
Mexico

e-mail: luis@matem.unam.mx

BOLYAI SOCIETY
MATHEMATICAL STUDIES, 24

Geometry –
Intuitive, Discrete, and Convex
pp. 219–257.

Survey on Decomposition of Multiple Coverings

JÁNOS PACH*, DÖMÖTÖR PÁLVÖLGYI† and GÉZA TÓTH‡

The study of multiple coverings was initiated by Davenport and L. Fejes Tóth more than 50 years ago. In 1980 and 1986, the first named author published the first papers about decomposability of multiple coverings. It was discovered much later that, besides its theoretical interest, this area has practical applications to sensor networks. Now there is a lot of activity in this field with several breakthrough results, although, many basic questions are still unsolved. In this survey, we outline the most important results, methods, and questions.

1. Cover-decomposability and the Sensor Cover Problem

Let $\mathcal{P} = \{P_i \mid i \in I\}$ be a collection of sets in $\mathbb{R}^d$. We say that $\mathcal{P}$ is an m-*fold covering* if every point of $\mathbb{R}^d$ is contained in at least m members of $\mathcal{P}$. The largest such m is called the *thickness* of the covering. A 1-fold covering is simply called a *covering*. To formulate the central question of this survey succinctly, we need a definition.

Definition 1.1. A planar set P is said to be *cover-decomposable* if there exists a (minimal) constant $m = m(P)$ such that every m-fold covering of the plane with translates of P can be decomposed into two coverings.

*Research partially supported by Swiss National Science Foundation Grants 200021-137574 and 200020-144531, by Hungarian Science Foundation Grant OTKA NN 102029 under the EuroGIGA programs ComPoSe and GraDR, and by NSF grant CCF-08-30272.

†Supported by János Bolyai Research Scholarship of the Hungarian Academy of Sciences, OTKA PD 104386 and OTKA NN 102029 under EUROGIGA project GraDR 10-EuroGIGA-OP-003. Part of this work was done in Lausanne and supported by Swiss National Science Foundation Grant 200021-125287/1.

‡Supported by OTKA K 83767 and by OTKA NN 102029 under EUROGIGA project GraDR 10-EuroGIGA-OP-003.

Note that the above term is slightly misleading: we decompose (partition) not the set P, but a collection $\mathcal{P}$ of its translates. Such a partition is sometimes regarded a *coloring* of the members of $\mathcal{P}$. The problem whether a set P is cover-decomposable is also referred to as the *cover decomposability problem* for P.

The problem of characterizing all cover-decomposable sets in the plane was proposed by Pach [17] in 1980. He made the following conjecture, which is still unsolved.

Conjecture 1.2 [17]. *Every plane convex set P is cover-decomposable.*

In the present survey, we concentrate on results and proof techniques related to this conjecture. Obviously, in addition to systems of *translates* of a set P, we could study the analogous questions for systems of *homothets* of P (that is, similar copies in parallel position) or for systems of *congruent* copies.

In [18], Conjecture 1.2 was proved for open centrally symmetric convex polygons. More than twenty years later the proof was extended by Tardos and Tóth [23] to open triangles and then by Pálvölgyi and Tóth [22] to any open convex polygon P. Sections 2 and 3 describe the basic ideas and techniques utilized in these proofs.

Theorem 1.3 [18]. *Every centrally symmetric open convex polygon is cover-decomposable.*

Theorem 1.4 [23]. *Every open triangle is cover-decomposable.*

Theorem 1.5 [22]. *Every open convex polygon is cover-decomposable.*

In fact, the proof gives a slightly stronger result: any set, which is the union of finitely many translates of the same open convex polygon is also cover-decomposable. See Section 4 for details.

Given a cover-decomposable set P, one can try to determine the exact value of $m(P)$, that is, the smallest integer m for which every m-fold covering of the plane with translates of P splits into 2 coverings (cf. Definition 1.1). For example, for any open triangle T, we have $12 \geq m(T) \geq 4$ [11]. However, in most of the cases, the best known upper and lower bounds are very far from each other.

One can further generalize the cover decomposability problem by asking whether a sufficiently thick multiple covering of the plane can be decomposed into k coverings, for a fixed $k \geq 2$. This question was raised in [18], and first addressed in detail in [20].

Definition 1.6. Given a set $P \subset \mathbb{R}^2$ and an integer $k \geq 2$, let $m_k(P)$ denote the smallest positive number m with the property that every m-fold covering of the plane with translates of P can be decomposed into k coverings. If such an integer m does not exist, we set $m_k(P) = \infty$.

We believe that $m_k(P)$ is finite for every cover-decomposable set P, but we cannot verify this conjecture in its full generality. However, the statement is true for all currently known families of cover-decomposable sets. In [18], it was shown that, for any centrally symmetric convex open polygon P, the parameter $m_k(P)$ exists and is bounded by an exponentially fast growing function of k. In [23], a similar result was established for open triangles, and in [22] for open convex polygons. However, all these results were improved to the optimal linear bound in a series of papers by Pach and Tóth [20], Aloupis et al. [2], and Gibson and Varadarajan [10].

Theorem 1.7 [20]. *For any open centrally symmetric convex polygon P, we have $m_k(P) = O(k^2)$.*

Theorem 1.8 [2]. *For any open centrally symmetric convex polygon P, we have $m_k(P) = O(k)$.*

Theorem 1.9 [10]. *For any open convex polygon P, we have $m_k(P) = O(k)$.*

The problem of determining $m_k(P)$ can be reformulated in a slightly different way: for a given m, try to decompose an m-fold covering into as many coverings as possible. This problem, more precisely, a slight generalization of this problem, is called the *sensor cover problem* in sensor network scheduling. Suppose that we have a finite number of sensors scattered in a region R, each monitoring some part of R, which is called the range of the sensor. Each sensor has a duration for which it can be active and once it is turned on, it has to remain active until this duration is over, after which it will stay inactive. The *load* of a point is the sum of the durations of all ranges that contain it, and the load of the arrangement of sensors is the minimum load of the points of R. A *schedule* for the sensors is a starting time for each sensor that determines when it starts to be active. The goal is to find a schedule to monitor the given area, R, for as long as we can. Clearly, the cover decomposability problem is a special case of the sensor cover problem, when the duration of each sensor is the same ("unit" time).

Buchsbaum et al. [3] and Gibson and Varadarajan [10] proved their result in this more general context. It was shown in [10] that for every open convex polygon P, there is a constant $c(P)$ such that for any instance of the sensor cover problem with load $c(P)k$, where the range of each sensor is a translate of P, there is a polynomial time computable schedule such that every point is monitored for k units of time.

Conjecture 1.2 cannot be extended to all (not necessarily convex) polygons.

Theorem 1.10 [19]. *No concave quadrilateral is cover-decomposable.*

In Section 5, following [19] and [21], we describe a large class of concave polygons that are not cover-decomposable.

The definition of cover-decomposability can be extended to higher dimensions in a natural way. It is interesting to note that most of the ideas presented in this survey fail to generalize to higher dimensions. The main reason for this is that the statement analogous to Conjecture 1.2 is false in higher dimensions.

Theorem 1.11 [16]. *For $d \geq 3$, the unit ball in $\mathbb{R}^d$ is not cover-decomposable.*

Theorem 1.12 [21]. *For $d \geq 3$, no convex polytope is cover-decomposable.*

However, there is a notable exception in 3-dimensions, albeit unbounded: the octant $\{(x, y, z) : x, y, z > 0\}$.

For the octant, even a 1-fold covering of the whole plane can be trivially decomposed into any number of coverings. We get a more interesting problem if we demand only a part of the plane to be covered.

Theorem 1.13 [11]. *Any 12-fold covering of a finite point set by octants can be decomposed into 2 coverings.*

This property established in the above theorem is called *finite-cover-decomposability;* see Definition 5.4. As an easy consequence, we obtain that any 12-fold covering of the plane with homothets of a fixed triangle can be split into two coverings.

As an easy consequence, we obtain that there is an integer m such that any m-fold covering of the plane with homothets of a fixed triangle can be split into two coverings. The statement holds with $m = 12$. In fact, Conjecture 1.2 can be (and was) formulated in the following more general form.

Conjecture 1.14. For every plane convex set P, there exists a positive integer $m = m(P)$ such that any m-fold covering of the plane with homothets of P can be split into two coverings.

The methods developed in the first substantial publication in this topic [18] were used in all later papers. Therefore, in the next two sections we concentrate on this paper and sketch the proof of Theorem 1.3. In Subsections 3.2 and 3.3, we establish Theorems 1.7 and 1.4. In Section 4, we outline the proofs of Theorem 1.5 and Theorem 1.9 for triangles. Section 5 contains constructions proving (an extension of) Theorem 1.10 and Theorem 1.12. We close this paper with some open problems.

2. Basic Tricks

A family of sets $\mathcal{P}$ is called *locally finite* if every point is contained in only finitely many members of $\mathcal{P}$. It follows by a standard compactness argument that any m-fold covering of the plane with translates of an open polygon P has a locally finite subfamily that forms an m-fold covering. Therefore, in the sequel we will assume without loss of generality that all coverings that we consider are locally finite.

In the next three subsections, we describe three basic tricks from [18] that enable us to reduce the cover decomposability problem to a finite combinatorial problem for hypergraphs.

2.1. Dualization

Let $\mathcal{P} = \{P_i \mid i \in I\}$ be a collection of translates of a finite polygon P in the plane, where I is a finite or infinite set. Let O_i denote the center of gravity of P_i. Obviously, $\mathcal{P}$ is an m-fold covering of the plane if and only if every translate of $\bar{P}$, the reflection of P through the origin, contains at least m elements of the point set $\mathcal{O} = \{O_i \mid i \in I\}$. Furthermore, $\mathcal{P} = \{P_i \mid i \in I\}$ can be decomposed into two coverings if and only if the point set $\mathcal{O} = \{O_i \mid i \in I\}$ can be colored with two colors such that every translate of $\bar{P}$ contains at least one point of each color.

Clearly, the reflected polygon $\bar{P}$ is cover-decomposable if and only if P is. Therefore, we have the following.

Lemma 2.1. *The polygon P is cover-decomposable if and only if there exists an integer m satisfying the following condition. Any point set S in the plane with the property that every translate of P contains at least m elements of S can be colored with two colors so that every translate of P contains at least one point of each color.*

The same argument applies if we want to decompose a covering into $k > 2$ coverings. Almost all later papers in the subject follow this "dual" approach. In the sequel, we also study this version of the problem.

2.2. Divide and conquer – Reduction to wedges

The second trick from [18] is to cut the plane and the set S in Lemma 2.1 into small regions so that with respect to each of them every translate of our polygon looks like an infinite "wedge".

We use the following terminology. Two half-lines (rays) emanating from the same point O divide the plane into two connected pieces, called *wedges*. A *closed wedge* contains its boundary, an *open wedge* does not. The point O is called the *apex* of the wedge. The *angle* of a wedge is the angle between its two boundary half-lines, measured inside the wedge.

Let P be an open or closed polygon of n vertices. Consider a multiple covering of the plane with translates of P. Then, the cover decomposition problem can be reduced to wedges as follows. Divide the plane into small regions, say squares, so that each of them intersects at most two consecutive sides of any translate of P. Every translate of P can intersect only a bounded number c of squares. If a translate of P contains at least cm points of a set S, then at least m of those will belong to one of the squares. Therefore, to find a coloring of the points of S meeting the requirements in Lemma 2.1, it is sufficient to focus on a fixed subset of $S' \subset S$, consisting of all points of S that lie in a single square. It is sufficient to 2-color the elements of S' so that no translate of P that covers at least m points of S' is monochromatic. Notice that, because of our assumption of local finiteness, each subset S' is finite. Moreover, from the point of view of S' any translate of P "looks like" a half-space or a wedge corresponding to one of the vertices of P. To make this statement more precise, denote by $v_1, \ldots, v_n$ the vertices of P in cyclic order, and denote by W_i the wedge bounded by the rays $\overrightarrow{v_i v_{i-1}}$ and $\overrightarrow{v_i v_{i+1}}$ which contains a piece of P in any small neighborhood of v_i. (The indices are taken mod n.) Now any subset of S' that can be cut off from S by a translate of P can also be cut off by a translate of one of $W_1, \ldots, W_n$.

Lemma 2.2. *Suppose that there is a positive integer m such that any finite point set S can be colored with two colors such that every translate of any wedge W_i of P that contains at least m elements of S, contains points of both colors. Then P is cover-decomposable.*

A straightforward generalization of the above argument can be applied when we want to decompose a covering into $k \geq 2$ coverings. Thus, from now on, to prove positive cover-decomposability results we will try to find colorings of finite point sets. However, it will turn out that coloring point sets with respect to wedges may also be very useful in proving negative results.

Observe that we can assume without loss of generality that our point set S is in general position with respect to P, that is, none of the lines determined by two points of S is parallel to a side of P. Indeed, if there is such a line, we can slightly perturb the point set such that any subset of S that can be cut off from S by a translate of P, can also be cut off from the perturbed point set S'.

2.3. Totalitarianism

So far we have only considered coverings of the *whole* plane. At this point it will be convenient to extend our definitions to coverings of subsets of the plane.

Definition 2.3. A set P is said to be *totally-cover-decomposable* if there exists a (minimal) constant $m^T = m^T(P)$ such that every m^T-fold covering of any (!) point set in the plane with translates of P can be decomposed into two coverings. More generally, for any fixed $k \geq 2$, let $m_k^T(P)$ denote the smallest number m^T with the property that every m^T-fold covering of *any* planar point set with translates of P can be decomposed into k coverings.

This notion was formally introduced only in [21], but, in view of Lemma 2.2, all proofs in earlier papers also work for this stronger version of decomposability. To avoid confusion with this notion, sometimes we will call cover-decomposable sets *plane-cover-decomposable*. By definition, every totally-cover-decomposable set is also plane-cover-decomposable. On the other hand, there exist sets (perhaps even open polygons) that are plane-cover-decomposable but not totally-cover-decomposable. For example, the disjoint union of a concave quadrilateral and a far enough half-plane is such a set. Using the notion of total decomposability, we obtain the following stronger version of Lemma 2.2.

Lemma 2.4. *A polygon P is totally-cover-decomposable if and only if there exists a positive integer m^T with the property that any finite point set S in the plane can be colored with two colors such that every translate of any wedge of P that contains at least m^T points of S, contains points of both colors.*

Note that if we want to show that a set P is not plane-cover-decomposable, then, using Lemma 2.4 with suitably chosen sets S, we can first show that it is not totally-cover-decomposable, and then we can add more points to S and apply Lemma 2.1. Of course, we have to be careful not to add any points to the translates that guarantee that P is not totally-cover-decomposable. This is the path followed in [16, 19] (and also in [21], but there the point set S cannot always be extended). These constructions will be discussed in detail in Section 5.

3. Boundary Methods

Let W be a wedge and s be a point in the plane. The translate of W with its apex at s is denoted by $W(s)$. More generally, given a *convex* wedge (whose angle is at most π) W, and points $s_1, s_2, \ldots, s_k$, let $W(s_1, s_2, \ldots, s_k)$ denote the *minimal* translate of W (for containment) whose closure contains $s_1, s_2, \ldots, s_k$.

Following [18], next we will define the boundary of a finite point set with respect to a collection of wedges. We establish and explore some basic combinatorial and geometric properties of the boundary, which will be the heart of the proofs of Theorems 1.3, 1.7, and 1.4. The details of these three proofs from [18], [20], and [23], respectively, will be sketched in the next three subsections.

3.1. Decomposition into two parts

In this subsection, we outline the proof of Theorem 1.3 in the special case when P is an axis-parallel square. This square has an *upper-left*, a *lower-left*, an *upper-right*, and a *lower-right* vertex. For each vertex v of the square, there is a corresponding convex wedge, whose apex is at v and whose boundary half-lines contain the sides of the square incident to v. Denote these wedges by W_{ul}, W_{ll}, W_{ur}, and W_{lr}, respectively. We refer to these four wedges as *P-wedges*.

Let S be a finite point set. By Lemma 2.2 it is sufficient to prove the following.

Lemma 3.1. *The set S can be colored with two colors such that any translate of a P-wedge which contains at least five points of S, contains points of both colors.*

At this point, we introduce the notion of the *boundary* of S with respect to the wedges of P. This notion will be similar to that of the boundary of the convex hull. A point s of S belongs to the boundary of the convex hull of S if there is a half-plane which contains s on its boundary, but none of the points of S in its interior. Similarly, a point s of S belongs to the boundary with respect to wedge W if $W(s)$ contains none of the points of S.

Definition 3.2. Let W be an open wedge. The *W-boundary* of S, that is, the boundary of S with respect to W is defined as $Bd^W(S) = \{s \in S : W(s) \cap S = \emptyset\}$. Two vertices, s and t, of the W-boundary are called *neighbors* if $W(s,t) \cap S = \emptyset$.

Obviously, one can define a natural ordering on the W-boundary points of S, according to which two vertices are consecutive if and only if they are neighbors. Observe that any translate of W intersects the W-boundary in an interval with respect to this ordering. The *boundary of S with respect to the four P-wedges* is the union of the W_{ul}-boundary, the W_{ll}-boundary, the W_{ur}-boundary, and the W_{lr}-boundary of S. All points of S that are not boundary vertices with respect to the P-wedges are called *interior* points.

The W_{lr}-boundary and the W_{ll}-boundary of S meet at the "highest" point of S, that is, at the point of maximum y-coordinate. (Assume, for simplicity that this point is unique). The W_{ll}-boundary and the W_{ul}-boundary meet at the rightmost point of S; the W_{ul}-boundary and the W_{ur}-boundary meet at the lowest point; and the W_{ur}-boundary and the W_{lr}-boundary meet at the leftmost point. See Figure 1. If it leads to no confusion, the translates of W_{ul}, W_{ll}, W_{ur}, W_{lr} will also be denoted by W_{ul}, W_{ll}, W_{ur}, W_{lr}.

If we link together the natural orderings of the boundary vertices of S corresponding to W_{ll}, W_{lr}, W_{ur}, and W_{ul}, in this cyclic order, then we obtain

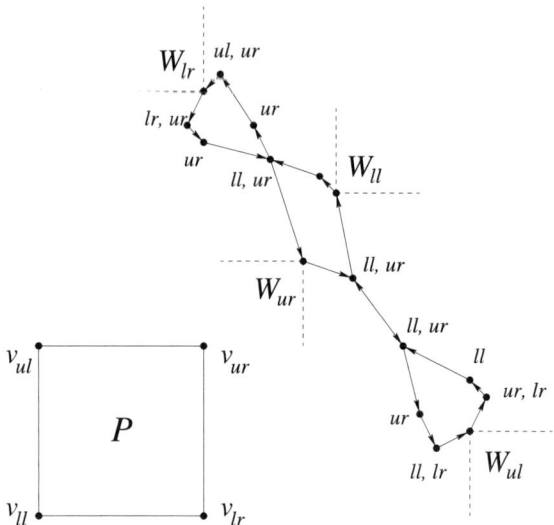

Fig. 1. The boundary of a point set

a counterclockwise cyclic enumeration of all boundary vertices. The main difference between the boundary of S with respect to P and the boundary of the convex hull of S is that in the cyclic enumeration of the boundary vertices some vertices may occur *twice*. These vertices are called *singular*, and all other vertices *regular*.

It is not hard to show, however, that no boundary vertex can appear *three* times in the cyclic enumeration. Moreover, all singular vertices must have the same type. In our case, all of them belong to both a W_{ul} and a W_{lr}, or all of them belong to a W_{ur} and a W_{ll}. This property generalizes to the case when P is any centrally symmetric convex polygon: all singular boundary vertices must belong to a pair of opposite P-wedges of the same type.

The most important observation is the following.

Observation 3.3. *If the intersection of S with a translate of some P-wedge, say, W_{ll}, is non-empty, then this set can be obtained as the union of three subsets:*

   (i) *an interval of consecutive elements in the cyclic enumeration of all vertices of the boundary of S, which contains at least one point from the W_{ll}-boundary;*

(ii) an interval of consecutive elements in the cyclic enumeration of all vertices of the boundary of S, which contains at least one point from the W_{ur}-boundary;

(iii) a set of interior points.

Note that while the subset in (i) contains at least one element, those in (ii) and (iii) may be empty. Analogous statements hold for the other three wedges, and also for other symmetric polygons.

A first naive attempt to find a suitable coloring of S is to color all boundary vertices blue and all interior vertices red. Unfortunately, it is possible that there is a P-wedge that contains lots of boundary vertices and no interior vertex, so this coloring is not necessarily good.

Another naive attempt is to color the boundary vertices alternately red and blue. Apart from the obvious problems that the size of the boundary may be odd and that the singular vertices are repeated in the cyclic order, there is a more serious difficulty with this approach: the translate of a wedge may contain just one boundary vertex and lots of interior vertices. Consequently, we have to be careful when we color the interior vertices, which may lead to further complications.

It turns out that a "mixture" of the above two naive approaches will work.

Definition 3.4. A boundary vertex $s \in S$ is called *m-rich* if there is a translate W of a P-wedge, such that s is the only W-boundary vertex in W, but W contains at least m points of S.[1]

This definition is used in different proofs with a different constant m, but when it leads to no confusion, we simply write "rich" instead of "m-rich." In this proof, "rich" means "5-rich," thus a boundary vertex s is rich if there is a wedge that intersects the W-boundary in s and contains at least four other points.[2]

Our general coloring rule will be the following.

---

[1] In [18] and [20] a slightly different definition was used: there s was required to be the only vertex from the whole boundary (and not only from the W-boundary) in the translate of W. For centrally symmetric polygons, both definitions work, but, for example, for triangles only the latter one does.

[2] Instead of $m = 5$, we could also choose $m = 4$ to define rich points in this proof. Only the last line of the argument would require a little more attention.

(1) Rich boundary vertices are blue.

(2) There are no two red neighbors along the boundary.

(3) Color as many points red as possible, that is, let the set of red points $R \subset S$ be *maximal* under condition (1) and (2).

Note that from (3) we can deduce

(4) Interior points are red.

A coloring that satisfies these conditions is called a *proper coloring*. The same point set may have many proper colorings. For centrally symmetric polygons, any proper coloring will be suitable for our purposes. In [18], an explicit proper coloring is described.

Now we are ready to sketch the proof of Lemma 3.1. Suppose that S is colored properly and W is a translate of a P-wedge such that it contains at least five points of S. We can assume without loss of generality that W contains exactly five points of S. By Observation 3.3, W intersects the W-boundary of S in an interval.

First, we find a blue point in W. If the above interval contains just one point then this point is rich, as the wedge contains at least five points, and rich points are blue according to (1). If the interval contains at least two points, then one of them should be blue, according to (2).

Now we show that W also has at least one red element. If W contains any interior point, then we are done, according to (4). Thus, we can assume by Observation 3.3 that $W \cap S$ is the union of two intervals and all points in W are blue. Since W has five points, at least one of them, say, x, is not the endpoint of any of the intervals. If x is not rich, then, according to (3), x or one of its neighbors is red. So, x must be rich. But then there is a translate W' of a P-wedge, W, or $-W$, which contains only x as a boundary vertex, and contains five points. Using that S is centrally symmetric, it can be shown that $S \cap W'$ is a proper subset of $S \cap W$, a contradiction, since both contain exactly five points. This concludes the proof of Lemma 3.1.

If we consider wedges with more points, we can guarantee more red points in them.

Lemma 3.5. *In a proper coloring of S, any translate of a P-wedge which contains at least $5i$ points of S contains at least one blue point and at least i red points $(i \geq 1)$.*

The proof is very similar to the proof of Lemma 3.1. The difference is that now we color $5i$-rich points red and we have to be a little more careful when counting red points, especially because of the possible singular points. If we delete the blue points (giving them color 1) and then recolor red points recursively by Lemma 3.5, we obtain an upper bound on $m_k(P)$, exponential in k. An analogous statement holds for any centrally symmetric open convex polygon, therefore, we have

Lemma 3.6. *For any centrally symmetric open convex polygon P, there is a constant c_P such that any c_P^k-fold covering of the plane with translates of P can be decomposed into k coverings.*

3.2. Decomposition to $\Omega(\sqrt{m})$ parts for symmetric polygons

Here we sketch the proof of Theorem 1.7, which is a modification of the argument described in the previous subsection. We continue to assume for simplicity that P is an axis-parallel square. Let $k \geq 2$. We will color the point set S with k colors such that any P-wedge that contains at least $m = 18k^2$ points has at least one point of each color. Recursively, we define k boundary layers and denote them by $B_1, B_2, \ldots, B_k$. Let B_1 denote the boundary of S, and let $S_2 = S \setminus B_1$. For any $i < k$, if the set $S_i \subset S$ has already been defined, let B_i be the boundary of S_i and let $S_{i+1} = S \setminus B_i$. The coloring of the boundary layer B_i will be "responsible" for color i. Color i takes the role of blue from the previous proof, while those points that were colored red there will be "uniformly" distributed among the other $k - 1$ colors.

Slightly more precisely, a vertex $v \in B_i$ is called *rich* if there is a translate of a P-wedge that intersects S_i in at least $18k^2 - 18ki$ points, and v is the only element of B_i in it. We color all rich vertices of B_i with color i, and color first the remaining singular, then the remaining regular points periodically: $1, i, 2, i, 3, i, \ldots, k, i, 1, i, \ldots$. The main observation is that, if a P-wedge intersects B_i (for any i) in at least $18k$ points, then it contains a long interval that contains a point of each color. Otherwise, it has to intersect each of the boundary layers B_i $(1 \leq i \leq k)$, but then for each i, its intersection with B_i contains a rich point of color i.

3.3. Triangles

The main difficulty with non-symmetric polygons is that Observation 3.3 does not hold here: the intersection with a translate of a P-wedge is not necessarily the union of two boundary intervals and some interior points. See Figure 2. In the case of triangles, Tardos and Tóth [23] managed to overcome this difficulty by defining a variant of proper colorings. In this subsection, we sketch their proof of Theorem 1.4. For other polygons, a different approach was needed (see Section 4.1).

Suppose that P is a triangle with vertices A, B, C. There are three P-wedges, W_A, W_B, and W_C. We define the boundary just like before. It consists of three parts, the *A-boundary, B-boundary,* and *C-boundary.* Each of them forms an interval in the cyclic enumeration of the boundary vertices. Here comes the first difficulty: there may exist a singular boundary vertex which appears *three* times in the cyclic enumeration of boundary vertices, once in each boundary. It is easy to see that there exists at most *one* such vertex, and we can get rid of it by decomposing our point set S into at most four subsets such that in each of them all singular boundary points belong to the same pair of boundaries, just like in the case of centrally symmetric polygons. For simplicity of the explanation, assume that S has no singular boundary vertex.

Again, we call a boundary vertex s *rich* if there is a translate W of a P-wedge, such that s is the only W-boundary vertex in W, but W contains at least *five* elements of S.

Our coloring will satisfy the following four conditions.

(1) Every rich boundary vertex is blue.

(2) There are no two red neighbors.

(3) Color as many points red as possible, that is, let the set of red points $R \subset S$ be *maximal* under condition (1) and (2).

(4) All interior points are red.

We describe explicitly how to find the set of red points using a greedy algorithm. Consider the linear order on the set of all lines of the plane parallel to the side BC, so that the line through A is *smaller* than the line BC. We define a partial order $<_A$ on our point set as follows. Let $x <_A y$

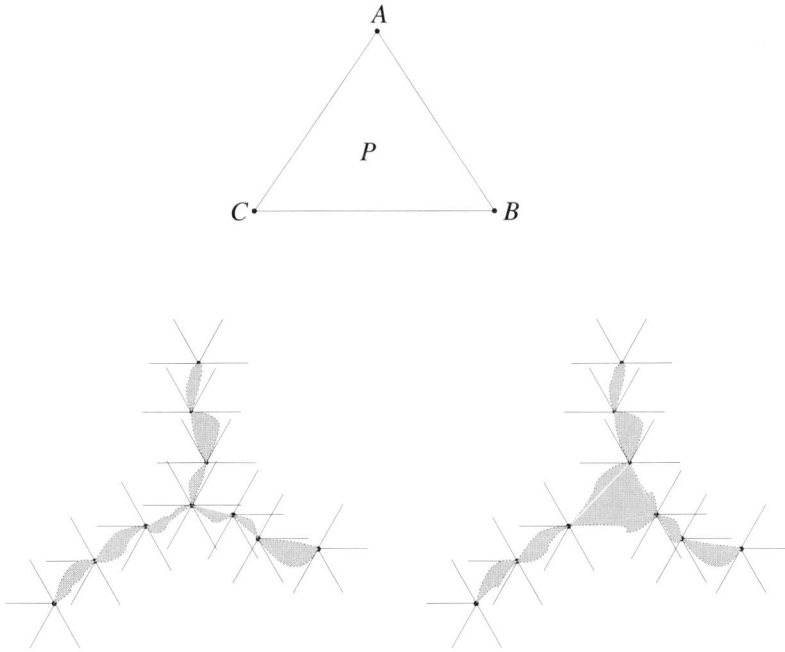

Fig. 2. Triangle P and the structure of the boundary

if the line through x is smaller than the line through y. We have $A <_A B$ and $A <_A C$. Analogously, define the partial order $<_B$ with respect to the side AC such that $B <_B C$ and $B <_B A$, and the partial order $<_C$ with respect to the side AB such that $C <_C A$ and $C <_C B$.

First, color all rich boundary vertices blue. Now take all A-boundary vertices of S and consider them in *increasing* order with respect to $<_A$. If we get to a point that is not colored, we color it red and its neighbors blue. Note that these neighbors may have already been colored blue (because they are rich, or because of an earlier red neighbor), but they were definitely not colored red, as any neighbor of any red point is immediately colored blue. Continue this procedure, until all points of the A-boundary are colored. Color the B-boundary and the C-boundary in a similar fashion, using the other two partial orders.

Suppose that W is a translate of a P-wedge covering *at least* five points of S. We can assume without loss of generality that W covers exactly five points of S. Assume that W is a translate of W_A. The other two cases can be treated similarly. To find a blue point in W, we proceed just like in the previous section; this works for any proper coloring. We know that W

intersects the A-boundary of S in an interval. If this interval contains just one point, then it must be rich and hence blue. It the interval contains at least two points, then one of them must be blue.

It remains to show that W also contains at least one red point. If W contains any interior point, then we are done. Therefore, we assume that all five points in W are boundary vertices. Since there are five points in W, one of them, say, x, is (i) not the first or last A-boundary vertex in W; (ii) not the $<_A$-minimal B-boundary point in W; and (iii) not the $<_A$-minimal C-boundary point in W.

Suppose that x is rich. Then there is a translate W' of a P-wedge, which contains only x as a boundary vertex, and contains five points. It can be shown by some simple geometric arguments that $S \cap W'$ is a proper subset of $S \cap W$, a contradiction, since both sets contain five points. So, x cannot be rich. But then why would it be blue? The only possible reason is that during the coloring process, one of its neighbors on the boundary, say, y, was colored red earlier. It can be shown that then $y \in W$, which implies that there is a red point in W.

The same idea works if there are some singular boundary vertices, but all of them belong to the A-boundary and the B-boundary, say. The only difference is that in this case we have to synchronize the coloring processes on the A-boundary and on the B-boundary, so that we arrive at the common vertices at the same time. This concludes the proof of Theorem 1.4. The original proof gave that every 43-fold covering with translates of a triangle splits into two coverings, but B. Ács [1] showed that the statement also holds for every 19-fold covering. Recently it was further improved to 12-fold coverings, by Keszegh and Pálvölgyi [11].

By a slightly more careful argument, we can establish

Lemma 3.7. *The points of S can be colored with red and blue such that any translate of a P-wedge which contains at least $5i + 3$ of the points, contains a blue point and at least i red points $(i \geq 1)$.*

If we apply Lemma 3.7 recursively, we obtain an bound on $m_k(P)$, exponential in k.

Lemma 3.8. *For any open triangle P, every $\frac{7 \cdot 5^k - 15}{20}$-fold covering of the plane with translates of P can be decomposed into k coverings.*

This result was later improved by the more general Theorem 1.9 of Gibson and Varadarajan.

4. PATH DECOMPOSITION AND LEVEL CURVES

In this section, we present three generalizations of the boundary method that can be used to establish cover-decomposability results.

4.1. Classification of wedges

Pálvölgyi and Tóth [22] developed some new ideas to establish Theorem 1.5 which states that all open convex polygons are cover-decomposable. In the previous section, we colored a point set with respect to P-wedges, for some fixed polygon P. Here we color point sets with respect to an *arbitrary* set of wedges.

Definition. A collection of wedges $\mathcal{W} = \{W_i \mid i \in I\}$ is said to be *non-conflicting* or, simply, *NC* if there is a constant m with the following property. Any finite set of points S in the plane can be colored with two colors so that any translate of a wedge $W \in \mathcal{W}$ that covers at least m points of S contains points of both colors.

It turns out that a single wedge is always NC. One can also characterize all pairs of wedges that are NC. Pálvölgyi and Tóth proved that a set of wedges is NC if and only if each pair is NC. It follows directly from this characterization that for any convex polygon P, the set of P-wedges is NC.

Lemma 4.1. *A single wedge is NC.*

An important tool in the proof of Lemma 4.1 and in the proof of the following lemmas is the *path decomposition,* which is a generalization of the concept of the boundary. To illustrate this technique, we present a proof of Lemma 4.1.

Proof of Lemma 4.1. Let S be a finite point set and let W be a wedge. We prove that the NC property holds with $m = 3$, that is, we show that S can be colored with two colors such that any translate of W that contains at least 3 points of S, contains a point of both colors. Suppose first that the angle of W is at least π. Then W is the union of two half-planes, A and B. Take the translate of A (resp. B) that contains exactly two points of S, say, A_1 and A_2 (resp. B_1 and B_2). There might be coincidences between A_1, A_2 and B_1, B_2, but still, we can color the set $\{A_1, A_2, B_1, B_2\}$ such that A_1 and A_2 (resp., B_1 and B_2) are of different colors. Now, if a translate of W

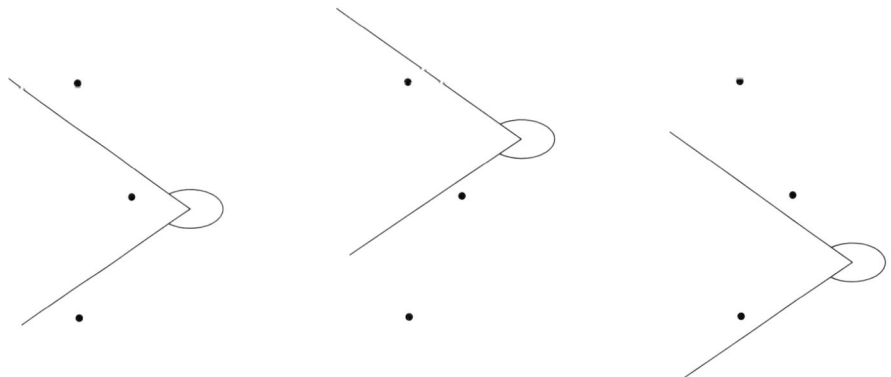

Fig. 3. A concave wedge and three points, any two of which can be cut off by a translate

contains three points, it contains either A_1 and A_2, or B_1 and B_2, and we are done. Note that three is optimal in this statement; see Figure 3.

Suppose now that the angle of W is less than π. We show that in this case the NC property holds with $m = 2$. We can assume that the positive x-axis is in W; this can be achieved by an appropriate rotation. For simplicity, also suppose that no line determined by a pair of points of S is parallel to the sides of W. This can be guaranteed by applying a suitable perturbation of the set S that does not effect which subsets of it can be cut off by a translate of W.

For any fixed y, let $W(2; y)$ be the translate of W which

(1) contains at most two points of S;

(2) the y-coordinate of its apex is y; and

(3) the x-coordinate of its apex is minimal.

For any y, the translate $W(2; y)$ is uniquely determined. Examine, how $W(2; y)$ varies as y runs over the real numbers. If y is very small (smaller than the y-coordinate of the points of S), then $W(2; y)$ contains two points, say X and Y, and one more, Z, on its boundary. As we increase y, the apex of $W(2; y)$ changes continuously. How can the set $\{X, Y\}$ of the two points in $W(2; y)$ change? For a certain value of y, one of them, say, X, moves to the boundary. At this point, Y is inside and two points, X and Z, are on the boundary. If we slightly further increase y, then Z *replaces* X, that is, Y and Z will be in $W(2; y)$ (see Figure 4). As y increases to infinity, the set $\{Z, Y\}$ could change several times, but each time it changes in the

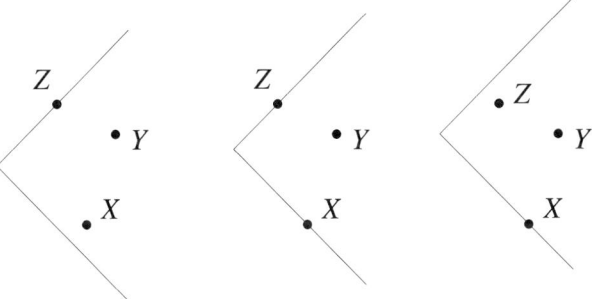

Fig. 4. Z replaces X in $W(2; y)$

above described manner. Define a directed graph whose vertices are the points of S, and there is an edge from u to v if v replaced u during the procedure. We get two paths, P_1 and P_2. The pair (P_1, P_2) is called the *path decomposition of S with respect to W, of order two* (see Figure 5).

Color the vertices of P_1 red, the vertices of P_2 blue. Observe that each translate of W that contains at least two points, contains at least one vertex of both P_1 and P_2. This completes the proof. ∎

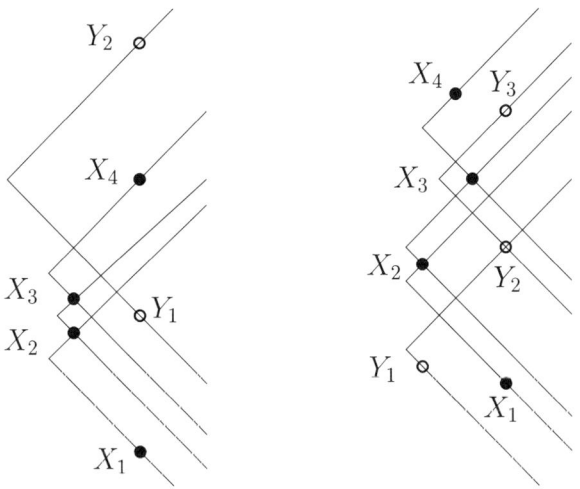

Fig. 5. Path decompositions of order two. $P_1 = X_1 X_2 \ldots$, $P_2 = Y_1 Y_2 \ldots$

The *path decomposition of S with respect to W, of order m* can be defined very similarly. Let $W(m; y)$ denote the translate of W which

(1) contains at most m points of S;

(2) the y-coordinate of its apex is y; and

(3) the x coordinate of its apex is minimal.

Suppose that, for a very small value of y, the set $W(m; y)$ contains the points $r_1, r_2, \ldots, r_m$, and at least one more point on its boundary. Just like in the proof above, as we increase the value of y, the set $\{r_1, r_2, \ldots, r_m\}$ changes several times. Every time one of the elements of this set is *replaced* by another point. Define a directed graph whose vertices are the points of S, and there is an edge from r to s if r is replaced by s at some point. This graph is the union of m directed paths, P_1^W, P_2^W, $\ldots$, P_m^W (and possibly some isolated vertices), which is called the *order m path decomposition of S with respect to W*. Note that the order 1 path decomposition is just the W-boundary of S, so this notion can be regarded as a generalization of the boundary. In general, in a higher order path decomposition, no path is identical to the boundary. The union of the paths, however, always contains the boundary.

Note that there is a hidden variable in this notation. When we write P_1^W, then it can mean the first path of the path decomposition of *any* order m, so it would be more precise to write $P_1^W(m)$. However, to ease readability, we use the (ambiguous) simpler notation as from the context the value of m will be always clear.

Lemma 4.2.

(i) *Any translate of W contains an interval of each of $P_1^W, P_2^W, \ldots, P_m^W$.*

(ii) *If a translate of W contains precisely m points of S, then it contains precisely one point from each of $P_1^W, P_2^W, \ldots, P_m^W$.*

Now we scrutinize the case when we have *two* wedges, V and W. We distinguish several cases according to their relative position.

Type 1 (Big): One of the wedges has angle at least π.

For the other cases, we can assume without loss of generality that W contains the positive x-axis. Extend the boundary half-lines of W to lines. They divide the plane into four parts: Upper, Lower, Left, and Right parts, the last of which is W itself. See Figure 6.

Type 2 (Half-plane): One side of V is in the Right part and the other one is in the Left one. That is, the union of the wedges cover a half-plane. See Figure 7.

Type 3 (Contain): One of the following three conditions is satisfied:

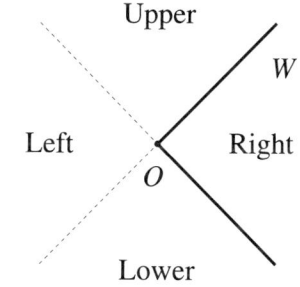

Fig. 6. Wedge W

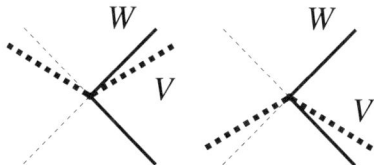

Fig. 7. Type 2 (Half-plane)

(i) one side of V is in the Upper part, the other is in the Lower part;

(ii) both sides are in the Right part;

(iii) both sides are in the Left part. See Figure 8.

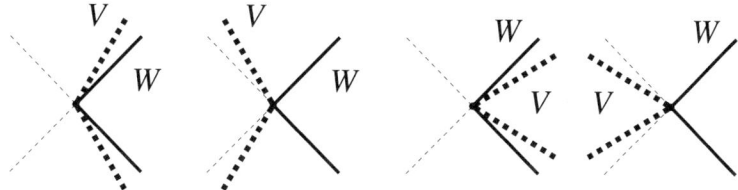

Fig. 8. Type 3 (Contain)

Type 4 (Hard): One side of V is in the Left part and the other side is either in the Upper part or in the Lower one. See Figure 9.

Type 5 (Special): One of the following three conditions is satisfied:

(i) one side of V is in the Right part and the other one is in the Upper or Lower part;

(ii) both sides of V are in the Upper part;

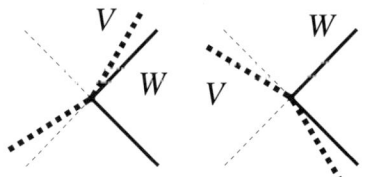

Fig. 9. Type 4 (Hard)

(iii) both sides are in the Lower part.

That is, the union of the wedges is in an open half-plane whose boundary contains the origin, but neither of them contains the other. See Figure 10.

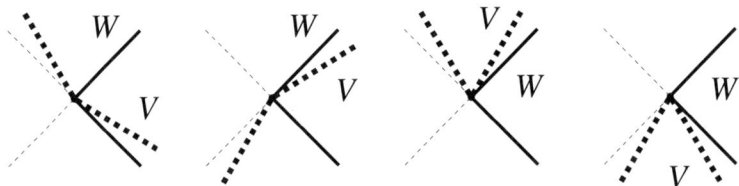

Fig. 10. Type 5 (Special)

It is not hard to see that there are no other possibilities.

Lemma 4.3. *Let* $\mathcal{W} = \{V, W\}$ *be a set of two wedges, of Type 1, 2, 3, or 4. Then* $\mathcal{W}$ *is NC.*

Here we omit the proof. It is different for each type, but in each case the basic idea is similar to that of the proof of Lemma 4.1. In the case of pairs of wedges of Type 4 (Hard), we have to take care of singular points in a somewhat similar way as in the previous section, in the proof for triangles. For pairs of wedges of Type 3 (Contain), we can apply an order 4 path decomposition.

Next, we turn to the case of several wedges.

Lemma 4.4. *A set of wedges* $\mathcal{W} = \{W_1, W_2, \ldots, W_t\}$ *is NC if and only if any pair* $\{W_i, W_j\}$ *is NC.*

It is obvious that if two wedges are not NC, then $\mathcal{W}$ cannot be NC. The proof in the other direction is more involved. It is based on a tricky application of path decompositions. In fact, it can be shown that if $\mathcal{W}$ is NC, then for any k there is an m_k such that any finite point set can be colored with k colors such that if a translate of a wedge from $\mathcal{W}$ contains at least m_k points, then it contains all k colors. However, the bound obtained in [22] grows very fast, the argument gives only $m_k \leq (8k)^{2^{t-1}}$.

To finish the proof of Theorem 1.5, observe that no two wedges corresponding to the vertices of a convex polygon can form a pair of Type 1 (Big) or of Type 5 (Special).

It is shown in [21] that if $\mathcal{W} = \{V, W\}$ is a set of two wedges of Type 5 (Special), then $\mathcal{W}$ is *not* NC. Therefore, a set of wedges is NC if and only if none of its pairs is of Type 5 (Special). For the construction and its consequences, see Section 5.

4.2. Level curves and decomposition to $\Omega(m)$ parts for symmetric polygons

The *level curve* method, which can be regarded as another extension of the boundary technique, was invented by Aloupis, Cardinal, Collette, Langerman, Orden, and Ramos [2] at about the same time, but independently from the introduction of path decompositions.

Suppose that W is an open wedge and its angle is less than π. The *level curve of depth l*, denoted by $\mathcal{C}(l)$, is defined as the boundary of the union of all translates of W that contain fewer than l points. If W contains the positive x-axis, then we can also define $\mathcal{C}(l)$ as the set of the apices of $W(l-1; y)$.

Note that this curve consists of straight-line segments that are parallel to the sides of W. See Figure 11. $\mathcal{C}(1)$ passes through all boundary points. If $p \in \mathcal{C}(l)$, then $|W(p) \cap S|$ is $l-1$, and $W(p)$ has one or two points of S on its boundary.

Consider all translates of W whose apices are on $\mathcal{C}(l)$. Call these translates $\mathcal{C}(l) - W$-*wedges*. Consider a point of S in a $\mathcal{C}(l) - W$-wedge. The apices of those $\mathcal{C}(l) - W$-wedges which contain this point form an interval on $\mathcal{C}(l)$. Therefore, each $\mathcal{C}(l) - W$-wedge corresponds to a point on $\mathcal{C}(l)$, and every point of S corresponds to an interval of $\mathcal{C}(l)$. The condition that each $\mathcal{C}(l) - W$-wedge contains at least $l - 1$ points translates to the condition that each of the points of $\mathcal{C}(l)$ is covered by at least $l - 1$ intervals. Here we want to color the *intervals* in such a way that each *point* is covered by intervals of all colors.

Now we sketch the proof of Theorem 1.8 given in [2], based on the level curve method.

$\mathcal{C}(l)$

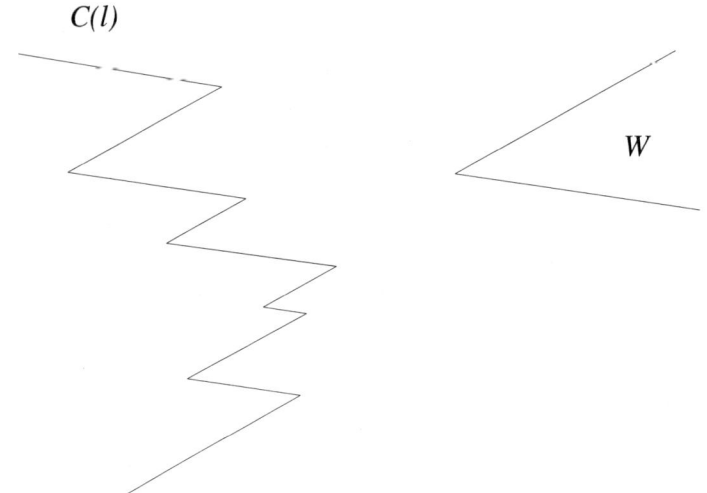

Fig. 11. The level curve $\mathcal{C}(l)$

Suppose that our symmetric polygon P has $2n$ vertices. Denote the wedges belonging to them by $W_0, \dots, W_{2n-1}$, in clockwise order. Throughout the proof, all the indices should be considered modulo $2n$. Two wedges, W_i and W_j, are called *antipodal* if $i + n \equiv j$ modulo $2n$, that is, if they belong to two opposite vertices of the polygon. A crucial observation, already used in Subsection 3.1 (more generally, in [18]), is that any two wedges that are not antipodal cover a half-space.

We want to color the points of the point set S with k colors such that every translate of W_i ($i = 0, \dots, 2n - 1$) that contains at least m'_k points, contains a point of each color. For any fixed l, the level curves $\mathcal{C}_i(l)$ that correspond to wedge W_i may cross each other in a complicated way. However, in the "middle" of S they form a structure similar to the boundary in Subsection 3.1. It turns out that it is enough to consider these parts of the level curves.

More precisely, let $l = 6k + 1$. For every side of P, take two lines parallel to it that cut off $2l + 3$ points from each side of S. Denote the intersection of the n strips formed by these lines by $\mathcal{T}$. For every i, let $\mathcal{C}'_i(l) = \mathcal{C}_i(l) \cap \mathcal{T}$. Call those translates of W_i whose apices are on $\mathcal{C}'_i(l)$ *witness W_i-wedges*. It is not hard to see that only level curves belonging to antipodal wedges may cross inside $\mathcal{T}$. Some further analysis shows that, in fact, there can be only at most one such pair. (Note the similarity to singular points in case of symmetric polygons.) This means that the regions cut off from $\mathcal{T}$

by the curves $C_i'(l)$ are all disjoint, with the possible exception of one pair. Without loss of generality we may assume that this pair is $C_i'(0)$ and $C_i'(n)$. It is not difficult to verify that any translate of W_i that contains at least $3l + 5$ points, must contain a witness W_i-wedge. Therefore, it is enough to concentrate on wedges with this property.

One can parameterize these witness wedges by $t \in [0, 2n)$ such that $W(t)$ is a translate of $W_{\lfloor t \rfloor}$. The most important geometric observation is that if $p \in W(\lfloor t \rfloor + x) \cap W(\lfloor t \rfloor + z)$, where $0 \le x \le 1$ and $0 \le z \le n$, then $p \in W(\lfloor t \rfloor + y)$ for all $x \le y \le z$.

If $p \in W(\lfloor t \rfloor + x) \cap W(\lfloor t \rfloor + z)$, where $0 \le x \le 1$ and $n \le z \le n+1$, then p is contained in two antipodal wedges, which implies that it is contained in translates of W_0 and W_n, but in no translates of any other wedge W_i. Therefore, every p corresponds to either an interval of the circle $[0, 2n)$ or to two intervals, one of which is a subinterval of $[0, 1]$, and the other a subinterval of $[n, n+1]$.

We can take care of these two cases separately, as any big wedge contains many points from one of these groups. The sets of the first type (intervals) form a circular interval graph. Using a simple greedy algorithm, we can partition the set of these circular intervals into k parts with the property that any point of the circle that is covered by at least $3k$ intervals will be covered by at least one interval in each part. For sets of the second type (unions of two intervals), we want to color points with respect to a wedge W and its antipodal pair $-W$. The greedy algorithm again gives a good partition of a $3k$-fold covering into k coverings. Since every witness wedge contains at least $6k$ points, we are done.

Combining these facts, we obtain that $m_k' \le 18k + 5$ for any system of wedges derived from a convex centrally symmetric polygon. This has to be multiplied by a constant depending on the shape of the polygon that comes from Lemma 2.2, to derive a bound for the multiple-cover-decomposability function m_k of the polygon.

4.3. Decomposition to $\Omega(m)$ parts for triangles

The case of not necessarily centrally symmetric polygons P was settled in [10]. In this subsection, we sketch the proof in the special case when P is a triangle, which already contains most of the key ideas of the general argument.

The first step is the usual dualization and reduction to wedges, therefore, it is enough to prove the following statement.

Lemma 4.5. *Let* W_A, W_B, W_C *be the wedges of a triangle* T, *and let* $k > 0$. *Then any point set* S *can be colored with* k *colors such that any translate of* W_A, W_B, *or* W_C *which covers at least* $14k$ *points of* S *contains at least one point in each color.*

Let S be a point set. Consider the level curve $\mathcal{C}_A = \mathcal{C}_A(14k + 1)$ of W_A of depth $14k + 1$. Again, for the coloring it is enough to consider those translates of W_A whose apices are on $\mathcal{C}_A$. As we have seen in the previous subsection, these wedges contain $14k$ points of S. Call these translates *witness A-wedges*. The *witness B-wedges* and *witness C-wedges* can be defined analogously.

The most important new idea is that first we *partially* color the points of S so that every witness A-wedge contains at least one point of each color, and all witness B-wedges and witness C-wedges have sufficiently many uncolored points. We proceed by extending this coloring in such a way that every witness B-wedge has a point of each color, and it is still true that every witness C-wedge has enough uncolored points. Finally, we take care of the witness C-wedges.

Lemma 4.6. *One can partially color the points of* S *with* k *colors such that*

(i) *each witness A-wedge contains all* k *colors, and*

(ii) *each witness B-wedge and C-wedge contains at least* $6k$ *uncolored points.*

Proof. We will again use the partial orders $<_A$, $<_B$, and $<_C$, defined in Subsection 3.3. First, we choose a subset $Q \subset S$ in the following way. Initially, set $Q = \emptyset$. Then, for each witness A-wedge W such that $|Q \cap W| < 2k$, we add the points of $S \cap W$ to Q, one by one, in decreasing order with respect to $<_A$, until $|Q \cap W| = 2k$. Then we proceed with another witness A-wedge. There are infinitely many witness A-wedges, but we have to consider only finitely many, since they can intersect S in only finitely many distinct subsets.

In the way described in the previous subsection, each witness A-wedge corresponds to a point on $\mathcal{C}_A$, and each point of Q corresponds to an interval. Thus, we obtain a system of intervals on $\mathcal{C}_A$ (or, equivalently, on a line) such that each point is covered at least $2k$ times. Take a *minimal* collection of these intervals that still form a covering. Is is easy to see that no point can be covered more than twice. Color these intervals with the first color, take another minimal cover for the second color, and continue until all colors are used. Since we started with a $2k$-fold covering and in each step the thickness decreased by at most two, we will be able to use all colors. This corresponds to a coloring of a subset $R \subset Q$. It is clear that each witness A-wedge contains at least one point of each color. Observe, that the intervals that correspond to R do not cover any point more than $2k$ times. That is, each witness A-wedge contains at most $2k$ points of R.

Now we prove (ii). By symmetry, it is enough to show that every witness B-wedge contains at least $6k$ uncolored points. Let W be a witness B-wedge, and let $p_1, p_2, \ldots$ be the points of $W \cap S$ in increasing order with respect to $<_B$. If none of them is in Q, then none of them is colored and we are done. Otherwise, let j be the largest number such that $p_j \in Q$. If $j < 8k$, then there are at least $6k$ uncolored points in W. Suppose that $j \geq 8k$. Point p_j was added to Q when we considered a certain witness A-wedge, say, V. Wedges W and V can have two types of intersection, since exactly one of them contains the apex of the other one.

Fig. 12. The two types of intersections of W and V

Case 1: V contains the apex of W. Consider the triangle $Z_1 = \{x \mid x \in W, p_j \not<_B x\}$. (See the left part of Figure 12.) It contains j points of S, but at most $2k$ of them are colored, so W contains at least $6k$ uncolored points.

Case 2: W contains the apex of V. Consider the triangle $Z_2 = \{x \mid x \in V, p_j \not<_A x\}$. (See the right part of Figure 12.) Since we added p_j to Q when we processed wedge V, there can be at most $2k - 1$ points p in V with $p_j <_A p$. Therefore, at least $12k$ points are in Z_2. Since we colored at most $2k$ of them, there must remain at least $10k$ uncolored points in $Z_2 \subset W$.

Now we run the same algorithm for the uncolored points and for the witness B-wedges. A very similar argument shows that there will still be at least $2k$ uncolored points in each witness C-wedge. We run the algorithm once more for the uncolored points and for the witness C-wedges. This concludes the proof.

5. Indecomposable Coverings

In this section, we describe some constructions of coverings with arbitrarily high multiplicity that cannot be decomposed into two coverings. The first such example was given by Mani and Pach [16], and it shows that the unit ball is not cover-decomposable. In other words, for any m, there exists a covering of $\mathbb{R}^3$ with unit balls such that every point is covered by at least m balls, but the covering cannot be decomposed into two coverings. Later in [19], several other constructions were given, all based on the geometric realization of the same m-uniform hypergraph (system of m-element sets) not having Property B.[3] The same hypergraph is used in the construction described in Subsection 5.1 below. It was shown by Erdős [8] that every m-uniform hypergraphs that does not have Property B has at least 2^{m-1} hyperedges, so any indecomposable construction must be exponentially large. As one of the first geometric applications of the Lovász Local Lemma [9], Pach showed that if a system of translates of a "nice" geometric set has the property that every point is covered by at least m and at most a subexponential (in m) number of sets, then the system is decomposable into two coverings.

First, we present the construction of [19] showing that no concave quadrilateral is cover-decomposable. In Subsection 5.2 (see also [21]), we show that general concave polygons are and polyhedra are not space-cover-decomposable. Finally, we discuss the difference between several variants of cover-decomposability.

5.1. Concave quadrilaterals—Proof of Theorem 1.10

We present the construction in the dual setting. Suppose that the vertices of the quadrilateral, Q, are A, B, C and D, in this order, the reflex angle being at D. This implies that W_A and W_C are of Type 5 (Special) (see

---

[3]We say that a hypergraph has Property B if the elements of its vertex set can be colored with two colors such that every hyperedge contains points of both colors.

Section 4.1 for the definition). Moreover, they belong to an even more special subclass, which we call *Very Special:* when we translate the wedges so that their apices are in the origin, then they are disjoint and their closures are both contained in the same open half-plane (see the two right examples in Figure 10).

First, for any m, we give a finite set of points and a finite number of translates of Q, each covering precisely m points, such that no matter how we color the points by two colors, at least one of the translates will be monochromatic. In the "primal" setting, this corresponds to a finite system of translates of Q with the property that no matter how we partition this system into two, we can find a point contained in precisely m translates, each of which belongs to the same part. Hence, Q is not totally-cover-decomposable. Finally, we show how this construction can be extended to an m-fold covering of the whole plane, which cannot be split into two coverings.

We use translates of the wedges W_A and W_C to realize the following m-uniform hypergraph $\mathcal{H}$, also used in [16]. The vertices of the hypergraph are the vertices of a rooted perfect m–ary tree of height $m - 1$. There are two types of hyperedges. To each vertex v which is not a leaf of the tree, we assign a hyperedge of the *first type*, formed by the children of v. To each leaf v, we assign a hyperedge of the *second type*, formed by the vertices along the path from the root to v. More precisely, the vertices of the hypergraph are sequences of length less than m, consisting of the integers from 1 through m: $V(\mathcal{H}) = [m]^{<m} = \cup_{i=0}^{m-1}[m]^i$. The hyperedges of the first kind consist of m-tuples of sequences of length l, for some l $(1 \leq l < m)$, such that removing their last elements, we obtain the same sequence of length $l - 1$. The hyperedges of the second kind consist of all initial segments of a sequence of length $m - 1$, where the empty sequence (corresponding to the root) is considered an initial segment of every sequence. Hence, $\mathcal{H}$ has $\sum_{i=0}^{m-1} m^i$ vertices and $\sum_{l=1}^{m-1} m^{l-1} + m^{m-1}$ hyperedges.

The hyperedges of the first kind are realized by translates of W_A, the hyperedges of the second kind by translates of W_C. For simplicity, suppose that W_A is a very thin wedge that contains the positive x-axis and W_C is a very thin wedge that contains the negative y-axis; although the construction would work for any pair of convex wedges that belong to opposite vertices of a concave quadrilateral. All vertices of $\mathcal{H}$ are very close to a vertical line. All vertices of a hyperedge of the first kind are on a horizontal line, for each edge on a different one (see Figure 13). It is easy to see that this is indeed a geometric realization of $\mathcal{H}$, so the points cannot be colored with two colors

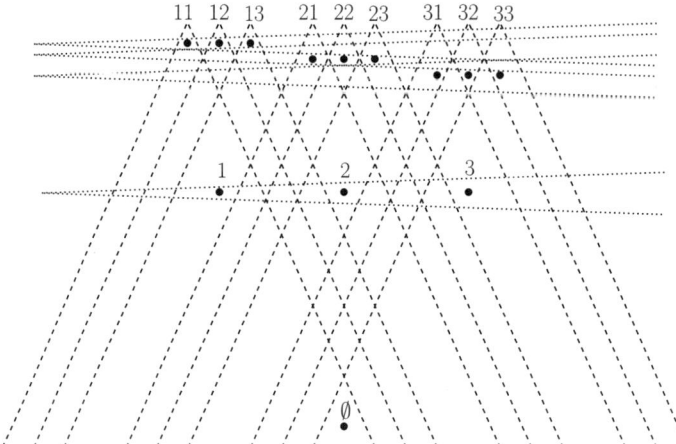

Fig. 13. Indecomposable covering with two special wedges of a concave quadrilateral

such that every translate of W_A and W_C of size m contains points of both colors.

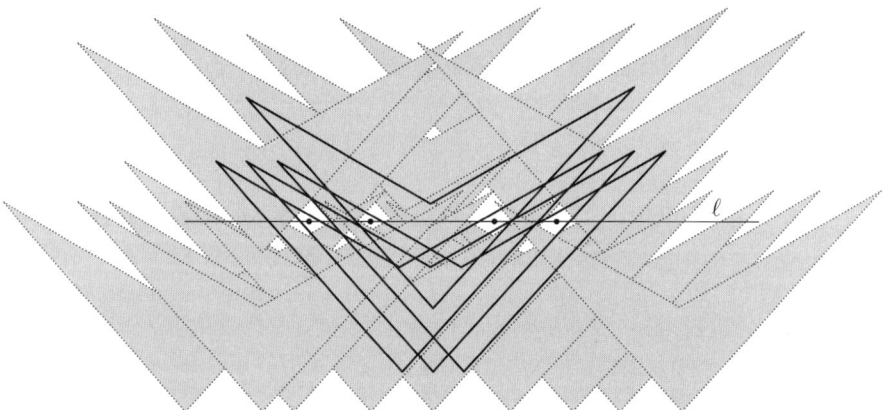

Fig. 14. Extending the original 2-fold covering of the four points by the solid quadrilaterals to a 2-fold covering of the whole plane by adding the dotted quadrilaterals

Now we switch back to the primal plane. We have a point set S, and a set $\mathcal{Q}$ of translates of Q. It remains to extend $\mathcal{Q}$ to an m-fold covering of the whole plane. Before doing so, notice that it can be achieved that all points of the set S are on a line ℓ, not parallel to the sides of Q. Add to this m-fold covering all translates of Q that are disjoint from S (see Figure 14). It is clear that the resulting arrangement remains indecomposable. The construction can be easily modified to obtain a "locally finite" covering, using a standard compactness argument. Note that the construction of [21] is not always extendable this way.

5.2. General concave polygons and polyhedra

The hypergraph $\mathcal{H}$ can be realized by two wedges that form a Very Special pair. Unfortunately, there are concave polygons that do not have two Very Special wedges (see, e.g., Figure 16). In fact, they might not even have two wedges that form a Type 5 (Special) wedge at all; e.g., in the case of the union of two axis-aligned rectangles. The cover-decomposability of such concave polygons follows from the proof of Theorem 1.5 (see Lemma 2.2, 4.3, 4.4). However, it can be shown that every concave polygon that has two wedges that form a pair of Type 5 (Special) is not totally-cover-decomposable. This includes all "typical" concave polygons, as any polygon that has no parallel sides has a Type 5 (Special) pair of wedges.

To prove indecomposability, we have to realize another hypergraph that does not have property B. This construction has fewer points than $\mathcal{H}$ (about 4^m). It is also more general, in the sense that it can be realized by any pair of Type 5 (Special) wedges. In fact, the following statement holds, which implies that no polygon with a Type 5 (Special) pair of wedges is totally-cover-decomposable [21].

Lemma 5.1 [21]. *For any pair of special wedges, V and W, and for any pair of positive integers, k and l, there is a point set P of size $\binom{k+l}{k} - 1$ such that for every coloring of P with red and blue, either there is a translate of V containing k red points and no blue points, or there is a translate of W containing l blue points and no red points.*

Proof. We proceed by induction on $k + l$. Denote by $P(k, l)$ a set of points that satisfy the conditions of the lemma for k and l. If k or l is equal to 1, then the statement is trivially true. In the induction step (see the left side of Figure 15), place a point p in the plane and a suitable small scaled down copy of $P(k - 1, l)$ with the property that any translate of V with its apex in the neighborhood of $P(k - 1, l)$ contains p, but none of the translates of W with its apex in the neighborhood of $P(k - 1, l)$ does. Analogously, place a scaled down copy of $P(k, l - 1)$ in such a way that any translate of W with its apex in the neighborhood of $P(k, l - 1)$ contains p, but none of the translates of V with its apex in the neighborhood of $P(k, l - 1)$ does.

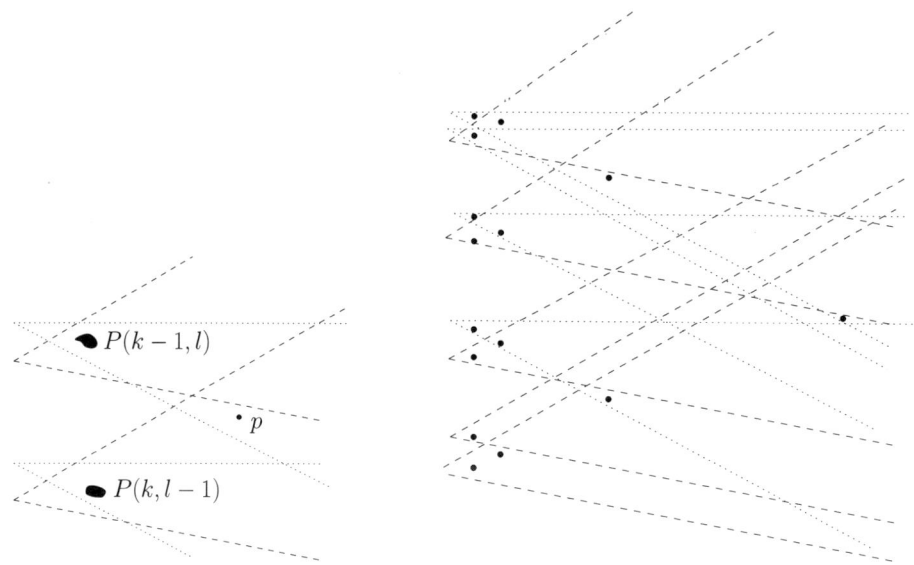

Fig. 15. Sketch of one step of the induction and iteration of some steps

If p is colored red, then either for the first part of the construction, similar to $P(k-1,l)$, there is a translate of V that covers point p together with $k-1$ other red points and no blue ones, or for the part similar to $P(k-1,l)$, there is a translate of W that covers l blue points, no red ones, and it does not contain p. In both cases, we are done. A similar argument works in the case when p is blue.

Remark 5.2. Instead of considering *all* translates of V and W, in order to find a wedge that meets the requirements of Lemma 5.1, it is sufficient to restrict our attention to a *finite set* of translates whose apices lie on the same line.

This construction, combined with Lemma 4.3 and 4.4, gives the following characterization of polygons.

Theorem 5.3 [22, 21]. *An open polygon P is totally-cover-decomposable if and only if none of the P-wedges form a pair of Type 5 (Special).*

Unfortunately, we still do not have a nice characterization for *plane-cover-decomposability*. The reason is that the above construction cannot always be extended to coverings of the whole plane. As pointed out in Remark 5.2, it is sufficient to consider a finite set of wedges whose apices lie on the same line. However, after dualization the centers of the translates will lie on *two* lines. An example of a polygon which is not totally-cover-decomposable but might be cover-decomposable is depicted in Figure 16. Some special cases when such an extension is always possible, were studied in [21].

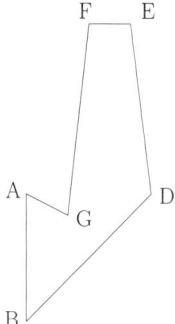

Fig. 16. Unknown hexagon: its only special pair of wedges are at A and E

In higher dimensions, the situation is completely different. According to Theorem 1.12 [21], for $d \geq 3$, no d-dimensional convex polytope is cover-decomposable.

The proof is based on the observation that for any polytope P, either there is a plane that intersects P in a concave polygon, which always has a special pair of wedges, or there are two parallel planes that intersect P in two polygons such that there is a special pair among their wedges. In both cases, we can take a plane in space and a family of translates of P that realize the above construction in this plane so that the intersection of the plane and the translates of P play the role of the wedges. Then we take the dual of this arrangement. To prove that this construction is extendable to an indecomposable covering of the entire space, observe that the centers of all the translates used in the construction lie in a plane, therefore, we can follow the same argument as for quadrilaterals in the plane.

5.3. Technical difficulties: closed polygons, finite covering

Notice that in all of our positive results (Theorems 1.3-1.9) we considered *open* polygons. This is due to the fact that at the very beginning of Section 2, based on a compactness argument, we restricted our attention to *locally finite* coverings. This does not work for closed polygons. The truth is that at the moment for not locally finite coverings with closed polygons, we cannot prove any positive result. (Our negative results, of course, remain valid for closed polygons as well.) In [21], we made an attempt to overcome this difficulty. To state the (rather weak) results obtained there, we need a definition.

Definition 5.4. A planar set P is said to be *finite-cover-decomposable (countable-cover-decomposable)* if there exists a constant $m \geq 2$ such that every m-fold covering of any point set with finitely (countably) many translates of P can be decomposed into two coverings.

By definition, we have: *P is totally-cover-decomposable $\Rightarrow$ P is countable-cover-decomposable $\Rightarrow$ P is finite-cover-decomposable.* But which of these implications can be reversed? In [21], it was proved that the first one can be for "nice" sets. The definition of nice includes all closed convex sets and polygons, but is much more general. The proof is based on the hereditary Lindelöf property of the plane.

Unfortunately, we have been unable to prove any such connection between finite-cover-decomposability and countable-cover-decomposability. Hence, the status of closed polygons is still undetermined. We believe, however, that using further geometric observations this problem can be settled.

6. OPEN QUESTIONS

The main unsolved problem in the field remains to verify (or refute) Conjecture 1.2 or, more generally Conjecture 1.14.

Problem 6.1. Is every plane convex set cover-decomposable?

Concerning coverings with homothetic copies of a set P, the first interesting special cases are when P is a disk or a square.

Problem 6.2. Does there exist a positive integer m such that every m-fold covering of the plane with open disks of arbitrary radii splits into two coverings?

Problem 6.3. Does there exist a positive integer m such that every m-fold covering of the plane with open squares of arbitrary side lengths, whose sides are parallel to the coordinate axes, splits into two coverings?

As we have seen in the Introduction, the answer to the corresponding question for triangles is affirmative [11].

In Subsection 2.3, we defined a notion somewhat stronger than cover-decomposability (see Definition 2.3).

Problem 6.4. Does there exist a bounded (convex) set P which is cover-decomposable, but not totally-cover-decomposable?

According to Theorem 1.5, every open convex polygon is cover-decomposable, that is, for every open convex polygon P, there is a positive constant $m(P)$ such that every $m(P)$-fold covering of the plane with translates of P splits into two coverings. The best known value of $m(P)$ depends on the shape of P.

Problem 6.5. Is it true that, for any integer $j \geq 3$, there is a positive constant m_j such that every m_j-fold covering of the plane with translates of any convex j-gon P splits into two coverings?

For open triangles the answer is yes with $m_3 \leq 12$. On the other hand, the same statement is not known for closed triangles, as we do not even know if closed triangles are cover decomposable.

It is possible that for any cover-decomposable set P, there exists a (smallest) positive integer $m = m_3(P)$ with the property that every m-fold covering of the plane with translates of P splits into *three* coverings. More generally, as in the Introduction, let $m_k(P)$ denote the smallest positive integer m such that every m-fold covering of the plane with translates of P splits into k coverings. If such an integer does not exist, let $m_k(P) = \infty$.

Problem 6.6. Is it true that if $m_2(P) < \infty$, then we also have $m_k(P) < \infty$, for every $k \geq 3$?

This may be true even in a very general combinatorial setting. Given a finite system of sets $\mathcal{F}$, a *multiset* of its members (with possible repetition!) is said to form a m-fold covering if every element of the underlying set is contained in at least m members of $\mathcal{F}$. For any positive integer k, let $m_k(\mathcal{F})$ denote the smallest number $m \geq 1$ such that every m-fold covering with members of $\mathcal{F}$ splits into k coverings. It is easy to see that this number is always finite: for example, we have $m_k(\mathcal{F}) \leq (k-1)|\mathcal{F}| + 1$.

Problem 6.7. Does there exist a function f such that, for every finite set system $\mathcal{F}$, we have $m_3(\mathcal{F}) < f(m_2(\mathcal{F}))$?

It is possible that the answer is yes even with the function $f(x) = O(x)$. As a matter of fact, the relation $m_k(\mathcal{F}) < Ckm_2(\mathcal{F})$ may also hold with an absolute constant $C > 0$.

In spite of substantial progress in this field, our knowledge on decomposability properties of multiple coverings is rather rudimentary. To our surprise, G. Tardos (personal communication) constructed a set system $\mathcal{F}$, which "almost" refutes Problem 6.7. This set system cannot be decomposed into 3 coverings, although every subsystem of it (with no repetition!) which forms a 2-fold covering splits into 2 coverings.

Finally, we mention another problem for finite set systems that has a strong connection to cover-decomposability.

For a subset $A \subset [n]$, let us denote by a_i the i-th smallest element of A. Given two k-element sets, $A, B \subset [n]$, we write $A \preceq B$ if $a_i \leq b_i$ for every i. A k-uniform hypergraph $\mathcal{H} \subset \binom{[n]}{k}$ is called a *shift-chain* if for any two hyperedges, $A, B \in \mathcal{H}$, we have $A \preceq B$ or $B \preceq A$. (So a shift-chain has at most $k(n-k) + 1$ hyperedges.)

Problem 6.8. Is it true that if k is sufficiently large, then every k-uniform shift-chain has Property B? In other words, is it true that for every shift-chain $\mathcal{H} \subset \binom{[n]}{k}$, one can color $[n]$ with two colors such that no hyperedge is monochromatic?

An affirmative answer would be a huge step towards Pach's conjecture that all planar convex sets are cover-decomposable. To see this, recall the following definition from Section 4.1. For a finite set of point S in the plane and for a plane convex set P, define $P(k; y)$ as the translate of P which

(1) contains exactly k points of S;

(2) the y-coordinate of its apex is y; and

(3) the x-coordinate of its apex is maximal,

if such a translate exists.

If we associate $i \in [n]$ with the element of S with the i-th smallest y-coordinate, then an easy geometric argument shows that $\mathcal{H} = \{P(k; y) \cap S \mid y \in \mathbb{R}\}$ is a shift-chain.

For $k = 2$, there is a trivial counterexample to the above problem: (12), (13), (23). For $k = 3$, a magic counterexample was found by a computer program written by Radoslav Fulek:

(123), (124), (125), (135), (145), (245), (345), (346), (347), (357),

(367), (467), (567), (568), (569), (579), (589), (689), (789).

If we allow the hypergraph to be the union of two shift-chains (with the same order), then the construction in Section 5.2 provides a counterexample for any k. Therefore, all arguments using that the average degree is small (like attempts based on Lovász Local Lemma) would probably fail.

Added in proof

Recently, several new related new results have been found. It was proved by I. Kovács and G. Tóth [14, 15], and, independently, by M. Vizer [24] that closed centrally symmetric polygons are cover-decomposable. In a series of papers, it was shown by J. Cardinal, K. Knauer, P. Micek and T. Ueckerdtand [5, 6], and by B. Keszegh and D. Pálvölgyi [12, 13] that $k^{O(1)}$-fold coverings by homothets of open triangles or by (finite collections of) octants are decomposable into k coverings.

References

[1] B. Ács: *Síkfedések szétbonthatósága*, Master Thesis, Eötvös University Budapest, 2010 (in Hungarian).

[2] G. Aloupis, J. Cardinal, S. Collette, S. Langerman, D. Orden, and P. Ramos: Decomposition of multiple coverings into more parts, *Discrete and Computational Geometry* **44** (2010), 706–723. Also in: *Proc. 20th Annual ACM-SIAM Symposium on Discrete Algorithms (SODA 09)*, ACM, New York, 2009, 302–310.

[3] A. L. Buchsbaum, A. Efrat, S. Jain, S. Venkatasubramanian, and K. Yi: Restricted strip covering and the sensor cover problem, in: *Proc. 18th Annual ACM SIAM Symposium on Discrete Algorithms (SODA 07)*, ACM, New York, 2007, 1056–1063.

[4] P. Brass, J. Pach, and W. Moser: *Research Problems in Discrete Geometry*, Springer, Berlin, 2005.

[5] J. Cardinal, K. Knauer, P. Micek and T. Ueckerdt: Making triangles colorful, arXiv:1212.2346.

[6] J. Cardinal, K. Knauer, P. Micek and T. Ueckerdt: Making octants colorful and related covering decomposition problems, arXiv:1307.2705.

[7] M. Elekes, T. Mátrai, and L. Soukup: On splitting infinite-fold covers, *Fundamenta Mathematicae* **212** (2011), 95–127.

[8] P. Erdős: On a combinatorial problem, *Nordisk Matematisk Tidskrift* **11** (1963), 5–10.

[9] P. Erdős and L. Lovász: Problems and results on 3-chromatic hypergraphs and some related questions, in: *Infinite and Finite Sets* (to Paul Erds on his 60th birthday), II. North-Holland, Amsterdam, 1975, 609–627.

[10] M. Gibson and K. Varadarajan: Optimally decomposing coverings with translates of a convex polygon, *Discrete & Computational Geometry* **46** (2011), 313–333. Also in: *Proc. 50th Annual IEEE Symposium on Foundations of Computer Science, (FOCS 09)*, IEEE Computer Soc., Los Alamitos, CA, 2009, 159–168.

[11] B. Keszegh and D. Pálvölgyi: Octants are cover decomposable, *Discrete & Computational Geometry*, DOI 10.1007/s00454-011-9377-1, to appear. Also in: *Proc. 7th Hungarian–Japanese Symposium on Discrete Mathematics and Its Applications, Kyoto, 2011,* 217–226.

[12] B. Keszegh and D. Pálvölgyi: Octants are cover-decomposable into many coverings, arXiv:1207.0672.

[13] B. Keszegh and D. Pálvölgyi: Convex Polygons are Self-Coverable, arXiv:1307.2411.

[14] I. Kovács: Többszörös fedések zárt sokszögekkel, *OTDK 2013* (in Hungarian), http://tdk.bme.hu/ttk/DownloadPaper/Tobbszoros-fedesek-zart-sokszogekkel.

[15] I. Kovács and G. Tóth: Multiple coverings with closed polygons, manuscript.

[16] P. Mani-Levitska, J. Pach: Decomposition problems for multiple coverings with unit balls, manuscript, 1986.

[17] J. Pach: Decomposition of multiple packing and covering, in: *Diskrete Geometrie, 2. Kolloq.* Math. Inst. Univ. Salzburg, 1980, 169–178.

[18] J. Pach: Covering the plane with convex polygons, *Discrete & Computational Geometry* **1** (1986), 73-81.

[19] J. Pach, G. Tardos, and G. Tóth: Indecomposable coverings, *Canadian Mathematical Bulletin* **52** (2009), 451–463. Also in: *The China–Japan Joint Conference on Discrete Geometry, Combinatorics and Graph Theory (CJCDGCGT 2005), Lecture Notes in Computer Science* **4381**, Springer, Berlin, 2007, 135–148.

[20] J. Pach and G. Tóth: Decomposition of multiple coverings into many parts, *Computational Geometry: Theory and Applications* **42** (2009), 127–133. Also in: *Proc. 23rd ACM Symposium on Computational Geometry, (SoCG07)* 2007, 133–137.

[21] D. Pálvölgyi: Indecomposable coverings with concave polygons, *Discrete & Computational Geometry* **44**, (2010), 577–588.

[22] D. Pálvölgyi and G. Tóth: Convex polygons are cover-decomposable, *Discrete & Computational Geometry* **43** (2010), 483–496.

[23] G. Tardos and G. Tóth: Multiple coverings of the plane with triangles, *Discrete & Computational Geometry* **38** (2007), 443–450.

[24] M. Vizer, personal communication.

János Pach

Ecole Polytechnique Fédérale de Lausanne
and
Alfréd Rényi Institute of Mathematics,
Budapest

e-mail: pach@cims.nyu.edu

Dömötör Pálvölgyi

Department of Mathematics,
Eötvös University,
Budapest

e-mail: dom@cs.elte.hu

Géza Tóth

Alfréd Rényi Institute of Mathematics,
Budapest,
Hungary

e-mail: geza@renyi.hu

BOLYAI SOCIETY
MATHEMATICAL STUDIES, 24

Geometry –
Intuitive, Discrete, and Convex
pp. 259–299.

Hanani–Tutte and Related Results

Marcus Schaefer

We are taking the view that crossings of adjacent edges are trivial, and easily got rid of.
                                                                    Bill Tutte

We interpret this sentence as a philosophical view and not a mathematical claim.
                                                                    László Székely

We investigate under what conditions crossings of adjacent edges and pairs of edges crossing an even number of times are unnecessary when drawing graphs. This leads us to explore the Hanani–Tutte theorem and its close relatives, emphasizing the intuitive geometric content of these results.

1. The Hanani–Tutte Theorem in The Plane

In 1934, Hanani [15] published a paper which—in passing—established the following result:

Any drawing of a K_5 or a $K_{3,3}$ contains two independent edges crossing each other oddly.[1]

Since by Kuratowski's theorem every non-planar graph contains a subdivision of K_5 or $K_{3,3}$, Hanani's observation implies that any drawing of a

---

[1] The result can be hard to find even if one reads German. It is stated as (1) on page 137 of the article and mainly an application of methods developed by Flores [20, 62. Kolloqium].

non-planar graph contains two vertex-disjoint paths that cross an odd number of times and therefore two independent edges that cross oddly—one in each path. This consequence was first explicitly stated by Tutte [49].[2]

Theorem 1.1 (The (Strong) Hanani–Tutte Theorem). *Any drawing of a non-planar graph contains two independent edges that cross oddly.*

Equivalently, but more spectacularly, the theorem can be phrased as saying that if we can draw a graph so that every two independent edges cross evenly, then the graph is planar (we study the algorithmic content of this statement in Section 1.4). Since the reverse direction is immediate, the Hanani–Tutte theorem can be viewed as a characterization of planarity.

In this paper we bring together different versions and applications of the Hanani–Tutte theorem to show that the result of Hanani–Tutte deserves the epithets "remarkable" and "beautiful" [27, 29]. We acknowledge the roots of Hanani–Tutte in the literature of algebraic topology, but this paper will take an intuitive, geometric approach which proves sufficient as long as we restrict ourselves to two-dimensional surfaces.

Some conventions used in this paper: When we speak of drawings of graphs we do not distinguish between an abstract edge and the arc representing it in the plane; or a vertex and the point it is located at. We will simply use "edge" and "vertex" for both concepts; we use *topological graph* when we want to emphasize that we are considering an abstract graph together with a drawing. We require drawings of graphs to fulfill the standard properties: there are only finitely many intersections, a vertex does not lie in the interior of an edge, no two vertices lie in the same location, and at most two edges intersect at any interior intersection point. Interior intersection points of edges come in two flavors: *crossings*, if the edges cross at that point, or *touching points* if the edges touch. A common endpoint of two edges is considered an intersection point, but it is neither a crossing nor a touching point.

---

[2]The theorem is generally known as the Hanani–Tutte theorem, though Levow [17] calls it the "van Kampen-Shapiro-Wu characterization of planar graphs" emphasizing the parallel history of the theorem in algebraic topology (ignoring Flores, however). In a recent paper [34] we introduced the name "strong Hanani–Tutte theorem" to distinguish it from a weaker version that is also often called the Hanani–Tutte theorem in the literature.

1.1. The Weak Hanani–Tutte Theorem, or, Even Crossings Don't Matter

The Hanani–Tutte theorem is often stated and used in a weak form. Call an edge in a drawing *even* if it crosses every other edge an even number of times (including 0 times). If a graph can be drawn so that all its edges are even, then the graph is planar. This weak version of the Hanani–Tutte theorem is easier to prove than the strong version, and yields a stronger conclusion: the graph can be embedded in the plane without changing its rotation system. The *rotation* at a vertex is the cyclic ordering of ends of edges at that vertex, the *rotation system* is the collection of rotations of all vertices.

Theorem 1.2 (Weak Hanani–Tutte). *If a graph can be drawn so that all its edges are even, then the graph is planar and can be embedded without changing the rotation system.*

The assumption of the weak Hanani–Tutte theorem can be weakened: it is enough to require that in the drawing every *even* subgraph of G, that is, a subgraph all of whose vertices have even degree, has an even number of self-crossings. This form of the theorem was suggested and proved by Loebl and Masbaum in their study of Norine's conjecture [18].

Theorem 1.3 (Loebl, Masbaum). *If a graph can be drawn so that every even subgraph has an even number of self-crossings, then the graph can be embedded in the plane without changing the rotation system.*

Or, as Loebl and Masbaum phrase it: "Even drawings don't help". Theorem 1.3 immediately implies Theorem 1.2. We include an easy proof of Theorem 1.3 using geometric rather than homological methods.

Proof. Suppose we are given a drawing of the graph so that (∗) *every even subgraph has an even number of self-crossings.*

We can assume that the graph is connected: Adding an edge between two components of the graph does not affect (∗) since that edge cannot be part of any even subgraph (it would, in each of the components, be incident to a subgraph of odd total degree, which contradicts the handshake lemma). Repeating this, we obtain a connected graph fulfilling (∗). If the new graph is embeddable without changing its rotation system, then the original graph can be embedded with its original rotation system (delete the additional edges). We prove the result for connected graphs by induction on the number of vertices and edges. To make the induction work, we allow multiple edges and loops.

If there is a non-loop edge $e = uv$ contract it by moving v along e towards u, eventually identifying u and v and merging the rotations of u and v as shown in Figure 1. We presently argue that ($*$) remains true, so by the inductive assumption, the new graph can be embedded without changing its rotation; but then we can split $u = v$ into two vertices again and move them apart slightly, recovering the original rotations of u and v and reinserting the edge $e = uv$ without introducing any crossings.

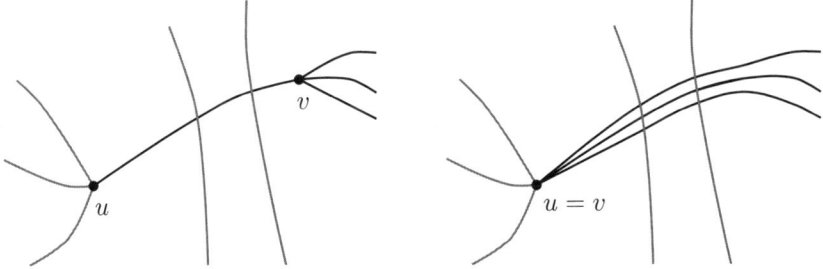

Fig. 1. Geometrically contracting edge $e = uv$ towards u

The contraction of e does not affect ($*$): let H be an arbitrary even subgraph before the contraction. If $e \notin E(H)$, then, since the degree of H at v is even, an even number of edges of H is pulled along e, so any crossing of e leads to an even number of crossing with H, so the number of self-crossings of H remains even. The same conclusion holds if $e \in E(H)$. In this case we are pulling an odd number of edges of H along e, but e itself belongs to H, so if v is pulled through a crossing with some edge $f \in E(H)$, the odd number of crossings added is balanced by the single crossing between e and f that is removed, so that the overall parity remains even.

Since we started with a connected graph, we are left with the case of a single vertex with loops. Any loop by itself is an even subgraph, so it has an even number of self-crossings. Since any two loops also form an even subgraph, any two loops must cross each other an even number of times. Pick any loop e whose ends are closest in the rotation; the ends of e must be consecutive (any loop starting between the two ends of e would also have to end between the two ends of e, but then we would have chosen it over e). Remove e and draw the remaining graph by induction (with the same rotation). We can then reinsert e at its original place in the rotation without introducing any crossings. This completes the proof. ■

The Loebl–Masbaum result no longer holds if we only require that every cycle has an even number of self-crossings: take two cycles sharing a single vertex so that the ends of the cycle alternate at the vertex. While this graph is planar it cannot be drawn without changing its rotation system. A simple modification of the theorem is true, however.

Theorem 1.4. *If a graph can be drawn so that every cycle has an even number of self-crossings, then the graph can be embedded in the plane without changing the rotation system of any 2-connected block of the graph. (Only the rotations at cut-vertices need to be adjusted: make the ends of edges belonging to the same 2-connected block consecutive in the rotation without otherwise changing the ordering of ends belonging to the same block.)*

Even though Theorem 1.4 leads to a change in rotation, it has the flavor of the weak Hanani–Tutte theorem; we can ask whether it is sufficient to assume that every cycle has an even number of *independent* crossings to guarantee planarity. Theorem 1.16 in the next section answers that question.

In the proof of the theorem, we will contract edges only partially, namely, up to a point where they are free of crossings.[3]

Proof of Theorem 1.4. We show that any 2-connected graph fulfilling the conditions of the theorem can be embedded without changing its rotation system. The general result then follows.

First note that contracting an edge, even partially, does not affect the parity of the number of crossings along any cycle, since a cycle always has even degree at every vertex. So we can pick a spanning tree T of the graph and partially contract edges in a breadth-first (or depth-first) order towards the root of the tree so that all edges of T are entirely free of crossings. This might introduce self-crossings along edges in $E(G) - E(T)$, but since each such edge forms a cycle with a path in T, it must have an even number of self-crossings, so we can remove all self-crossings of edges without changing the parity of crossings along any cycle. At this point the only crossings are between distinct edges in $E(G) - E(T)$. We claim that any two such edges $e, f \in E(G) - E(T)$ cross evenly: for a contradiction assume that e and f cross oddly. Let P_e and P_f be the sub-paths of T connecting the endpoints of e and f and $C_e = P_e \cup \{e\}$, $C_f = P_f \cup \{f\}$. First consider the case that e

---

[3]Partial contractions were used by Černý [9] in his proof of the weak Hanani–Tutte theorem in the plane.

and f are independent. Since e and f cross oddly, but two closed curves in the plane always cross evenly, C_e and C_f must have at least a vertex in common. If C_e and C_f share an edge, we argue as follows: the symmetric difference $C' = C_e \triangle C_f$ is a cycle consisting of e, f and crossing-free edges from T. Since by assumption C' has an even number of self-crossings, e and f must cross evenly. If C_e and C_f share only a vertex, say v, then $C_e - v$ and $C_f - v$ are disjoint, so since G is 2-connected, there must be an edge $g \notin E(T)$ connecting them. Consider the following cycles pictured in Figure 2:

$C_{e,g}$: start at the endpoint of g lying on C_e, follow C_e to e, traverse e, follow C_e to v, follow C_f to the other endpoint of g, traverse g,

$C_{f,g}$: start at the endpoint of g lying on C_f, follow C_f to f, traverse f, follow C_f to v, follow C_e to the other endpoint of g, traverse g,

C': start at the endpoints of g lying on C_e and C_f, follow C_e to v traversing e and follow C_f to v traversing f; add g.

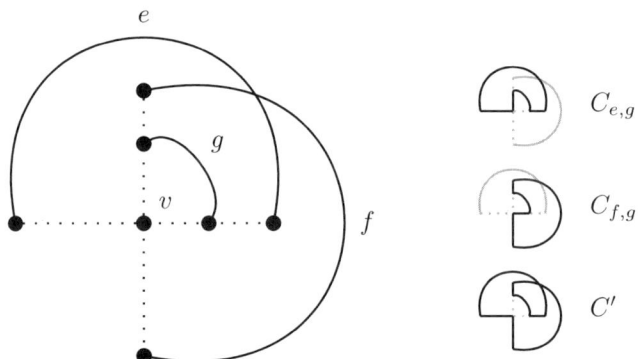

Fig. 2. Cycles $C_{e,g}$, $C_{f,g}$ and C' in case e and f are independent

Now $C_{e,g}$ and $C_{f,g}$ are even, so both e and g as well as f and g cross evenly (all other edges in these cycles are crossing-free). Since C' is even, this means that e and f also have to cross evenly.

In case e and f share an endpoint v, let T_e and T_f be the components of $T - \{v\}$ containing the other endpoints of e and f. If $T_e = T_f$, then $T - \{v\}$ contains a path P connecting the two endpoints of e and f which are different from v. Now the cycle $P \cup \{e\} \cup \{f\}$ is even; since all edges of P are free of crossings, this implies that e and f cross evenly.

So we can assume that $T_e \neq T_f$. Since G is 2-connected, there must be an edge g between T_e and T_f. With g we can construct three cycles as shown in Figure 3:

$C_{e,g}$: start at v, follow e, follow T_e to g, follow g, follow T_f to v,

$C_{f,g}$: start at v, follow f, follow T_f to g, follow g, follow T_e to v,

C': start at v, follow e, follow T_e to g, follow g, follow T_f to f, follow f back to v.

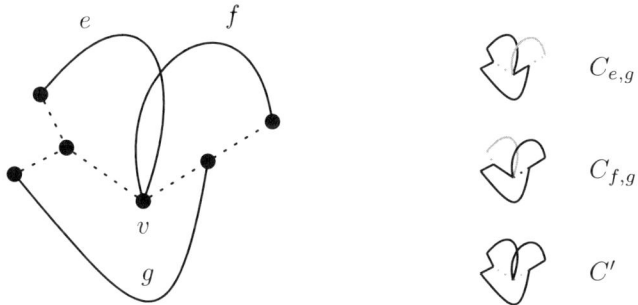

Fig. 3. Cycles $C_{e,g}$, $C_{f,g}$ and C' in case e and f are adjacent

Since $C_{e,g}$ is even, e and g cross evenly (by an argument similar to the one above); by the same token, the evenness of $C_{f,g}$ implies that f and g cross evenly. Finally, since C' is even, and, we assumed e and f cross oddly, e or f must cross oddly with g, but we saw that this is not the case, so e and f cross evenly.

In other words, all edges in the current drawing of G are even. By the weak Hanani–Tutte theorem, we can then embed G in the plane without changing its rotation system. ∎

Theorem 1.4 suggests a more general family of Hanani–Tutte type results: if we know that all subgraphs belonging to some family of graphs have an even number of self-crossings, what does this tell us about the graph? To guarantee planarity we saw that it is enough to look at pairs of edges (weak Hanani–Tutte), pairs of independent edges (strong Hanani–Tutte), even subgraphs (Loebl–Masbaum), cycles (Theorem 1.4). What other families of graphs guarantee planarity?

Paths furnish a trivial example: if all paths have an even number of self-crossings, then all edges are even: consider two edges e and f, and let P be a shortest path containing both e and f (so they must be the first and last edge of the path). Since $P - \{e, f\}$, $P - \{e\}$, $P - \{f\}$ as well as P are even, e and f have to cross evenly. Hence all edges are even, and the graph is planar.

Stars on the other hand, do not appear very promising at first: any drawing of a graph which minimizes the number of crossings has no crossings between adjacent edges, so for any graph all stars can be made free of self-crossings. Looking at pairs of stars trivializes the problem: any two edges are a pair of stars, so by the weak Hanani–Tutte theorem, the graph is planar. However, there is an interesting variant hiding here: what happens if we consider pairs of maximal stars?

Question 1.5. If G is drawn so that every union of two maximal stars in G has an even number of self-crossings, is G planar?

Note that this fails for the torus: take a toroidal grid $C_n \,\square\, C_n$ and add a pair of diagonal edges to each square. Then the union of two maximal stars will always contain 0 or 2 self-crossings.

Other families of graphs worth exploring might be triples of (independent) edges and Θ-graphs, that is, graphs consisting of three internally disjoint paths connecting the same pair of vertices.

Even if we do not get planarity, we can still ask whether requiring certain subgraphs in the drawing to have an even number of self-crossings allows us to draw any conclusions about the graph. One might, for example, ask extremal questions. We are not aware of any extremal results of this particular form, however, there are several very similar extremal results we will discuss in Section 3.2. There is a result by Pach and Tóth [31] worth mentioning in this context: a topological graph for which any set of $k \geq 2$ independent edges contains two edges that cross evenly has at most $O(n \log^{4k-8} n)$ edges. For $k = 2$ this is a consequence of the strong Hanani–Tutte theorem.[4]

On the structural side there is Norine's fine characterization of Pfaffian graphs [22]: he shows that a graph is Pfaffian if and only if it has a drawing in which every perfect matching has an even number of self-crossings (edges are not allowed to self-intersect in this characterization).

One might ask which natural properties of graphs are invariant under weakening crossing-free to even or independently even, where an edge is *independently even* if it crosses every non-adjacent edge an even number of times. Suppose, for example, that we have a drawing of a graph in which all vertices lie on the boundary of the same region, and all edges are independently even. Then the graph is outerplanar. So outerplanarity survives the weakening of crossing-free to independently even. (The proof is simple: to the region containing all vertices on the boundary, add a new vertex and connect it to all other vertices by crossing-free edges. All edges in the resulting drawing are independently even, so by the strong

---

[4]This result, with a slightly weaker bound, was rediscovered and used in [6].

Hanani–Tutte theorem the graph is planar. Removing the new vertex from the planar drawing yields an outerplanar drawing of the graph.) On the other hand, the notion of crossing number is not invariant under replacing crossing-free with even or independently even as we will see in Section 3.4.

Another example is furnished by a result of Pach and Tóth's on x-monotone drawings; call a drawing x-*monotone* if all its edges are x-monotone, that is, functions on an interval. It is known that every x-monotone embedding of a graph can be turned into a straight-line embedding without changing the x-coordinates of any vertex [11, 30].

Theorem 1.6 (Pach, Tóth [30]). *If all edges in an x-monotone drawing of a graph are even, then the graph has a straight-line embedding in which every vertex keeps its x-coordinate.*

Does the result remain true if we only require edges to be independently even rather than even? Since x-monotone embeddings can be turned into straight-line embeddings without changing x-coordinates, it would be sufficient to establish the following conjecture.

Conjecture 1.7. If all edges in an x-monotone drawing of a graph are independently even, then there is an x-monotone embedding of the graph in which every vertex keeps its x-coordinate.

The proof of Theorem 1.6 uses the Cairns-Nikolayevsky proof of Theorem 1.2; it is not immediately clear whether the redrawing techniques we have used in this section can be adapted to establish Theorem 1.6 or the conjecture.

The weak version of Hanani–Tutte is a simple but popular form of the theorem. As such it has been discovered independently a couple of times. Cairns and Nikolayevsky used homology theory and intersection forms to prove the result for arbitrary surfaces [7]; an intuitive geometric proof, again for arbitrary surfaces, can be found in [36]. The proof for the plane was independently found by Černý [9]. It also follows from a redrawing result of Pach and Tóth [29] which we discuss in Section 3.4.

1.2. Planarity Criteria and Weak Hanani–Tutte

In a way all versions of Hanani–Tutte are planarity criteria, but the two variants we explore in this section are special in that they depend on the rotation system of the graph only, and not on the particular drawing. The first criterion is due to Cairns and Nikolayevsky [7] and is implicit in their proof of the weak Hanani–Tutte theorem. To state the result we need to define a new notion of the number of crossings between two cycles of a graph: Consider two cycles C_1 and C_2 and let P be a maximal path in $C_1 \cap C_2$; contract P to a single vertex. If the ends of C_1 and C_2 alternate at that vertex, we say the cycles *cross* in P and count this as one crossing; otherwise C_1 and C_2 *touch* in P and we do not count this as a crossing. Let $\sigma(C_1, C_2)$ be the total number of crossings—in this sense—between C_1 and C_2. Note that we do not count crossings between edges from C_1 and C_2, so $\sigma(C_1, C_2)$ is completely determined by a rotation system of the graph.

Theorem 1.8 (Cairns, Nikolayevsky [7])**.** *If a graph can be drawn so that $\sigma(C_1, C_2)$ is even for every two cycles C_1, C_2 in the graph, then the graph can be embedded in the plane without changing the rotation system.*

Theorem 1.8 is easily seen to imply the weak Hanani–Tutte theorem. A proof of the theorem can proceed along the same lines as the proof of Theorem 1.3: The parity of $\sigma(C_1, C_2)$ is not affected by contractions of edges, so we can contract the graph to a single vertex with loops; now by assumption any two loops cross an even number of times (in the traditional sense of crossings); but then we can embed the graph as we did in the proof of Theorem 1.3.

The second planarity criterion is due to Lovász, Pach and Szegedy [19] and, like the Cairns-Nikolayevsky criterion, arose in the study of thrackles (of which more in Section 3.3). Recall that a Θ-*graph* is a pair of vertices connected by three internally disjoint paths. In a drawing of a Θ-graph the cyclic clockwise ordering in which the three paths end at the two vertices is either the same or reversed; if the order is reversed, we call the Θ-graph a *converter*. A plane Θ-graph is always a converter.

Theorem 1.9 (Lovász, Pach, and Szegedy [19])**.** *A graph is planar if and only if it can be drawn so that every Θ-subgraph is a converter. In that case, the graph can be embedded in the plane without changing the rotation system of any 2-connected block of the graph (only the rotations at cut-vertices need to be changed).*

Rather than establishing the planarity criterion from scratch (which isn't very hard assuming Kuratowski's theorem), we will show that it is really an incarnation of Theorem 1.2. Indeed, this is how we obtain the conclusion about the rotation system, which is not part of the original result of Lovász, Pach and Szegedy (using Kuratowski's theorem doesn't allow any conclusions about the rotation). In the proof we use a characterization of Θ-graphs in terms of self-crossings.

Lemma 1.10. *A Θ-graph is a converter if and only if it has an even number of cycles with an odd number of self-crossings.*

The lemma could be established using exhaustive case-analysis; instead we opt to obtain it as a consequence of Theorem 1.4.

Proof. If a Θ-graph is drawn so that it has an even number of cycles with an odd number of self-crossings, there are either no or two cycles with an odd number of self-crossings. In case there are two cycles with an odd number of crossings, the two cycles share one of the three paths; introduce a self-crossing (within one edge) along that path. Hence, we can assume that all cycles have an even number of self-crossings. By Theorem 1.4, the graph can be embedded in the plane with the same rotation system, so it must be a converter.

To establish the other direction, swap the ends of two edges at one of the two vertices defining the Θ-graph; this changes the parity of the number of cycles with an odd number of self-crossings, so it becomes even. Then by the argument we made in the first paragraph, the modified graph is a converter, which means that the original graph (before swapping the ends) is not. ■

Θ-graphs are not closed under contracting edges, leading to problems with proofs centered around contraction. To address this issue, we introduce *ϑ-graphs* as graphs resulting from completely contracting one of the paths of a Θ-graph; in other words, a ϑ-graph is a vertex with two closed, internally disjoint paths starting and ending at that vertex. We call a ϑ-graph a *converter* if the ends of the two closed paths do not alternate at the shared vertex. (If we contract one of the paths of a Θ-graph which is a converter, the resulting ϑ-graph is a converter.) It is easy to see that a ϑ-graph is a converter if and only if its two closed paths cross an even number of times (not counting the shared vertex v).

Lemma 1.11. *If every Θ-subgraph in a drawing of a 2-connected graph (without loops) is a converter, then every ϑ-subgraph in the drawing is a converter.*

Proof. Fix a drawing of the graph and assume that every Θ-subgraph is a converter. Pick a ϑ-subgraph at some vertex v formed by two closed paths P and Q. Since the graph is 2-connected, there must be a path connecting $P - v$ and $Q - v$ (each of these paths contains at least one vertex, since the graph contains no loops). Let R be a shortest such path with endpoints $p \in P$ and $q \in Q$. Now P consists of two (proper) v, p-paths P_1, P_2 and Q of two v, q-paths Q_1, Q_2. With these pieces we can build two Θ-graphs between v and p: P_1, P_2, $R + Q_1$ and P_1, P_2, $R + Q_2$. By assumption, both of the Θ-graphs are converters, so the cyclic clockwise orderings of the three paths at v and p must be reversed. Let us assume that paths P_1, R, P_2 occur in this clockwise, cyclic ordering at p (the other case P_2, R, P_1 is analogous). But then the two continuations of R (Q_1 and Q_2) must each occur between P_2 and P_1 in the clockwise cyclic ordering at v since the two Θ-graphs are converters. Hence, Q_1 and Q_2 are consecutive at v, which implies that the ϑ-subgraph formed by P and Q is a converter. ∎

With these results we are in a position to show that Theorem 1.9 is a variant of the weak Hanani–Tutte theorem.

Proof of Equivalence of Theorem 1.9 and Theorem 1.2. In a drawing in which every two edges cross an even number of times, every Θ-subgraph is a converter by Lemma 1.10 (every cycle has an even number of self-crossings and we can assume that edges are free of self-crossings). This shows that Theorem 1.9 implies Theorem 1.2.

To see the other direction, let the graph be drawn so that every Θ-subgraph is a converter. It is sufficient to prove the result for 2-connected graphs, and then recombine the drawings at the cut-vertices. As we did in the proof of Theorem 1.4, we can partially contract the edges of a spanning tree T so that all edges of T are free of crossings. Let e and f be any two edges of the graph. We show that e and f cross an even number of times. This is obvious if either one of them belongs to $E(T)$, so we can assume that $e, f \notin E(T)$ and there are cycles $C_e \subseteq T + e$ and $C_f \subseteq T + f$. If C_e and C_f do not share a vertex, they must cross an even number of times in the plane, so e and f cross evenly; if C_e and C_f share a vertex, but not an edge, then $C_e \cup C_f$ is a ϑ-graph, so by Lemma 1.11 it is a converter, and e and f cross evenly. Finally, if C_e and C_f share more than one vertex, they share a non-empty path, and $C_e \cup C_f$ is a Θ-graph which, by assumption, is a converter. But then e and f cross evenly by Lemma 1.10. ∎

1.3. Strong Hanani–Tutte and Cycles

Maybe the shortest and most elegant proof of the Hanani–Tutte theorem is due to Kleitman [16]. He essentially establishes Hanani's original result using an intuitive geometric approach by showing that all drawings of K_{2i+1} and $K_{2i+1,2j+1}$ have an odd number of independent crossings.[5] Together with Kuratowski's theorem this yields the strong Hanani–Tutte theorem in the same way that Hanani's result did.

Archdeacon and Richter later showed that K_{2i+1} and $K_{2i+1,2j+1}$ are the only graphs for which the parity of independent crossings is the same in all drawings [2].

We do not include Kleitman's proof since we will show how to obtain an even stronger version of the Hanani–Tutte theorem using ideas similar to his. The following lemma has been used and stated in many forms, but its core ideas really go all the way back to van Kampen [54]. An (e, v)-*move* consists of deforming a small part of e, moving it close to v and then pulling it over v. It changes the parity of crossing of e with every edge incident to v, but with no other edge, see Figure 4. A *rotation swap* consists of swapping the order of two consecutive ends at a vertex.

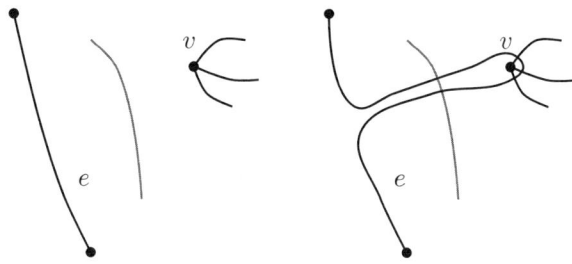

Fig. 4. Performing an (e, v)-move

Lemma 1.12. *Given two drawings D_1 and D_2 of the same graph, there is a set of (e, v)-moves and rotation swaps that can be applied to D_1 so that the resulting drawing D_1' has the same parity of crossing between every pair of edges as D_2.*

Proof. We follow Kleitman's argument [16]: start with D_1. By deforming the plane, we can assume that each vertex has the same location in D_1

---

[5]The slightly weaker result that this result is true if one restricts oneself to drawings in which adjacent edges do not cross, is already contained in Levow [17].

and D_2. Now continuously deform each edge e from its drawing in D_1 to its drawing in D_2. The only ways the parity of crossing between e and another edge f can change is by e moving over an endpoint v of f—corresponding to an (e, v)-move, or by the consecutive ends of e and f swapping order at a shared vertex—corresponding to a rotation swap. ∎

Lemma 1.12 simplifies reasoning about drawings, by turning it into an algebraic problem. Our first use of the lemma is to establish a strengthening of Kleitman's observation. Let a *principal k-cycle* of a subdivision of a graph G be a cycle that contracts to a k-cycle of G.

Lemma 1.13. *Given a drawing of a subdivision of $K_{3,3}$, the number of principal 4-cycles with an odd number of independent self-crossings is odd.*

Proof. The claim is true for the standard straight-line drawings of $K_{3,3}$ and its subdivisions, so we only have to show that the number of principal 4-cycles with an odd number of independent self-crossings does not change under (e, v)-moves.

First observe that an edge and a vertex in a $K_{3,3}$ always determine an even number of 4-cycles (namely 2 or 4) that use the edge and the vertex; on the other hand two disjoint edges determine a unique 4-cycle.

Fix a drawing of a $K_{3,3}$-subdivision, and select any vertex v and edge e of the graph. If v is a degree-2 vertex in a subdivided $K_{3,3}$ edge f, and e occurs in the subdivision of a $K_{3,3}$-edge that is not adjacent to f, then the two edges incident to v cannot be adjacent to e, so the (e, v)-move flips the parity of crossing of e with both edges incident to v. Since either both or neither of those two edges belong to a cycle, the number of independent self-crossings along any cycle cannot change in this case.

Otherwise, v is a degree-2 vertex in the same subdivided edge that e belongs to, or v is a vertex of the original K_5. In either case, by the observation, an (e, v)-move affects an even number of principal 4-cycles (by either changing the parity of independent self-crossings of all of them or none of them), so the total number of principal 4-cycles with an odd number of independent self-crossings does not change parity. ∎

As often, K_5 turns out to be the harder case.

Lemma 1.14. *Any drawing of a K_5 contains a cycle with an odd number of independent self-crossings.*

Proof. Let an *i-odd pair* be a pair of independent edges that crosses an odd number of times. First note that in any drawing of a K_5 the number of i-odd pairs is odd. This is true for the standard (convex) straight-line drawing of K_5, which contains five i-odd pairs; so it is sufficient to show that the parity of i-odd pairs does not change under (e, v)-moves; but this is clear, since any (e, v)-move either does not change the parity of crossings between independent edges at all (if v is an endpoint of e), or changes the parity of crossing between e and each of the two edges incident to v that are not adjacent to e. In either case, the parity of i-odd pairs does not change.

Now assume, for a contradiction, that K_5 can be drawn so that every cycle has an even number of independent self-crossings. Every 4-cycle is made up of two independent pairs of edges that, by assumption, must have the same parity of crossing. But then the three pairs of independent edges that make up a K_4-subgraph must also all have the same parity of crossing. So each K_4 has either three or no i-odd pairs. Since the number of i-odd pairs in the drawing is odd, this implies that of the five K_4s into which the independent pairs in K_5 can be partitioned, an odd number of them will contain three i-odd pairs. Now consider any C_5; it shares exactly one i-odd pair with each of the five K_4s so it has an odd number of independent self-crossings, contradicting the assumption. ∎

With Lemma 1.14 as the base case we are ready to deal with subdivisions of K_5.

Lemma 1.15. *Any drawing of a subdivision of K_5 contains a cycle with an odd number of independent self-crossings.*

Proof. For the purposes of this proof only, the *parity* of a cycle is the parity of the number of pairs of independent edges that belong to the cycle and cross oddly. Suppose then that there is a subdivision G of K_5 which can be drawn so that all cycles have even parity.

By Lemma 1.14, G cannot be K_5, so G must contain a degree-2 vertex v. We show that in that case we can redraw G so that v can be contracted away without changing the parity of any cycle. This gives us an inductive proof of the lemma. For the redrawing it will be useful to understand the effects of (e, v)-moves on the parity of cycles: if v is one of the endpoints of e, then an (e, v)-move has no effect on the parity of any pairs of edges crossing, and so it does not affect the parity of any cycles. If, on the other hand, v is not a neighbor of either endpoint of e, then an (e, v)-move also has no effect on the parity of any cycle: each cycle has even degree at v, so an (e, v)-move will change the parity of crossing between e and the two cycle edges incident to v, so the parity of the cycle does not change. Finally, we are in the case that there is an edge f which shares one endpoint with e and has v as its other endpoint. Then an (e, v)-move will change the parity of every cycle that contains both e and f excepting, if it exists, a C_3 containing both e and f; a C_3 always has even parity.

First suppose that G contains a path $uvwx$ where v and w have degree 2 (so $uvwx$ is part of a subdivided edge). There are two possible obstacles to merging vw and wx into a single edge vx without changing the parity of any cycle: some edge incident to x crosses vw oddly or uv crosses wx oddly. If uv crosses wx oddly, then perform an (uv, x)-move; since there cannot be an edge between u and x, this does not change the parity of any cycle, and makes uv cross wx evenly. If there is an edge f incident to x that crosses vw oddly, perform an (f, v)-move. As above, f cannot have u as an endpoint, so no cycle changes parity, and f crosses vw evenly. Hence, all edges incident to v and x (excepting vw and wx) cross both vw and wx evenly, so we can replace vw and wx by a single edge vx without changing the parity of any cycle.

If G does not contain a path $uvwx$ as above and is not a K_5, it must contain a path uvw so that both u and w have degree 4 and v has degree 2. If all edges (other than uv and vw) incident to u or w cross both uv and vw evenly, then we can merge uv and vw into a single edge uw without changing the parity of any cycle in G. Hence, there must be some edge e incident to u or w that crosses uv or vw oddly. Without loss of generality, e is incident to u. If e crosses uv oddly, then we can perform an (e, u)-move. This makes e cross uv evenly, and it does not change the parity of any cycle (since e is incident to u). Hence, we can assume that e crosses vw oddly. Let x be the other endpoint of e. If xw is not an edge in G, then we can perform an (e, w)-move without changing the parity of any cycle, and making e cross vw evenly. Hence $f = xw$ must be an edge of G. But now uv, vw, f, e form a 4-cycle in G which, by assumption, has an even number of independent crossings. However, there are only two independent pairs: e, vw and f, uv. Since e and vw cross oddly, so must f and uv. Now perform both an (e, w)- and a (f, u)-move. This reduces the number of odd crossings along uvw and does not change the parity of any cycle (each move by itself changes the parity of all cycles containing both e and f; performed together, the parity of each cycle is unaffected). In summary, we can ensure that all edges incident to u or w cross both uv and vw evenly, so uv and vw can be merged into a single edge uw. ∎

Since by Kuratowski's theorem every non-planar graph contains a subdivision of K_5 or $K_{3,3}$, Lemmas 1.13 and 1.15 imply the following theorem. The traditional strong Hanani–Tutte theorem is an immediate consequence.

Theorem 1.16. *If a graph can be drawn so that all its cycles have an even number of independent self-crossings, then the graph is planar.*

One might ask, whether it is possible to prove the strong Hanani–Tutte theorem or even Theorem 1.16 without taking recourse to Kuratowski's theorem? This is not an idle question as we will see in Section 1.4. For Theorem 1.16 we have to leave the question open, but there is an elementary proof of the Hanani–Tutte theorem—in the style of the proof of Theorem 1.3—that does not use Kuratowski's theorem [34].[6] This, in turn, leads to the question of whether Kuratowski's theorem can be obtained from the Hanani–Tutte theorem. The answer is yes, as shown by van der Holst [51].[7]

1.4. Algorithmic and Algebraic Aspects

While Kuratowski's theorem gives us a characterization of planar graphs, it does not directly lead to either an efficient planarity test or an efficient embedding technique for planar graphs. These problems were first addressed in the sixties, culminating in the linear-time algorithm by Hopcroft and Tarjan. The Hanani–Tutte theorem offers an alternative algorithmic approach to planarity testing along two separate routes: one practical, the tree approach, based on work of de Fraysseix and Rosenstiehl [10] and the more theoretical algebraic approach, first suggested by Wu [56, 57].[8]

1.4.1. Trémaux Orders. A *Trémaux tree* is a (rooted) depth-first search tree of a graph; it defines a partial order on the vertices of the tree, a

---

[6]Sarkaria [41] in 1991 claimed the same result. His proof contains several flaws: The redrawing suggested in his Figure 4 (page 82) introduces odd crossings between β and edges that end between α and β. This not only changes the parity of crossings between two edges, but it may also introduce crossings with edges that have previously been cleared of crossings. Both problems can be dealt with, but, as far as we know, not by locally working with a single vertex. This is why we believe that Sarkaria's proof cannot be fixed along the lines described in his paper.

[7]Sarkaria [41] also claims this result, however we were not able to verify it.

[8]Wu's 1985 papers are translations of work originally published in the 1970s in Chinese.

Trémaux order, with the root as the smallest element. In a Trémaux order any non-tree edge of a graph has two endpoints that are comparable. Given a spanning tree T of a graph G, a *T-embedding* of G is a drawing of G in which the edges of T are crossing-free; let $T[e]$ denote the unique path in T that connects the endpoints of e. Liu established the following characterization of planarity [10].

Theorem 1.17 (Trémaux Crossing Theorem (Liu)). *If T is a Trémaux tree of a non-planar graph G, then any T-embedding of G contains two edges e and f that cross oddly, and so that $T[e]$ and $T[f]$ have an edge in common.*

The Trémaux Crossing Theorem follows from the strong Hanani–Tutte theorem: in a T-embedding there are two independent edges e and f that cross oddly by Hanani–Tutte; but then $T[e]$ and $T[f]$ must share an edge: if they only shared a vertex, the endpoints of e and f together with the shared vertex do not form a Trémaux order.

The Trémaux Crossing Theorem is at the root of de Fraysseix–Rosenstiehl's planarity criterion which has been used to justify the correctness of linear time planarity algorithms including Hopcroft-Tarjan and the Left-Right algorithm of de Fraysseix–Rosenstiehl [10].

1.4.2. Algebraic Characterizations of Planarity. Call a drawing of a graph *i-even* if all pairs of independent edges cross evenly. To find a planar embedding of a graph, it is enough to, (i), find an i-even drawing of the graph (if this fails, the graph is not planar), and, (ii), convert the i-even drawing into a planar drawing.

For step (i) we can exploit the algebraic characterization of planarity suggested by the Hanani–Tutte theorem. For a given graph G, let $U(G)$ be the GF(2)-vector space over the basis $[e, f]$, where e and f range over all independent pairs of edges of G and $e < f$ in some ordering of the edges; to simplify notation, we allow $[f, e]$ for $[e, f]$ and let $[e, f] = 0$ if e and f are not independent.

Consider a drawing D of $G = (V, E)$. With D we associate the vector

$$\mathbf{x}_D := \sum_{e<f}(\mathrm{cr}_D(e, f) \mod 2)[e, f],$$

where $\mathrm{cr}_D(e, f)$ is the number of crossings between e and f in D. Let $X(G) := \{\mathbf{x}_D : D$ is a drawing of $G\}$. Also, we define $\mathbf{w}_{e,v} := \sum_{f=(u,v)}[e, f]$

(this vector corresponds to the effect of an (e, v)-move). Let $W(G)$ be the GF(2)-vector space in $U(G)$ spanned by the $\mathbf{w}_{e,v}$, $e \in E$, $v \in V$.

Theorem 1.18 (Van Kampen, Tutte, Wu, Levow). *The difference between any two vectors in $X(G)$ lies in $W(G)$ and if $\mathbf{x} \in X(G)$ and $\mathbf{w} \in W(G)$, then $\mathbf{v} + \mathbf{w} \in X(G)$. In other words, $X(G)$ is a coset of $W(G)$ in $U(G)$.*

Who proved this result? Van Kampen [54] proved one direction, $X(G) - X(G) \subseteq W(G)$, but could only prove the other direction for higher dimensions; the first explicit statements are in Wu [56] and Levow [17]. Tutte essentially proved the same result over a different vector space, as we will see below; Levow was aware of Tutte's work, Wu was not.

By Theorem 1.18 planarity is equivalent to $\mathbf{0} \in X(G)$, which means that planarity can be phrased as system of linear equations over the field GF(2) which can be solved by Gaussian elimination in cubic time (even less). Unfortunately, the number of variables and equations is $\Theta(|E|^2)$, which yields a planarity algorithm running in worst-case time $O(|E|^6)$; it is not clear whether the structure of the problem can be used to obtain a practical algorithm.

Let $\|\mathbf{x}\|_1$ denote the 1-norm of $\mathbf{x}$, that is, the sum of the absolute values of the entries of $\mathbf{x}$. Then $\min_{\mathbf{x} \in X(G)} \|\mathbf{x}\|_1$ is known as the *independent odd crossing number* of G written as iocr(G) (see Section 3.4 for more on crossing numbers). Thus, computing iocr(G) can be expressed as a minimization problem over a vector space. More precisely, let $\mathbf{s} \in X(G)$ be an arbitrary vector (for example, position the vertices of G on the boundary of a disk in arbitrary order and consider the resulting straight-line drawing). Then iocr(G) is the minimum of $\|\mathbf{s} + \mathbf{x}\|_1$ over all $\mathbf{x} \in W(G)$. In other words, we are looking for $\mathbf{x} \in W(G)$ that is closest, in the 1-norm, to $\mathbf{s}$. This is a special case of the nearest vector problem, which is **NP**-hard to approximate to within any constant [3]. Hence, the algebraic approach does not seem to offer any help in the efficient computation or approximation of crossing numbers.

We have counted crossings along edges modulo 2, corresponding to the traditional Hanani–Tutte theorem; however, there is another way to count crossings that might be worth exploring, and that goes back at least as far as Whitney's 1944 paper [55], though Florcs already hints at the possibility [20, 62. Kolloquium §12]; it's first explicitly worked out by Tutte. Form $U_{\mathbb{Z}}(G)$ like $U(G)$, but as a $\mathbb{Z}$-vector space. If we orient all the edges in G, we can assign $+1$ and -1 to each crossing between two edges depending on the direction of the crossing; let $\operatorname{acr}_D(e, f)$ be the sum of these values along e and f, and $\mathbf{a}_D := \sum_{e<f} \operatorname{acr}_D(e, f)[e, f]$, and $A(G) := \{\mathbf{a}_D : D \text{ is a drawing of } G\}$. For an (e, v)-move we now have two vectors $\mathbf{w}_{e,v}$ depending on the direction in which we pull e over v (and the entries are $+1$ and -1 depending on the direction of crossing). Without spelling out the details, let $W_{\mathbb{Z}}(G)$ be the $\mathbb{Z}$-vector space generated by all these vectors.

Theorem 1.19 (Tutte). *$A(G)$ is a coset of $W_{\mathbb{Z}}(G)$ in $U_{\mathbb{Z}}(G)$.*

As above, planarity is equivalent to $\mathbf{0} \in A(G)$. Tutte also showed that if $\mathbf{x} \in A(G)$, then $2\mathbf{x} \in W_{\mathbb{Z}}(G)$.[9]

As with $X(G)$, we can make $A(G)$ the basis of a crossing number definition: $\min_{\mathbf{x} \in A(G)} \|\mathbf{x}\|_1$ is the *independent algebraic crossing number* of G, or $\operatorname{iacr}(G)$, for short. On the question of whether $\operatorname{iacr} = \operatorname{cr}$, where cr is the traditional crossing number (see Section 3.4), Levow [17] writes "it seems reasonable to hope that equality holds for all graphs"; as we will see in Section 3.4 this is not the case, the two notions of crossing numbers differ. Interestingly, Whitney came close to asking the same question 30 years earlier [55]. Levow continues "whether or not equality holds, the algebraic setting may be useful in helping to compute crossing numbers, for it leads to a lower bound for the crossing number given in terms of the solution to an integer or Boolean minimization problem."

Step (ii) requires an effective version of the strong Hanani–Tutte theorem. From that perspective all the proofs based on Kuratowski's theorem fail. The first proof of strong Hanani–Tutte that does not appeal to Kuratowski's theorem and constructs an embedding starting with an i-even drawing is from [34] (a straightforward implementation of the algorithm will run in quadratic time, better bounds might be possible). The approach has a flavor similar to the proof shown in Theorem 1.3. For a variant see Theorem 3.14, for a strengthening, Theorem 3.17 in Section 3.4.[10]

As an immediate consequence of steps (i) and (ii) we get that planarity testing can be performed, and the planar graph embedded, in time $O(|E|^6)$,

---

[9]In his terminology: "A crossing chain is half a cross-coboundary."

[10]We note that there were two previous claims for algorithms solving (ii). Wu extends the system of linear equations by a set of quadratic equations whose solution will describe an embedding. However, solving quadratic systems of equations is **NP**-complete, so Wu's approach does not lead to an efficient solution. Sarkaria [41] claims that there is a "one-dimensional version of Whitney's trick by means of which any graph [G which has an i-even drawing] can be, step by step embedded in $\mathbb{R}^2$." Unfortunately, his version of the Whitney trick for $n = 1$ is fatally flawed as explained in an earlier footnote.

so the algebraic approach, at least at this point, does not seem to offer any algorithmic advantages over the graph-theoretical approach. However, on the theoretical side, the algebraic point of view has led to interesting research, including several recent papers by van der Holst giving purely algebraic characterizations of planarity, outerplanarity, and linkless embeddability [51, 52].

2. SURFACES

Different from many other planarity criteria—such as Kuratowski or MacLane's—Hanani–Tutte can easily be restated for arbitrary surfaces.[11] Take the weak Hanani–Tutte theorem:

Theorem 2.1 (Weak Hanani–Tutte for Surfaces [7, 36]). *If a graph can be drawn in a surface S so that every pair of edges crosses an even number of times, then the graph can be embedded in S without changing the embedding scheme.*[12]

For orientable surfaces this was established by Cairns and Nikolayevsky [7] using homology theory; the result can also be established using a geometric proof which also works for non-orientable surfaces as shown by Pelsmajer, Schaefer, and Štefankovič [36]. Cairns and Nikolayevsky established the slightly stronger Theorem 1.8 for orientable surfaces. The parity of crossing between two closed curves in a surface depends only on their isotopy classes (we assume the two curves have a finite number of crossings and don't touch at any point); for example, in the plane any two closed curves cross an even number of times, and any two generators of the torus will cross an odd number of times. Given two cycles C_1 and C_2 in a graph drawn in surface S, let c_1 and c_2 be two curves isotopic to the drawings of C_1 and C_2 that cross finitely and don't touch. Let $\Omega_S(C_1, C_2)$ denote the parity of crossing between c_1 and c_2. We use $\sigma_S(C_1, C_2)$ for the notion

---

[11]Both Kuratowski and MacLane can be restated for arbitrary surfaces, but in the case of Kuratowski we do not know the list of excluded topological minors except for the plane and the projective plane [21], and in the case of MacLane's criterion the generalization is far from obvious [5].

[12]Embedding schemes generalize the notion of rotation system to arbitrary surfaces, including non-orientable ones. We do not include a formal definition, but refer the reader to [21] or [36] for details.

of crossing number defined at the beginning of Section 1.2 generalized to surface S.

Theorem 2.2 (Cairns, Nikolayevsky [7]). *If a graph can be drawn in a surface S so that $\sigma_S(C_1, C_2) \equiv \Omega_S(C_1, C_2) \bmod 2$ for every two cycles in the graph, then the graph can be embedded in S without changing the embedding scheme.*

In the plane, $\Omega(C_1, C_2) = 0$ for all pairs of cycles and so the planar version, Theorem 1.8, is a special case. The proof we sketched of Theorem 1.8 generalizes to arbitrary surfaces, including non-orientable surfaces: The parity of $\sigma(C_1, C_2)$ is not affected by contractions of edges, so we can contract the graph to a single vertex with loops; now the number of crossing between two loops e and f equals $\sigma_S(e, f) + \Omega_S(e, f) \equiv 0 \bmod 2$, so e and f cross an even number of times. By Theorem 2.1 that graph can be embedded without changing its embedding scheme.

The Loebl–Masbaum result, Theorem 1.3 also generalizes to arbitrary surfaces.

Theorem 2.3. *If a graph can be drawn in surface S so that every even subgraph has an even number of self-crossings, then G can be embedded in S without changing the embedding scheme.*

Proof. The proof of Theorem 1.3 does not use the fact that the ambient surface is a plane until it deals with the one-vertex case. However, in the case of a single vertex the assumption of the theorem implies that any two loops cross evenly. This allows us to apply Theorem 2.1 (which, by its proof, is true for graphs with loops) to redraw the one-vertex graph without crossings and without changing the embedding scheme. ∎

Similarly, a closer look at the proof of Theorem 1.4 shows that for 2-connected graphs it does not use planarity at all, but only relied on the weak Hanani–Tutte theorem for the plane. Since we just saw that that theorem can be lifted to arbitrary surfaces, the cycle version of weak Hanani–Tutte is true for arbitrary surfaces:

Theorem 2.4. *If a 2-connected graph can be drawn in surface S so that every cycle has an even number of self-crossings, then the graph can be embedded in S without changing the embedding scheme.*

The result fails if the graph is not 2-connected; on the torus, for example, we can take two K_7 and overlap them in one vertex. Each K_7 by itself can be embedded on the torus without self-crossings, so all cycles in the drawing are free of self-crossings.

The planarity criterion by Lovász, Pach, and Szegedy, Theorem 1.9, does not seem to generalize in a straight-forward manner. The notion of converters seems too closely bound to the plane, but a careful analysis of the proof might yield an equivalent form for higher-order surfaces, at least for orientable surfaces.

The story of strong versions of Hanani–Tutte is, unfortunately, much shorter. We currently only know that strong Hanani–Tutte is true on the projective plane.

Theorem 2.5 (Pelsmajer, Schaefer, Stasi [32]). *If a graph can be drawn in the projective plane so that every pair of independent edges crosses an even number of times, then the graph can be embedded in the projective plane.*

The proof of Theorem 2.5 relies on the excluded minors for embeddability in the projective plane, so while some of ideas of the proof might be useful, it will not guide the way to establishing the Hanani–Tutte theorem for other surfaces, like the Klein bottle or the torus (for which the list of excluded minors is not known). It also means, that, at least with its present proof, Theorem 2.5 does not lead to an algorithm for embeddability in the projective plane.[13]

We know that in the plane every graph which can be drawn so that all its cycles have an even number of independent self-crossings is planar (Theorem 1.16). The example after Theorem 2.4 shows that on surfaces other than the plane we need to require 2-connectedness.

Conjecture 2.6. If a 2-connected graph can be drawn in the projective plane so that all its cycles have an even number of independent self-crossings, then the graph can be embedded in the projective plane.

If all edges in a graph are independently even, all cycles in the graph have an even number of independent self-crossings. Therefore Conjecture 2.6 implies the strong Hanani–Tutte theorem on the projective plane; analogously, a version of Conjecture 2.6 for surface S implies strong Hanani–Tutte of surface S. The reverse implication is not clear.

---

[13] In the terminology of Section 1.4, step (ii) fails. However, step (i) also seems to fail for the projective plane: the natural way of adding the cross-cap into the system of equations will lead to a quadratic system which, in general, is **NP**-complete to solve. Levow [17] shows how to extend Tutte's algebraic characterization of planarity to arbitrary surfaces.

3. Applications of Hanani–Tutte

We survey some of the applications of Hanani–Tutte, both weak and strong; we do not claim completeness; for example, we do not pursue the alternative history of Hanani–Tutte in the literature of algebraic topology.

3.1. Arrangements of Geometric Objects

A *collection of pseudo-disks* is a collection of simply connected regions (bounded by simple, closed curves) so that the boundary curves of any two regions intersect at most twice. In their extension of a point-selection result from disks to pseudo-disks, Smorodinsky and Sharir established the following theorem:

Theorem 3.1 (Smorodinsky, Sharir [45]). *Let P be a collection of n points and C a collection of m pseudo-disks in the plane so that the boundary of every pseudo-disk passes through a distinct pair of points in P and so that no pseudo-disk contains a point of P in its interior. Then $m \leq 3n - 6$.*

To see that the theorem is true, construct a multi-graph on the vertices of P with edges formed by the two boundary arcs of each pseudo-disk (formed by the two points of P on the boundary). We argue that any two independent edges e and f of the graph cross an even number of times; suppose, for a contradiction, that e and f cross oddly. Since e and f are independent, they belong to two different pseudo-disks bounded by e, e' and f, f'. Since e and f cross oddly, they must cross once (being part of the boundaries of two pseudo-disks, they can cross at most twice). If e' also crosses f, then neither can cross f or f', so the two endpoints of e are on different sides of the pseudo-circle formed by f, f', which is a contradiction, so we conclude that e' does not cross f. But then the endpoints of f are on opposite sides of e, e', again a contradiction. Hence, any two independent edges cross evenly, and, thus, by Hanani–Tutte, the multi-graph is planar. By construction, each edge in the multi-graph is doubled, so we conclude that $m \leq 3n - 6$.[14]

The proof needs the strong Hanani–Tutte theorem, since adjacent edges might very well cross oddly in the setting of Theorem 3.1. This makes generalizations of Theorem 3.1 to other surfaces hard, since we do not have strong Hanani–Tutte for

---

[14]The argument closely follows the proof given in [45].

arbitrary surfaces; however, we can extend Theorem 3.1 to the projective plane: Note that apart from the application of Hanani–Tutte, the proof of Theorem 3.1 works for arbitrary surfaces, since the union of two pseudo-disks (even in surfaces other than the plane) is a planar region, so the argument can proceed as is; the only difference is that we need to replace the Euler bound on m by the corresponding bound for the given surface; in the case of the projective plane, the proof outline above, combined with Theorem 2.5 establishes the following result:

Theorem 3.2. *Let P be a collection of n points and C a collection of m pseudo-disks in the projective plane so that the boundary of every pseudo-disk passes through a distinct pair of points in P and so that no pseudo-disk contains a point of P in its interior. Then $m \leq 3n - 3$.*

To generalize Theorem 3.1 to arbitrary surfaces we do not actually need the full strong Hanani–Tutte theorem, since we can assume that any pair of edges crosses at most twice. This suggests the following parameterized form of the Hanani–Tutte theorem for arbitrary surfaces:

Conjecture 3.3. *If a graph can be drawn in a surface S so that any pair of independent edges crosses evenly and every pair of edges crosses at most t times, then the graph can be embedded in the surface.*

To establish Theorem 3.1 for surface S (with adjusted Euler bound), it would be sufficient to prove the conjecture for $t = 2$, but even $t = 1$ does not appear to be obvious.

The idea of using Hanani–Tutte with unions of objects seems to have been first used by Pach and Sharir [25] in a new proof of an earlier result of Whitesides and Zhao. Call a collection of simply connected regions k-*admissible* if no region disconnects another, the boundaries of regions don't touch and they cross at most k times (for example, collections of pseudo-disks that don't touch are 2-admissible).

Theorem 3.4 (Whitesides, Zhao). *The boundary of the union of a k-admissible family of size $n \geq 3$ contains at most $k(3n - 6)$ arcs.*

As in the case of Theorem 3.1, the proof does not rely on properties of the plane other than the application of the Hanani–Tutte theorem (establishing Conjecture 3.3 for $t = 2k$ would be sufficient), so the theorem can be established for the projective plane with a bound of $k(3n-3)$; other surfaces remain open, as does the question of whether results based on Theorems 3.1 and 3.4 can be extended to surfaces other than the plane.

It appears that the Hanani–Tutte theorem can play a role in extending results about geometric disks to pseudo-disks; indeed, another example due to Buzaglo, Pinchasi, and Rote [6] concerns the Vapnik–Červonenkis-dimension of pseudo-disks. The *Vapnik–Červonenkis-dimension* of a collection $\mathcal{C}$ of sets is the largest number of points such that every subset of the points can be obtained as an intersection of the set of points with a set in $\mathcal{C}$. It is well-known that the Vapnik–Červonenkis-dimension of disks is 3 (for any set of four points, there always is some subset of the points that cannot be obtained by intersecting the four points with a disk).

Theorem 3.5 (Buzaglo, Pinchasi, and Rote [6]). *The Vapnik–Červonenkis-dimension of any collection of pseudo-disks is at most 3.*

The proof does not require the Hanani–Tutte theorem, but it is based on studying drawings of K_4 in which edges are allowed to cross evenly.

A *collection of pseudo-parabolas* is a collection of functions from $\mathbb{R} \mapsto \mathbb{R}$ so that any two functions cross twice or share one point of tangency (and no point lies on more than two pseudo-parabolas). Then the *tangency graph* in which each pseudo-parabola is represented by a vertex, and an edge represents tangency between two pseudo-parabolas is a biparite, planar graph and thus has at most $2n - 4$ edges, if $n \geq 3$ is the number of pseudo-parabolas. This result from [1] is based on strong Hanani–Tutte and it has several interesting consequences, for example, that the number of empty bigons (or lenses) in a collection of pairwise intersecting pseudo-circles is linear in the number of pseudo-circles [1], also see the exposition in [26, Section 5.2]. Ezra and Sharir [12] use the same approach to bound the complexity of the lower envelope of n functions in $\mathbb{R}^3$ (under certain conditions), a much more complicated situation.

3.2. Excluded Subgraphs

Traditionally, $\mathrm{ex}(n, G)$ is the largest number of edges of a (simple) graph on n vertices without a subgraph isomorphic to G; in topological graph theory the corresponding notion is $\mathrm{ex_{cr}}(n, G)$, asking for the largest number of edges in a topological graph that contains no self-intersecting G (equivalently: all copies of G are crossing free). Pinchasi and Radoičić introduced a parity version of this they called $\mathrm{ex_{ocr}}(n, G)$ which is the largest number of edges of a topological graph on n vertices for which every two edges in the same copy of G cross evenly [40]. In the same spirit we can define $\mathrm{ex_{iocr}}(n, G)$ in which we only require independent edges of G to cross evenly (the reasons for the names will become clear in Section 3.4 on crossing numbers). By the definition we have $\mathrm{ex}(n, G) \leq \mathrm{ex}_{\mathrm{cr}}(n, G) \leq \mathrm{ex}_{\mathrm{ocr}}(n, G) \leq \mathrm{ex}_{\mathrm{iocr}}(n, G)$.

Let us consider two trivial cases. If $G = P_3$, the path of length 3, then $\mathrm{ex}(n, P_3) = \lfloor n/2 \rfloor$, the size of a largest matching on n vertices. On the other hand, in a convex straight-line drawing of K_n no two adjacent edges cross, so $\mathrm{ex}_{\mathrm{cr}}(n, P_3) = \mathrm{ex}_{\mathrm{ocr}}(n, P_3) = \mathrm{ex}_{\mathrm{iocr}}(n, P_3) = \binom{n}{2}$; for the same reason,

$$\mathrm{ex}_{\mathrm{cr}}(n, K_{1,m}) = \mathrm{ex}_{\mathrm{ocr}}(n, K_{1,m}) = \mathrm{ex}_{\mathrm{iocr}}(n, K_{1,m}) = \binom{n}{2}$$

so stars are not of interest. If $G = 2K_2$, two independent edges, then $\mathrm{ex}(n, 2K_2) = n - 1$, for $n \geq 4$, and $\mathrm{ex}_{\mathrm{cr}}(n, 2K_2) \leq \mathrm{ex}_{\mathrm{ocr}}(n, 2K_2) \leq \mathrm{ex}_{\mathrm{iocr}}(n, 2K_2) = 3n - 6$, using the strong Hanani–Tutte theorem. From these examples, it is clear that ex differs from all the other variants, but we do not know of any examples separating $\mathrm{ex}_{\mathrm{cr}}$, $\mathrm{ex}_{\mathrm{ocr}}$ and $\mathrm{ex}_{\mathrm{iocr}}$.

As Pach, Pinchasi, Tardos and Tóth [23] point out any of these notions are only interesting for planar bipartite graphs G; if G is non-planar, then $\mathrm{ex}(n, G) = \mathrm{ex}_{\mathrm{cr}}(n, G) = \mathrm{ex}_{\mathrm{ocr}}(n, G) = \mathrm{ex}_{\mathrm{iocr}}(n, G)$ (using the Hanani–Tutte theorem), and if G is not bipartite, $\mathrm{ex}(n, G)$ is already $\Omega(n^2)$ so there are no interesting asymptotic results.

So the smallest interesting cases to consider are P_4 and C_4 and there are bounds for both. Pach, Pinchasi, Tardos and Tóth show that both $\mathrm{ex}_{\mathrm{cr}}(n, P_4)$ and $\mathrm{ex}_{\mathrm{iocr}}(n, P_4)$ are of order $\Theta(n^{3/2})$ [23] compared to the traditional $\mathrm{ex}(n, P_4) = n$. Pinchasi and Radoičić show that $\mathrm{ex}_{\mathrm{ocr}}(n, C_4) = O(n^{8/5})$ [40], while $\mathrm{ex}(n, C_4) = \Theta(n^{3/2})$ which also is the best current lower bound on $\mathrm{ex}_{\mathrm{ocr}}(n, C_4)$. The C_4 problem is particularly interesting, since it has implications for the number of cuts needed to turn an arrangement of pseudo-parabolas into pseudo-segments [40].

Other variants of these problems are possible; for example, Pach, Pinchasi, Tardos and Tóth study the geometric version $\mathrm{ex}_{\mathrm{rcr}}(n, P_4)$ of $\mathrm{ex}_{\mathrm{cr}}(n, P_4)$ (in which all edges are line segments). For geometric versions, the $\mathrm{ex}_{\mathrm{cr}}$, $\mathrm{ex}_{\mathrm{ocr}}$ and $\mathrm{ex}_{\mathrm{iocr}}$ versions collapse. One can also consider $\mathrm{ex}_{\mathrm{cr}(2)}(n, G)$ ($\mathrm{ex}_{\mathrm{cr}-(2)}(n, G)$) in which we require that every copy of G in the topological graph has an even number of (independent) self-crossings. This has the flavor of Theorem 1.4 and the problems suggested in the subsequent remark. As far as we know, nothing is known about $\mathrm{ex}_{\mathrm{cr}(2)}$ and $\mathrm{ex}_{\mathrm{cr}-(2)}$ or how they relate to $\mathrm{ex}_{\mathrm{ocr}}$ or $\mathrm{ex}_{\mathrm{iocr}}$.

3.3. Thrackles

John Conway has a penchant for asking simple questions that are hard to answer. His Devil and Angel problem had to wait more than 20 years for a solution, and his even older thrackle conjecture is still unsettled. It forcefully drives home the point how little we really know about drawings of graphs.

Conway defined a *thrackle* as a graph that can be drawn so that every pair of edges intersects exactly once. A common endpoint of two edges counts as an intersection, so if we rephrase this condition in terms of crossings it requires that the graph can be drawn so that every pair of independent edges crosses exactly once and adjacent edges do not cross.

Conway conjectured that the number of edges of a thrackle is at most the number of its vertices [4, Section 9.5]. While this conjecture is open, we do know that $|E(G)| = O(|V(G)|)$ for thrackles G; this was first shown by Lovász, Pach and Szegedy [19].[15] Their proof uses the notion of a *generalized thrackle,* a graph which can be drawn so that every two edges cross an odd number of times.

The traditional definition of a generalized thrackle requires that every two edges *intersect* an odd number of times. The two definitions are equivalent [36, Remark 4.2]: one can flip the rotation at each vertex changing the parity of crossing between any pair of adjacent edges to move back and forth between the two variants. For traditional thrackles this implies that an intersection-thrackle is always a crossing-thrackle, but the reverse is not true: C_4 is known not to have a (intersection)-thrackle drawing, but it can easily be drawn so that every pair of edges crosses exactly once. We are not aware of any research specifically on crossing-thrackles.

Theorem 3.6 (Lovász, Pach and Szegedy [19]). *A bipartite graph is a generalized thrackle if and only if it is planar.*

Let G be a thrackle; split $V(G)$ into V_1 and V_2 so as to maximize the number of edges between V_1 and V_2. Then every vertex has at least as many neighbors in the other partition as it has in its own partition (otherwise we would move it to the other partition), so we can remove at most half the edges of G to turn it into a bipartite graph G'. By Theorem 1.9, G' is planar and thus has at most $2|V(G')| - 4$ edges (using Euler); but then $|E(G)| \leq 4|V(G')| - 8 \leq 4|V(G)|$, so $|E(G)| = O(|V(G)|)$ for thrackles G. In fact, this bound can be improved by sharpening the reasoning:

---

[15]The best current upper bound of, approximately, $|E| < 1.428|V|$ is due to Fulek and Pach [13] using computational methods.

Corollary 3.7 (Lovász, Pach and Szegedy [19]). *If G is a thrackle, then $|E(G)| \leq 2|V(G)| - 3$.*

We include a very simple proof of Theorem 3.6 based on a proof from [36]. Recall that an (e, v)-move pulls e over v changing the parity of crossing between e and every edge incident to v as shown in Figure 4.

Proof of Theorem 3.6. Let G be a bipartite graph and U one of its partitions. Fix an order of the vertices in U and say that e *precedes* v, if e's endpoint in U precedes v in the ordering of U. Now, for any pair $e \in E(G)$ and $u \in U$ such that e precedes u perform an (e, u)-move. The result of these moves is that the parity of any pair of independent edges changes, whereas the parity of any pair of adjacent edges remains unaffected. Reversing the rotation of every vertex in U and redrawing the edges incident to it in a small neighborhood of the vertex changes the parity of every pair of adjacent edges. In summary, the parity of every pair of edges changed. This means that for a bipartite graph a drawing in which every pair of edges crosses oddly can be turned into a drawing in which every pair of edges crosses evenly and vice versa. This immediately implies that a planar bipartite graph is a generalized thrackle, and, in the reverse direction, that a generalized bipartite thrackle is planar by the weak Hanani–Tutte theorem. ∎

Cairns and Nikolayevsky showed that Theorem 1.9 remains true on any orientable surface. Indeed, the proof of Theorem 3.6 we just gave works for arbitrary surfaces if we replace the application of the weak Hanani–Tutte theorem for the plane with the version for an arbitrary surface.

Corollary 3.8 (Cairns, Nikolayevsky [7]). *A bipartite graph is a generalized thrackle in a surface if and only if it can be embedded in that surface.*

Cairns and Nikolayevsky managed to find a pleasant generalization of this result to non-bipartite graphs using the notion of a *parity embedding* which is is an embedding of a graph on a non-orientable surface so that even cycles are two-sided curves and odd cycles are one-sided curves.

Theorem 3.9 (Cairns, Nikolayevsky [8]). *G is a generalized thrackle on an orientable surface S if and only if G has a parity embedding on the (nonorientable) surface obtained by adding a crosscap to S.*

A short proof-sketch of the easy direction: Given G in S push each edge across the new crosscap as shown in Figure 5. Every pair of edges will then cross evenly, and thus, by Theorem 2.1 be embeddable in the new surface.

Fig. 5. Pushing edges over the crosscap

This result can be extended to non-orientable surfaces using the notion of an X-parity embedding. For a non-orientable surface, let X be a particular crosscap of the surface; an *X-parity embedding* is an embedding in which a cycle is odd if and only if it passes through X an odd number of times. (If S is orientable, then a parity embedding on $S + X$ is the same as an X-parity embedding, so Theorem 3.9 is subsumed.)

Theorem 3.10 (Pelsmajer, Schaefer, Štefankovič [36]). *G is a generalized thrackle on a surface S if and only if G has an X-parity embedding on the surface obtained by adding a crosscap X to S, with the same embedding scheme. In that case, we can assume that every edge passes through X an odd number of times.*

As a consequence, we can recover a result shown by Perlstein and Pinchasi in their study of Vázsonyi's conjecture [39]. A *centrally symmetric S^2-lifting* of a graph G is a bipartite graph G' embedded on the sphere so that G' is centrally symmetric and every vertex of G corresponds to two antipodal points of G' that belong to different partitions of G' and every edge of G corresponds to two edges of G' so that the endpoints of each edge belong to different partitions.

Theorem 3.11 (Perlstein, Pinchasi [39]). *A graph is a generalized thrackle if and only if it has a centrally symmetric S^2-lifting.*

Proof sketch: A centrally symmetric S^2-lifting is really a double-cover of an embedding of G in the projective plane, so one direction is obvious. In the other direction, Theorem 3.10 tells us that a generalized thrackle can be embedded in the projective plane so that every edge crosses through the crosscap an odd number of times. If we think of the projective plane as a disk with the crosscap as its boundary, then the natural S^2-double-cover of this embedding is a centrally symmetric S^2-lifting.

A *geometric graph* is a graph with a straight-line embedding. With Theorem 3.11 Perlstein and Pinchasi are able to show that every geometric graph in $\mathbb{R}^3$ in which every two edges are *strongly avoiding*—they can be projected onto some 2-dimensional plane so that they belong to two distinct rays that form a non-acute angle between them—is a generalized thrackle (in the plane).

3.4. Crossing Numbers

The crossing number, $\mathrm{cr}(G)$, of a graph G is the smallest number of crossings necessary to draw the graph in the plane. We do not allow edges to pass through vertices or more than two edges to cross in a point.

The Hanani–Tutte theorem has been closely linked to the study of the crossing number and many of its variants. Kleitman's parity result, for example, was part of his proof that Zarankiewicz's conjecture holds for the crossing numbers of $K_{5,n}$ and $K_{6,n}$. Tutte's paper tried to establish an algebraic theory of crossing numbers, see Section 1.4 for details on Tutte's approach.

Do we really need to count all the crossings? The (weak) Hanani–Tutte theorem seems to suggest that it should be sufficient to count crossings only modulo 2: let $\mathrm{ocr}(G)$, the *odd crossing number* of G be the smallest number of pairs of edges that cross oddly in a drawing of G [29]. The odd crossing number in many ways behaves like the standard crossing number, for example, the famous crossing lemma, $\mathrm{cr}(G) \geq 1/64 |E(G)|^3/|V(G)|^2$, remains true for the odd crossing number, with the original proof, though some recent strengthening of the constant factor apparently do not carry over to the odd crossing number [24].

The weak Hanani–Tutte theorem can now be stated as saying that $\mathrm{ocr}(G) = 0$ implies that $\mathrm{cr}(G) = 0$. This might suggest that $\mathrm{ocr}(G) = \mathrm{cr}(G)$ for all G, but equality does not hold between the two crossing numbers: One can construct a graph for which $\mathrm{ocr}(G) < \mathrm{cr}(G) \leq 10$ [35, 47][16]. However, it is true that $\mathrm{ocr}(G) = \mathrm{cr}(G)$ as long as $\mathrm{ocr}(G) \leq 3$ [34], that is, if a graph can be drawn so that at most $k \leq 3$ edges cross oddly, then the graph can be drawn with at most k crossings. To establish a result like this, one must be able to remove crossings along edges that are not involved in odd crossings. The first such "removing crossings" result is due to Pach and Tóth.

Lemma 3.12 (Pach, Tóth [29]). *If D is a drawing of G in the plane, and E_0 is the set of even edges in D, then G can be drawn in the plane so that no edge in E_0 is involved in any crossings.*

---

[16]The original separation is from [35]; Tóth's approach leads to a separation of $\mathrm{ocr}(G)$ and $\mathrm{cr}(G)$ with $\mathrm{cr}(G) \leq 10$ [48].

Lemma 3.12 implies that cr cannot be arbitrary larger than ocr (clear all even edges of crossings using the lemma and then draw the remaining edges in their faces so they cross each other at most once), resulting in the pairwise crossing of at most $2\,\mathrm{ocr}(G)$ edges. This was observed by Pach and Tóth.

Corollary 3.13 (Pach, Tóth [29]). $\mathrm{cr}(G) \leq \binom{2\,\mathrm{ocr}(G)}{2}$.

Lemma 3.12 has the disadvantage that it may introduce new odd pairs, that is, pairs of edges that cross oddly, so it cannot be used to show, for example, that $\mathrm{ocr}(G) = \mathrm{cr}(G)$ for small values. This issue is addressed in the following variant.

Lemma 3.14 (Pelsmajer, Schaefer, Štefankovič [34]). *If D is a drawing of G in the plane, and E_0 is the set of even edges in D, then G can be drawn in the plane so that no edge in E_0 is involved in any crossings and there are no new pairs of edges that cross an odd number of times.*

Lemma 3.14 can be used to prove the strong Hanani–Tutte theorem without appealing to Kuratowski's theorem: Pick a cycle in the graph, make its edges even (since there are no odd independent crossings this can be done by locally modifying rotations of vertices on the cycle), use Lemma 3.14 to remove crossings with the cycle, and induct. The induction has to be set up carefully. This approach yields an effective procedure for constructing the embedding from the original drawing.

Lemma 3.14 should be useful as a first step in improving the upper bound of Corollary 3.13. The general feeling is that $\mathrm{cr}(G) \leq O(\mathrm{ocr}(G))$, but there is no hard evidence for this. Lemma 3.14 unfortunately fails on surfaces other than the plane (there are counterexamples for projective plane and torus that show that the pairs of edges that cross oddly may have to change [36]). It is possible that a surface version of the lemma can be proved which only concludes that the odd crossing number of the drawing does not increase. Meanwhile, the following weaker version (which in the plane is the same as Pach-Tóth's result, Lemma 3.12) is true:

Lemma 3.15 (Pelsmajer, Schaefer, Štefankovič [36]). *If D is a drawing of a graph G on some surface S, and E_0 is the set of even edges in D, then G can be drawn in S so that no edge in E_0 is involved in any crossings.*

One concludes, as in the planar case, that cr_S, the crossing number on surface S is bounded in terms of ocr_S, the odd crossing number on S:

Corollary 3.16 (Pelsmajer, Schaefer, Štefankovič [36]).

$$\text{cr}_S(G) \leq \binom{2\,\text{ocr}_S(G)}{2}$$

for any surface S, orientable or non-orientable.

The *independent odd crossing number,* $\text{iocr}(G)$, is the smallest number of independent pairs of edges that cross in a drawing of G. The (strong) Hanani–Tutte theorem translates into "$\text{iocr}(G) = 0$ implies $\text{cr}(G) = 0$". Since, by definition, $\text{iocr}(G) \leq \text{ocr}(G) \leq \text{cr}(G)$, we already know that equality does not hold between $\text{iocr}(G)$ and $\text{cr}(G)$ (as ocr and cr can be separated), however, it is entirely open whether $\text{iocr}(G) = \text{ocr}(G)$. One might again ask, whether $\text{iocr}(G) = \text{cr}(G)$ for small values and whether cr can be bounded in terms of iocr. The situation is more difficult than cr versus ocr, since the number of odd crossings in a drawing can be arbitrarily large even if iocr is bounded. However, we could recently establish the following redrawing result:

Lemma 3.17 (Pelsmajer, Schaefer, Štefankovič [37]). *If D is a drawing of a graph G in the plane and E_0 is the set of independently even edges in D, then G can be redrawn so that no edge in E_0 is involved in any crossings and every pair of edges crosses at most once.*

As an immediate consequence one obtains, as earlier:

Corollary 3.18 (Pelsmajer, Schaefer, Štefankovič [37]). $\text{cr}(G) \leq \binom{2\,\text{iocr}(G)}{2}$.

It is open whether the conclusion of Lemma 3.17 can be strengthened to say that there are no new pairs of independent edges crossing oddly. The following result is another consequence of Lemma 3.17; the proof in this case is rather intricate though.

Corollary 3.19 (Pelsmajer, Schaefer, Štefankovič [37]). $\text{iocr}(G) = \text{cr}(G)$ *for graphs G with $\text{iocr}(G) \leq 2$.*

We do not know whether any of the last three results hold for surfaces other than the plane.

We started with the odd and independent odd crossing numbers, since they are most closely related to the Hanani–Tutte results, however, there are two other crossing number variants worth mentioning in this context: pair and algebraic crossing number (the latter we saw before in Section 1.4).

With this, our—still incomplete—list of basic crossing number variants becomes:

crossing number: $\mathrm{cr}(G)$, the smallest number of crossings in a drawing of G,

pair crossing number: $\mathrm{pcr}(G)$, the smallest number of pairs of edges crossing in a drawing of G,

algebraic crossing number: $\mathrm{acr}(G)$, orient all the edges in a drawing and distinguish between positive and negative crossings along an edge, counting them as $+1$ and -1; minimize the sum of the absolute values of these counts for each edge,

odd crossing number: $\mathrm{ocr}(G)$, the smallest number of pairs of edges crossing oddly in a drawing of G.

We can modify each of these notions by two rules, suggested by Pach and Tóth [28]:

"Rule $+$": restrict the drawings to drawings in which adjacent edges are not allowed to cross.

"Rule $-$": allow crossings of adjacent edges, but does not count them towards the crossing

We add $+$ and $-$ as a subscript to the crossing number to denote that we are following that particular rule. Rule $+$ is inspired by the observation that crossing-number minimal drawings fulfill it, that is, $\mathrm{cr} = \mathrm{cr}_+$, but it is not clear whether this holds for any other crossing number variant. Of the twelve possible combinations of Rule $+$ and Rule $-$ with the four crossing numbers, these are the only two that are known to coincide. Rule $-$ is what turns ocr into iocr, namely, $\mathrm{ocr}_- = \mathrm{iocr}$.

This leaves us with eleven, potentially different, notions of crossing number:

Rule $+$	ocr_+	acr_+	pcr_+	
	ocr	acr	pcr	cr
Rule $-$	$\mathrm{iocr} = \mathrm{ocr}_-$	$\mathrm{iacr} = \mathrm{acr}_-$	pcr_-	cr_-

Little is known about the relationship between these crossing numbers. The variants are monotone in the sense that going from bottom to top in the table does not decrease the value and neither does going from left to right as long as we drop either the acr-column or the pcr-column. So $\mathrm{iocr}(G)$ is the smallest and $\mathrm{cr}(G)$ the largest value, but we do not know whether $\mathrm{acr}(G) \leq \mathrm{pcr}(G)$. We are aware of only two separations among all these variants: one can construct families of graphs for which $\mathrm{ocr}(G) < \lambda\, \mathrm{cr}(G)$ for some $\lambda < 1$ (originally [35], improved λ in [47]). The original examples realize $\mathrm{ocr}(G) < \mathrm{acr}(G) = \mathrm{pcr}(G) = \mathrm{cr}(G)$ [35] and the new examples $\mathrm{ocr}(G) = \mathrm{acr}(G) < \mathrm{pcr}(G) = \mathrm{cr}(G)$ [47]. Combining these two types of examples, one can build a graph G for which $\mathrm{ocr}(G) < \mathrm{acr}(G) < \mathrm{pcr}(G)$. We are not aware of any other separations. Corollary 3.18 shows that all the crossing number variants listed here are within a square of each other; for cr versus pcr this bound can be improved: Valtr [50] showed that $\mathrm{cr}(G) = O(\mathrm{pcr}^2(G)/\log \mathrm{pcr}(G))$, which Tóth [47] improved to $\mathrm{cr}(G) = O(\mathrm{pcr}^2(G)/\log^2 \mathrm{pcr}(G))$. Using a separator theorem for string graphs due to Pach and Fox, Tóth has recently been able to lower this bound to $\mathrm{cr}(G) = O(\mathrm{pcr}^{7/4}(G)/\log^{3/2} \mathrm{pcr}(G))$ [46]. These are the only non-quadratic upper bounds between crossing numbers we are aware of.

We conjecture that $\mathrm{pcr} = \mathrm{cr}$ and $\mathrm{cr}_- = \mathrm{cr}$. Evidence for the first conjecture is purely computational: for two vertex multi-graphs with rotations it appears to be true, according to computer searches performed in connection with [35]. A first step towards the second conjecture is the proof that crossings with adjacent edges can be removed if they are the only crossings along an edge:

Theorem 3.20 (Schaefer [42]). *If D is a drawing of G in the plane, and E_0 is the set of edges in D that have no independent crossings, then G can be drawn in the plane so that no edge in E_0 is involved in any crossings and there are no new independent crossings.*

This brings us only slightly closer to proving $\mathrm{cr}_- = \mathrm{cr}$, but it is a first, necessary, step to showing even $\mathrm{cr} = O(\mathrm{cr}_-)$. The extent of our ignorance about cr_- is captured in the following conjecture.

Conjecture 3.21. *If a graph can be drawn on a surface S so that no two independent edges cross, then the graph can be embedded in S.*

By the strong Hanani–Tutte theorem we know that the conjecture is true for the plane and the projective plane. Beyond that we know nothing. So we can only agree with the first half of Tutte's sentiment that "crossings of adjacent edges are trivial, and easily got rid of" [49].

What about the computational complexity of all of these crossing numbers? Obviously, they are all **NP**-complete? Well, yes, and no. Most of them are **NP**-

complete, but not always for obvious reasons. Garey and Johnson showed that the crossing number problem is **NP**-complete [14]. This proof also shows that pcr is **NP**-hard, but it does not imply that the problem lies in **NP**; that was established later in connection with the string graph problem [43]. The variant ocr is **NP**-complete as shown by Pach and Tóth [29]; **NP**-hardness is a modification of the Garey-Johnson proof and containment in **NP** relies on a refinement of Theorem 1.18 which takes into account the rotation system. For acr, Pach and Tóth's hardness proof for ocr still works and acr $\in$ **NP**, since $\mathrm{acr}(G) \leq k$ can be rephrased as an integer linear program (along the lines of Theorem 1.19). (This means that drawings of G with $\mathrm{acr}(G) \leq k$ may require an exponential number of crossings, since this is the best bound known for integer linear programming. We leave it open whether this bound can be improved for acr.)

For the Rule − variants, we know that iocr, pcr_ and cr_ are **NP**-complete [38]. In all three cases, **NP**-hardness follows from showing that the underlying crossing number concept ocr, pcr and cr remains **NP**-hard if the graph is given with a rotation system. (This turns out to be highly non-trivial in the case of ocr.) All three problems lie in **NP**; for iocr this follows directly from Theorem 1.18, for pcr_ and cr_ the situation is more complicated, since there is no immediate bound on the number of crossings that do not count; using techniques from [43] and [44] the problems can be placed in **NP** (the upper bounds on the uncounted crossings are exponential). We do not know the complexity of iacr, though it is quite possible that the iocr-hardness proof can be adapted.

Finally, ocr_+ and pcr_+ are **NP**-hard; we know that ocr and pcr remain hard if the rotation system of the graph is specified. So let (G, R) be a graph with rotation system; from it construct G' by replacing each vertex v of degree d with a wheel W_d and attach edges originally connected to v to the d outer vertices of W_d. Finally, replace edges of W_d with multiple, parallel P_3s to ensure that the W_d are embedded. Then the pair or odd crossing number of G with rotation R is at most k if and only if pcr_+ or ocr_+ of G' is at most k. Showing that ocr_+ and pcr_+ lie in **NP** can be done using the approach from [43] and [44]. Again we leave open the complexity of the algebraic variant, acr_+.

There are also results on the parameterized complexity of pcr and ocr [33].

4. In Place of a Conclusion

We have sprinkled open problems and conjectures liberally throughout the survey and there are many obvious questions one can ask (is there a Hanani–Tutte theorem for hypergraphs? For matroids?), so instead of reiterating this material, let us mention one more tempting direction that one can take the Hanani–Tutte theorem. We restricted ourselves to surfaces following the graph-theoretical tradition; the algebraic topology literature went a different route. The Hanani–Tutte theorem is there known as a version of the Flores-van Kampen theorem, which has many generalizations and variants, typically in higher-dimensional spaces. Closer to our versions of the Hanani–Tutte theorem is a recent result of van der Holst and Pendavingh [53]: imagine a graph embedded in $\mathbb{R}^3$. An embedding is called *flat* if one can attach an open disk to each cycle of the graph so that the boundary of the disk is the cycle and the disk is disjoint from the graph. If a graph can be embedded in $\mathbb{R}^3$ so that each disk crosses each non-incident edge of the graph an even number of times, then the graph has a flat embedding. The proof uses methods from algebraic topology; can it be shown using an intuitive geometric argument?

Acknowledgment. I would like to thank Martin Loebl, Hein van der Holst and the anonymous referee for helpful comments. And Daniel and Michael for being there. Without them there would have been less to survey.

REFERENCES

[1] Pankaj K. Agarwal, Eran Nevo, János Pach, Rom Pinchasi, Micha Sharir, and Shakhar Smorodinsky, Lenses in arrangements of pseudo-circles and their applications, *J. ACM,* 51(2):139–186 (electronic), 2004.

[2] Dan Archdeacon and R. Bruce Richter, On the parity of crossing numbers, *J. Graph Theory,* 12(3):307–310, 1988.

[3] Sanjeev Arora, László Babai, Jacques Stern, and Z. Sweedyk, The hardness of approximate optima in lattices, codes, and systems of linear equations, *J. Comput. System Sci.,* 54(2, part 2):317–331, 1997. 34th Annual Symposium on Foundations of Computer Science (Palo Alto, CA, 1993).

[4] Peter Brass, William Moser, and János Pach, *Research Problems in Discrete Geometry.* Springer, New York, 2005.

[5] Henning Bruhn and Reinhard Diestel, Maclane's theorem for arbitrary surfaces, *Journal of Combinatorial Theory, Series B,* 99(2):275 – 286, 2009.

[6] Sarit Buzaglo, Rom Pinchasi, and Günter Rote, Topological hyper-graphs. Unpublished manuscript, 2007.

[7] Grant Cairns and Yury Nikolayevsky, Bounds for generalized thrackles, *Discrete Comput. Geom.,* 23(2):191–206, 2000.

[8] Grant Cairns and Yury Nikolayevsky, Generalized thrackle drawings of non-bipartite graphs, *Discrete Comput. Geom.*, 41(1):119–134, 2009.

[9] Jakub Černý, *Combinatorial and Computational Geometry*. PhD thesis, Charles University, Prague, 2008.

[10] Hubert de Fraysseix and Pierre Rosenstiehl, A characterization of planar graphs by Trémaux orders, *Combinatorica*, 5(2):127–135, 1985.

[11] Peter Eades, Qing-Wen Feng, and Xuemin Lin, Straight-line drawing algorithms for hierarchical graphs and clustered graphs. In *Graph drawing*, volume 1190 of *Lecture Notes in Comput. Sci.*, pages 113–128, London, UK, 1997. Springer.

[12] Esther Ezra and Micha Sharir, Lower envelopes of 3-intersecting surfaces in $\mathbb{R}^3$. Unpublished manuscript, 2007.

[13] Radoslav Fulek and János Pach, A computational approach to conway's thrackle conjecture, *CoRR*, abs/1002.3904, Feb 2010.

[14] Michael R. Garey and David S. Johnson, Crossing number is NP-complete, *SIAM Journal on Algebraic and Discrete Methods*, 4(3):312–316, 1983.

[15] Chaim Chojnacki (Haim Hanani), Über wesentlich unplättbare Kurven im drei-dimensionalen Raume, *Fundamenta Mathematicae*, 23:135–142, 1934.

[16] Daniel J. Kleitman. A note on the parity of the number of crossings of a graph, *J. Combinatorial Theory Ser. B*, 21(1):88–89, 1976.

[17] Roy B. Levow, On Tutte's algebraic approach to the theory of crossing numbers. In *Proceedings of the Third Southeastern Conference on Combinatorics, Graph Theory, and Computing (Florida Atlantic Univ., Boca Raton, Fla., 1972)*, pages 315–314, Boca Raton, Fla., 1972. Florida Atlantic Univ.

[18] Martin Loebl and Gregor Masbaum, On the optimality of the arf invariant formula for graph polynomials, *CoRR*, abs/0908.2925, 2009.

[19] Laszlo Lovász, János Pach, and Mario Szegedy, On Conway's thrackle conjecture, *Discrete Comput. Geom.*, 18(4):369–376, 1997.

[20] Karl Menger, *Ergebnisse eines mathematischen Kolloquiums*. Springer-Verlag, Vienna, 1998.

[21] Bojan Mohar and Carsten Thomassen, *Graphs on surfaces*. Johns Hopkins Studies in the Mathematical Sciences. Johns Hopkins University Press, Baltimore, MD, 2001.

[22] Serguei Norine, Pfaffian graphs, T-joins and crossing numbers, *Combinatorica*, 28(1):89–98, 2008.

[23] János Pach, Rom Pinchasi, Gábor Tardos, and Géza Tóth, Geometric graphs with no self-intersecting path of length three, *European J. Combin.*, 25(6):793–811, 2004.

[24] János Pach, Radoš Radoicić, Gábor Tardos, and Géza Tóth, Improving the crossing lemma by finding more crossings in sparse graphs: [extended abstract]. In *SCG '04: Proceedings of the twentieth annual symposium on Computational geometry*, pages 68–75, New York, NY, USA, 2004. ACM.

[25] János Pach and Micha Sharir, On the boundary of the union of planar convex sets, *Discrete Comput. Geom.*, 21(3):321–328, 1999.

[26] János Pach and Micha Sharir, Geometric incidences. In *Towards a theory of geometric graphs*, volume 342 of *Contemp. Math.*, pages 185–223. Amer. Math. Soc., Providence, RI, 2004.

[27] János Pach and Micha Sharir, *Combinatorial geometry and its algorithmic applications: The Alcalá lectures*, volume 152 of *Mathematical Surveys and Monographs*. American Mathematical Society, Providence, RI, 2009.

[28] János Pach and Géza Tóth, Thirteen problems on crossing numbers, *Geombinatorics*, 9(4):194–207, 2000.

[29] János Pach and Géza Tóth, Which crossing number is it anyway? *J. Combin. Theory Ser. B*, 80(2):225–246, 2000.

[30] János Pach and Géza Tóth, Monotone drawings of planar graphs, *J. Graph Theory*, 46(1):39–47, 2004.

[31] János Pach and Géza Tóth. Disjoint edges in topological graphs, In *Combinatorial geometry and graph theory*, volume 3330 of *Lecture Notes in Comput. Sci.*, pages 133–140. Springer, Berlin, 2005.

[32] Michael J. Pelsmajer, Marcus Schaefer, and Despina Stasi, Strong Hanani–Tutte on the projective plane, *SIAM Journal on Discrete Mathematics*, 23(3):1317–1323, 2009.

[33] Michael J. Pelsmajer, Marcus Schaefer, and Štefankovič, Crossing numbers and parameterized complexity. In Seok-Hee Hong, Takao Nishizeki, and Wu Quan, editors, *Graph Drawing*, volume 4875 of *Lecture Notes in Computer Science*, pages 31–36. Springer, 2007.

[34] Michael J. Pelsmajer, Marcus Schaefer, and Daniel Štefankovič, Removing even crossings, *J. Combin. Theory Ser. B*, 97(4):489–500, 2007.

[35] Michael J. Pelsmajer, Marcus Schaefer, and Daniel Štefankovič, Odd crossing number and crossing number are not the same, *Discrete Comput. Geom.*, 39(1):442–454, 2008.

[36] Michael J. Pelsmajer, Marcus Schaefer, and Daniel Štefankovič, Removing even crossings on surfaces, *European J. Combin.*, 30(7):1704–1717, 2009.

[37] Michael J. Pelsmajer, Marcus Schaefer, and Daniel Štefankovič, Removing independently even crossings, *SIAM Journal on Discrete Mathematics*, 24(2):379–393, 2010.

[38] Michael J. Pelsmajer, Marcus Schaefer, and Daniel Štefankovič, Crossing numbers of graphs with rotation systems, *Algorithmica*, 60:679–702, 2011. 10.1007/s00453-009-9343-y.

[39] Amitai Perlstein and Rom Pinchasi, Generalized thrackles and geometric graphs in $\mathbb{R}^3$ with no pair of strongly avoiding edges, *Graphs Combin.*, 24(4):373–389, 2008.

[40] Rom Pinchasi and Radoš Radoičić, Topological graphs with no self-intersecting cycle of length 4. In *Towards a theory of geometric graphs*, volume 342 of *Contemp. Math.*, pages 233–243. Amer. Math. Soc., Providence, RI, 2004.

[41] K. S. Sarkaria, A one-dimensional Whitney trick and Kuratowski's graph planarity criterion, *Israel J. Math.*, 73(1):79–89, 1991.

[42] Marcus Schaefer, Removing incident crossings. Unpublished Manuscript, 2010.

[43] Marcus Schaefer, Eric Sedgwick, and Daniel Štefankovič, Recognizing string graphs in NP, *J. Comput. System Sci.*, 67(2):365–380, 2003. Special issue on STOC2002 (Montreal, QC).

[44] Marcus Schaefer, Eric Sedgwick, and Daniel Štefankovič, Computing Dehn twists and geometric intersection numbers in polynomial time. In *Proceedings of the 20th Canadian Conference on Computational Geometry*, pages 111–114, 2008.

[45] Shakhar Smorodinsky and Micha Sharir, Selecting points that are heavily covered by pseudo-circles, spheres or rectangles, *Combin. Probab. Comput.*, 13(3):389–411, 2004.

[46] Géza Tóth, A better bound for the pair-crossing number. Presentation at the *Conference on Geoemtric Graph Theory*, Lausanne, Switzerland, 2010.

[47] Géza Tóth, Note on the pair-crossing number and the odd-crossing number, *Discrete Comput. Geom.*, 39(4):791–799, 2008.

[48] Géza Tóth, 2010. Personal communication.

[49] William T. Tutte, Toward a theory of crossing numbers, *J. Combinatorial Theory*, 8:45–53, 1970.

[50] Pavel Valtr, On the pair-crossing number. In *Combinatorial and Computational Geometry*, volume 52 of *Math. Sci. Res. Inst. Publ.*, pages 569–575. Cambridge University Press, Cambridge, 2005.

[51] Hein van der Holst, Algebraic characterizations of outerplanar and planar graphs, *European J. Combin.*, 28(8):2156–2166, 2007.

[52] Hein van der Holst, A polynomial-time algorithm to find a linkless embedding of a graph, *J. Combin. Theory Ser. B*, 99(2):512–530, 2009.

[53] Hein van der Holst and Rudi Pendavingh, On a graph property generalizing planarity and flatness, *Combinatorica*, 29(3):337–361, 2009.

[54] E. R. van Kampen, Komplexe in euklidischen Räumen, *Abhandlungen aus dem Mathematischen Seminar der Universität Hamburg*, 9(1):72–78, December 1933.

[55] Hassler Whitney, The self-intersections of a smooth n-manifold in 2n-space, *The Annals of Mathematics*, 45(2):220–246, 1944.

[56] Wen Jun Wu, On the planar imbedding of linear graphs. I, *J. Systems Sci. Math. Sci.*, 5(4):290–302, 1985.

[57] Wen Jun Wu, On the planar imbedding of linear graphs (continued), *J. Systems Sci. Math. Sci.*, 6(1):23–35, 1986.

Marcus Schaefer

School of Computing,
DePaul University,
Chicago, Illinois 60604,
USA

e-mail: mschaefer@cdm.depaul.edu

BOLYAI SOCIETY
MATHEMATICAL STUDIES, 24

Geometry –
Intuitive, Discrete, and Convex
pp. 301–330.

EXTREMAL PROPERTIES OF RANDOM MOSAICS

ROLF SCHNEIDER

László Fejes Tóth's fascinating book [2] demonstrates in many ways the phenomenon that figures of discrete or convex geometry that are very economical, namely solving an extremal problem of isoperimetric type, often show a high degree of symmetry. Among the examples are also planar mosaics where, for instance, an extremal property leads to the hexagonal pattern. Mosaics, or tessellations, have become increasingly important for applications. Random tessellations in two or three dimensions have been suggested as models for various real structures. We refer, e.g., to chapter 10 of the book by Stoyan, Kendall, Mecke [35] and to the book by Okabe, Boots, Sugihara, and Chiu [29] on Voronoi tessellations, which also contains a chapter on random mosaics. Apart from possible applications, random mosaics are also an interesting object of study from a purely geometric point of view. Among the results of geometric appeal that have been obtained, some concern extremal problems for (roughly speaking) expected sizes of average cells under some side condition, leading to random mosaics with high symmetry or of a very simple type, namely made up of parallelepipeds only. High symmetry here means that the distribution of the random mosaic, which is usually assumed to be translation invariant, is also invariant under rotations. Extremal problems for the sizes of average cells (not taking expectations) seem senseless at first, since extrema cannot be attained. Nevertheless, in many problems, average cells of large size approximate certain definite shapes, for example balls, segments, regular simplices, with high probability.

The purpose of the following is a survey over results, older and more recent, that have been obtained on extremal properties of random mosaics. These random tessellations are mostly of special types, namely hyperplane, Voronoi or Delaunay mosaics generated by Poisson processes, and satisfying an assumption of homogeneity, also called stationarity.

1. Explanations

We work in Euclidean space $\mathbb{R}^d$, with scalar product $\langle \cdot, \cdot \rangle$ and norm $\| \cdot \|$. Its unit ball and unit sphere are denoted, respectively, by B^d and S^{d-1}. The set $\mathcal{K}$ of convex bodies (nonempty, compact, convex subsets) in $\mathbb{R}^d$ and its subset $\mathcal{P}$ of polytopes are equipped with the topology induced by the Hausdorff metric. Lebesgue measure on $\mathbb{R}^d$ is denoted by λ, and m-dimensional Hausdorff measure by $\mathcal{H}^m$. We write $\lambda(B^d) =: \kappa_d = \pi^{d/2}/\Gamma(1 + d/2)$. For a topological space T, the σ-algebra of its Borel sets is denoted by $\mathcal{B}(T)$. A 'measure' on a topological space in the following is always a measure on its Borel σ-algebra.

The mosaics to be considered will be 'face-to-face'. Therefore, by a **mosaic** in $\mathbb{R}^d$, or a **tessellation** of $\mathbb{R}^d$, we understand here a locally finite set m of d-dimensional polytopes in $\mathbb{R}^d$ with the following properties: the polytopes of m cover $\mathbb{R}^d$, and the intersection of any two different polytopes of m is either empty or a face of both polytopes. The polytopes of m are called its **cells,** and every k-dimensional face of some cell is, by definition, a k**-face** of m $(k = 0, \ldots, d)$. We denote by $\mathcal{F}_k(\mathsf{m})$ the set of all k-faces of m and by skel_k m the union of these k-faces.

Since random mosaics will be modeled as special particle processes, we must explain these first. For a detailed introduction we refer to chapters 3 and 4 of the book [34]. Here we restrict ourselves to convex particles and simple processes. Thus, by a **particle process** X in $\mathbb{R}^d$ we understand a simple point process in $\mathcal{K}$, that is, a measurable mapping from some probability space $(\Omega, \mathsf{A}, \mathbb{P})$ into the space $\mathsf{N}_s(\mathcal{K})$ of simple, locally finite counting measures on $\mathcal{K}$, equipped with the usual σ-algebra. We identify a simple counting measure η with its support and often write $x \in \eta$ for $\eta(\{x\}) = 1$. Then we can view a realization of the particle process X as a locally finite system of (generally overlapping) convex bodies in $\mathbb{R}^d$. The particle process X is **stationary** (or homogeneous) if for any $t \in \mathbb{R}^d$ the process X and its translate $X + t$ (defined in the obvious way) have the same distribution. The **intensity measure** Θ of X is defined by $\Theta(A) := \mathbb{E}X(A)$ for $A \in \mathcal{B}(\mathcal{K})$, where $\mathbb{E}$ denotes mathematical expectation. We assume (this is part of the definition of a particle process) that Θ is locally finite with respect to the hit-or-miss topology, which means that $\Theta(\{K \in \mathcal{K} : K \cap C \neq \emptyset\}) < \infty$ for each $C \in \mathcal{K}$. From now on we assume that X is stationary and that Θ is not the zero measure; then the intensity measure has a decomposition

$$\int_K f \, \mathrm{d}\Theta = \gamma \int_{\mathcal{K}_0} \int_{\mathbb{R}^d} f(K + x) \, \lambda(\mathrm{d}x) \, \mathbb{Q}(\mathrm{d}K),$$

for any Θ-integrable function f on $\mathcal{K}$. Here, γ is a positive number, the **intensity** of X, and $\mathbb{Q}$, the **grain distribution** of X, is a probability measure on $\mathcal{K}_0$, the space of convex bodies K with Steiner point $s(K)$ at the origin. (Other than in [34], we use here the Steiner point as center function, which is possible and convenient in the case of convex bodies.) If $A \in \mathcal{B}(\mathcal{K})$ and $B \in \mathcal{B}(\mathbb{R}^d)$ is a set with $\lambda(B) = 1$, then (denoting by $\mathbf{1}_A$ the indicator function of A)

$$\gamma \mathbb{Q}(A) = \mathbb{E} \sum_{K \in X, \, s(K) \in B} \mathbf{1}_A\big(K - s(K)\big),$$

which reveals the intuitive meaning of the intensity γ and the grain distribution $\mathbb{Q}$ (and incidentally shows that they are uniquely determined by X).

A random convex body with distribution $\mathbb{Q}$ is called the **typical grain** of X.

With a stationary particle process X one can associate two body-valued parameters, which comprise much information about the process. Let $S_{d-1}(K, \cdot)$ denote the surface area measure of the convex body K. The **Blaschke body** $B(X)$ of X is the unique body in $\mathcal{K}_0$ with

$$S_{d-1}\big(B(X), \cdot\big) = \gamma \int_{\mathcal{K}_0} S_{d-1}(K, \cdot) \, \mathbb{Q}(\mathrm{d}K);$$

it exists by a theorem going back to Minkowski. The projection body, $\Pi_{B(X)}$, of $B(X)$ is called the **associated zonoid** of X and is denoted by Π_X. Thus, its support function, denoted by h, is given by

$$h(\Pi_X, u) = \frac{1}{2} \int_{S^{d-1}} \big| \langle u, v \rangle \big| \, S_{d-1}\big(B(X), \mathrm{d}v\big)$$

$$= \frac{\gamma}{2} \int_{\mathcal{K}_0} \int_{S^{d-1}} \big| \langle u, v \rangle \big| \, S_{d-1}(K, \mathrm{d}v) \, \mathbb{Q}(\mathrm{d}K)$$

$$= \gamma \int_{\mathcal{K}_0} h(\Pi_K, u) \, \mathbb{Q}(\mathrm{d}K).$$

The formula

$$h(\Pi_X, u) - \frac{1}{2} \mathbb{E} \sum_{K \in X} \mathrm{card}\left([0, u] \cap \mathrm{bd}\, K\right), \qquad u \in \mathbb{R}^d,$$

where $[0, u]$ denotes the line segment with endpoints 0 and u, reveals the intuitive meaning of the associated zonoid (see [34], in particular section 4.6).

A further body-valued parameter is the set-valued expectation of the typical grain. Let Z be the typical grain of X. Its **set-valued expectation** (or Aumann expectation, also called selection expectation) is defined by

$$\mathbb{E}\, Z := \left\{ \mathbb{E}\, \xi : \xi : \Omega \to \mathbb{R}^d \text{ is measurable and } \xi \in Z \text{ a.s.} \right\}$$

(a.s. stands for 'almost surely'). It is a convex body, and $h(\mathbb{E}\, Z, \cdot) = \mathbb{E}\, h(Z, \cdot)$, hence

$$h(\mathbb{E}\, Z, \cdot) = \gamma \int_{\mathcal{K}_0} h(K, \cdot)\, \mathbb{Q}(\mathrm{d}K).$$

In the plane, one has $S_{d-1}(K + M, \cdot) = S_{d-1}(K, \cdot) + S_{d-1}(M, \cdot)$ for $K, M \in \mathcal{K}$, from which one can deduce that

(1) $$B(X) = \gamma \mathbb{E}\, Z \qquad \text{for}\ \ d = 2.$$

A **random mosaic** is now defined as a particle process which is almost surely a mosaic. Let X be a stationary mosaic in $\mathbb{R}^d$. For $k = 0, \dots, d$, the set $\mathcal{F}_k(X)$ defines a stationary particle process, the process of k-faces of X, denoted by $X^{(k)}$. Assuming that it has locally finite intensity measure (this will be satisfied in the examples considered in later sections), we denote its intensity by $\gamma^{(k)}$ and its grain distribution by $\mathbb{Q}^{(k)}$.

There are several natural ways of defining 'average faces' of a stationary random mosaic. The **typical k-face** of X is, by definition, the random polytope $Z^{(k)}$ (unique up to stochastic equivalence) with distribution $\mathbb{Q}^{(k)}$. The intuitive idea behind this is that in every realization of the random mosaic one picks out a k-face at random, with equal chances for all the k-faces (which is, of course, only possible in a bounded region), and translates it to bring its Steiner point to the origin, to obtain a realization of $Z^{(k)}$. A more precise manifestation of this idea is given by the formula

$$\mathbb{P}\left\{ Z^{(k)} \in A \right\} = \lim_{r \to \infty} \frac{\mathbb{E} \sum_{F \in X^{(k)},\, F \subset rW} \mathbf{1}_A\left(F - s(F)\right)}{\mathbb{E} \sum_{F \in X^{(k)},\, F \subset rW} 1},$$

which holds for $A \in \mathcal{B}(\mathcal{P})$ and any $W \in \mathcal{K}$ with $\lambda(W) > 0$ (this follows from [34, Th. 4.1.3]). We write $Z^{(d)} =: Z$ and call this the **typical cell** of X.

Another way of defining an average cell of X consists in choosing the (almost surely unique) cell that contains a given point. By stationarity, it is inessential which point we choose; we choose 0 and call the cell containing this point the **zero cell** of X and denote it by Z_0. Up to translations, the distribution of the zero cell is the volume-weighted distribution of the typical cell. In fact, for every translation invariant, nonnegative, measurable function f of $\mathcal{P}$ we have $\mathbb{E}f(Z_0) = \mathbb{E}\left[f(Z)\lambda(Z)\right] / \mathbb{E}\lambda(Z)$ (see [34, Th. 10.4.1]).

For a stationary random mosaic X, the Blaschke body $B(X)$ is always centrally symmetric with respect to 0, and the support function of the associated zonoid is given by

$$h(\Pi_X, u) = \mathbb{E} \operatorname{card}\left([0, u] \cap \operatorname{skel}_{d-1} X\right).$$

For $d = 2$, $B(X) = \gamma \mathbb{E}Z$ and

(2) $$\Pi_X = 2\vartheta_{\pi/2} B(X),$$

where $\vartheta_{\pi/2}$ denotes a rotation by $\pi/2$ (observe that $B(X) = -B(X)$). Thus, for a planar random mosaic, the three introduced parameter bodies differ from each other only by elementary transformations.

2. Poisson Hyperplane Tessellations

A particularly accessible class of random mosaics are those generated by Poisson hyperplane processes. They were already an essential topic in the early work on stochastic geometry by Miles and Matheron. In this section, we give a brief survey of the older extremal properties that have been shown for these random mosaics, and we mention a few new results. The subsequent sections are then devoted to a broader survey of more recent results.

We denote by $G(d, k)$ the Grassmannian of k-dimensional linear subspaces and by $A(d, k)$ the affine Grassmannian of k-flats in $\mathbb{R}^d$, both with their usual topologies ($k = 1, \ldots, d - 1$). In particular, $A(d, d - 1)$ is the space of hyperplanes. A hyperplane is often written in the form

$$H(u, \tau) = \left\{x \in \mathbb{R}^d : \langle x, u \rangle = \tau\right\}$$

with $u \in S^{d-1}$ and $\tau \in \mathbb{R}$. Note that $H(u, \tau) = H(-u, -\tau)$ (which will not cause ambiguities).

Let $\widehat{X}$ be a stationary Poisson hyperplane process in $\mathbb{R}^d$ with intensity measure $\widehat{\Theta} \neq 0$. This means that $\widehat{\Theta}$ is a translation invariant, locally finite measure on $A(d, d-1)$ and $\widehat{X}$ is a measurable mapping from some probability space $(\Omega, \mathsf{A}, \mathbb{P})$ into the measurable space $\mathsf{N}_s\big(A(d, d-1)\big)$ of simple, locally finite counting measures on $A(d, d-1)$ with its usual σ-algebra, such that

$$\mathbb{P}\big\{\widehat{X}(A) = k\big\} = e^{-\widehat{\Theta}(A)}\frac{\widehat{\Theta}(A)^k}{k!}, \qquad k = 0, 1, 2, \ldots,$$

for $A \in \mathcal{B}\big(A(d, d-1)\big)$ with $\widehat{\Theta}(A) < \infty$. In particular, $\widehat{\Theta}(A) = \mathbb{E}\,\widehat{X}(A)$. The measure $\widehat{\Theta}$ has a unique decomposition of the form

$$\int_{A(d,d-1)} f\,\mathrm{d}\widehat{\Theta} = \widehat{\gamma}\int_{S^{d-1}}\int_{-\infty}^{\infty} f\big(H(u, \tau)\big)\,\mathrm{d}\tau\,\widehat{\varphi}(\mathrm{d}u)$$

for every nonnegative, measurable function f on $A(d, d-1)$, with a number $\widehat{\gamma} > 0$ and an even probability measure $\widehat{\varphi}$ on S^{d-1}. The number $\widehat{\gamma}$ is the **intensity** and $\widehat{\varphi}$ is the **spherical directional distribution** of $\widehat{X}$. We assume that $\widehat{X}$ is **nondegenerate**, which means that the measure $\widehat{\varphi}$ is not concentrated on a great subsphere. Again, simple counting measures are identified with their supports. Then a.s. every realization $\widehat{X}(\omega)$, $\omega \in \Omega$, can be viewed as a locally finite system of hyperplanes (i.e., every compact set is met by only finitely many of the hyperplanes). It defines a tessellation $X(\omega)$ of $\mathbb{R}^d$, the cells of which are the closures of the connected components of $\mathbb{R}^d \setminus X(\omega)$. Since $\widehat{X}$ is nondegenerate, the cells are a.s. bounded, hence in this way we define a random mosaic X. It is called the **Poisson hyperplane mosaic** induced by $\widehat{X}$. The random mosaic X is stationary.

For $k = 0, \ldots, d$, the process $X^{(k)}$ of k-faces of X has a locally finite intensity measure (see [34, sect. 10.3]) and hence a positive, finite intensity $\gamma^{(k)}$ and a grain distribution $\mathbb{Q}^{(k)}$.

Further natural geometric parameters for the description of $\widehat{X}$ and hence X are obtained as follows. For $k \in \{0, \ldots, d-1\}$, let $\widehat{X}_{d-k}$ denote the intersection process of order $d - k$ of $\widehat{X}$ (see [34, sect. 4.4]). It is a stationary process in $A(d, k)$, defined by the intersections of any $d - k$ hyperplanes of $\widehat{X}$ which are in general position. We denote the intensity of $\widehat{X}_{d-k}$ by $\widehat{\gamma}_{d-k}$ (hence $\widehat{\gamma}_1 = \widehat{\gamma}$) and its directional distribution, which is a measure on $G(d,k)$, by $\widehat{\mathbb{Q}}_{d-k}$. The intuitive meaning of these parameters can be read off from

$$\widehat{\gamma}_{d-k}\widehat{\mathbb{Q}}_{d-k}(A) = \frac{1}{\kappa_{d-k}} \mathbb{E}\,\text{card}\,\{E \in \widehat{X}_{d-k} \,:\, E \cap B^d \neq \emptyset,\ E_0 \in A\}$$

for $A \in \mathcal{B}\big(G(d,k)\big)$, where $E_0 \in G(d,k)$ denotes the linear subspace parallel to the flat E.

The typical cell Z and the zero cell Z_0 of the stationary random mosaic X are random polytopes, which can be considered as 'average' cells of X, in different ways. We shall be interested in estimating the 'size' of these average cells, metrically or combinatorially. A quite general class of functionals for measuring the size of a d-polytope P (already considered by Miles [27] for typical cells) is given, for $0 \le r \le s \le d$, by

$$Y_{r,s}(P) := \sum_{F \in \mathcal{F}_s(P)} V_r(F),$$

where $\mathcal{F}_s(P)$ is the set of s-faces of P and V_r denotes the rth intrinsic volume. This comprises the following special cases.

- $Y_{r,d}(P) = V_r(P)$ is the rth intrinsic volume of P; in particular, $V_d(P)$ is the volume, $2V_{d-1}(P)$ is the surface area, $(2\kappa_{d-1}/d\kappa_d)V_1(P)$ is the mean width of P, and $V_0(P) = 1$.
- $Y_{s,s}(P) =: L_s(P)$ is the total s-dimensional volume of the s-faces of P.
- $Y_{0,s}(P) =: f_s(P)$ is the number of s-faces of P.

For $0 \le r \le s \le k$, the functional $Y_{r,s}$ is defined for k-polytopes. We want to evaluate $\mathbb{E}\,Y_{r,s}\big(Z^{(k)}\big)$, its expectation for the typical k-face of the random mosaic X. Define the densities

$$(3) \qquad d_r^{(k,s)} := \gamma^{(k)}\mathbb{E}\,Y_{r,s}\big(Z^{(k)}\big), \qquad d_r^{(k)} := d_r^{(k,k)} = \gamma^{(k)}\mathbb{E}\,V_r\big(Z^{(k)}\big).$$

It follows from [34], Theorem 10.1.2 and (10.9), that

$$(4) \qquad d_r^{(k,s)} = 2^{k-s}\binom{d-s}{d-k}d_r^{(s)},$$

where we have used that a.s. every s-face of X with $s \le k$ lies in precisely $2^{k-s}\binom{d-s}{d-k}$ k-faces of X. By [34, Th. 10.3.1],

$$(5) \qquad d_r^{(s)} = \binom{d-r}{d-s}d_r^{(r)},$$

in particular (case $j = 0$),

$$(6) \qquad \gamma^{(s)} = \binom{d}{s} \gamma^{(0)} = \binom{d}{s} \widehat{\gamma}_d.$$

From (3)–(6) we get

$$(7) \qquad \mathbb{E}\, Y_{r,s}\big(Z^{(k)}\big) = 2^{k-s} \binom{k}{r} \binom{k-r}{k-s} \mathbb{E}\, V_r\big(Z^{(r)}\big).$$

The case $r = 0$ gives

$$(8) \qquad \mathbb{E}\, f_s\big(Z^{(k)}\big) = 2^{k-s} \binom{k}{s}.$$

So far, Poisson assumptions were not required, only some finiteness assumptions, which are satisfied in the Poisson case.

For a stationary Poisson hyperplane process $\widehat{X}$, the remaining essential parameters can be expressed as intrinsic volumes of an associated zonoid. This surprising and very useful fact was discovered by Matheron.

Since X is a stationary particle process, its Blaschke body $B(X)$ and associated zonoid Π_X are defined as in Section 1. They can now be expressed more directly in terms of the data $\widehat{\gamma}$ and $\widehat{\varphi}$ of the hyperplane process $\widehat{X}$. Since $\widehat{\varphi}$ is an even measure on S^{d-1} and not concentrated on a great subsphere, there exists a unique 0-symmetric convex body $B(\widehat{X})$ with $S_{d-1}\big(B(\widehat{X}),\cdot\big) = \widehat{\gamma}\widehat{\varphi}$, the **Blaschke body** of the hyperplane process $\widehat{X}$. Its projection body $\Pi_{B(\widehat{X})} =: \Pi_{\widehat{X}}$, and thus the body with support function

$$h(\Pi_{\widehat{X}}, \cdot) = \frac{\widehat{\gamma}}{2} \int_{S^{d-1}} |\langle \cdot, v \rangle| \, \widehat{\varphi}(\mathrm{d}v),$$

is the **associated zonoid** of $\widehat{X}$. It turns out ([34, p. 489]) that

$$(9) \qquad B(X) = 2^{\frac{1}{d-1}} B(\widehat{X}), \qquad \Pi_X = 2\Pi_{\widehat{X}}.$$

We remark already here that the Blaschke body $B(\widehat{X})$ is closely related to the shape of (weighted) typical cells. One instance is the following result ([34], Theorem 10.4.11) (another one is (34) and the subsequent remark). Among all convex bodies $K \in \mathcal{K}$ with $0 \in K$ and given positive volume, precisely the homothets of the Blaschke body $B(\widehat{X})$ maximize the probability that the zero cell Z_0 of the tessellation X contains K. A similar result (but requiring a different proof) was obtained in Hug and Schneider [12]: Among all convex bodies $K \in \mathcal{K}$ of given positive volume, precisely the homothets of the Blaschke body maximize the probability that the typical cell Z of the tessellation X contains a translate of K.

The following remarkable relations can be stated (see [34], (4.63), (10.43), (10.44) for proofs and references). For $0 \le r \le d$,

$$(10) \qquad \widehat{\gamma}_{d-r} = d_r^{(r)} = V_{d-r}(\Pi_{\widehat{X}}),$$

in particular,

$$(11) \qquad \widehat{\gamma} = \widehat{\gamma}_1 = V_1(\Pi_{\widehat{X}}).$$

From (3), (6), (10) we obtain

$$(12) \qquad \mathbb{E}\, V_r\big(Z^{(r)}\big) = \frac{V_{d-r}(\Pi_{\widehat{X}})}{\binom{d}{r} V_d(\Pi_{\widehat{X}})}.$$

In the plane, we can consider the set-valued expectation $\mathbb{E}\, Z$ of the typical cell Z. From (1), (2), (9) we get the nice formula

$$V_2(\mathbb{E}\, Z) = \mathbb{E}\, V_2(Z).$$

Since also

$$V_1(\mathbb{E}\, Z) = \mathbb{E}\, V_1(Z)$$

by linearity, the isoperimetric inequality gives

$$\big[\mathbb{E}\, V_1(Z)\big]^2 \ge \pi \mathbb{E}\, V_2(Z),$$

with equality if and only if $\widehat{X}$ is isotropic.

In higher dimensions, the Aleksandrov–Fenchel inequalities for intrinsic volumes can be used to obtain inequalities of isoperimetric type for the considered mosaics. A first example is a sharp inequality for the intersection densities (the intensities of the intersection processes), given the intensity, namely

$$(13) \qquad \frac{\widehat{\gamma}_k}{\widehat{\gamma}^k} \le \frac{\binom{d}{k} \kappa_{d-1}^k}{d^k \kappa_{d-k} \kappa_d^{k-1}}$$

for $k \in \{2, \ldots, d\}$. Equality holds if and only if $\widehat{X}$ is isotropic (see [34], Section 4.6, also for references). Thus, *for given intensity, precisely the most symmetric, namely isotropic, hyperplane processes yield the highest kth intersection density.* The case $k = d$ together with (6) gives

$$
(14) \qquad \frac{\gamma^{(k)}}{\widehat{\gamma}^d} \leq \frac{\binom{d}{k}\kappa_{d-1}^d}{d^d\kappa_d^{d-1}},
$$

where again equality holds precisely in the isotropic case. This shows that *for given intensity $\widehat{\gamma}$, the intensity of the k-faces of the mosaic X is maximal in the isotropic case.*

As simple examples show, there is no counterpart to (13) in the opposite direction: if $\widehat{\gamma} > 0$ is given, $\widehat{\gamma}_k$ can be arbitarily close to zero. However, a sharp estimate in the other direction is possible if the intersection density is measured in an affine-invariant way. For a linear transformation $\Lambda \in \mathrm{GL}\,(d)$, the hyperplane process $\Lambda\widehat{X}$ is obtained from $\widehat{X}$ by applying Λ to each hyperplane of $\widehat{X}$. Then the inequality

$$
(15) \qquad \sup_{\Lambda \in \mathrm{GL}\,(d)} \frac{\widehat{\gamma}_k(\Lambda\widehat{X})}{\widehat{\gamma}(\Lambda\widehat{X})^k} \geq \frac{1}{d^k}\binom{d}{k}
$$

holds for $k = 2, \ldots, d$, with equality if and only if the hyperplanes of $\widehat{X}$ attain almost surely only d fixed directions. This was proved by Hug and Schneider [12], also using the associated zonoid.

In contrast to the inequalities (13), the following one was not mentioned in [25] or [34]. Let $0 < k < d - 1$. From the inequalities (6.4.6) in [30], where we replace i, j, k by $0, k, d - 1$ and observe that

$$
W_r = \frac{\kappa_r}{\binom{d}{r}}V_{d-r},
$$

we obtain

$$
(16) \qquad \left(\frac{V_{d-k}(\Pi_{\widehat{X}})}{\binom{d}{k}V_d(\Pi_{\widehat{X}})} \right)^{d-1} \geq \frac{\kappa_{d-1}^k}{\kappa_k^{d-1}} \left(\frac{V_1(\Pi_{\widehat{X}})}{dV_d(\Pi_{\widehat{X}})} \right)^k
$$

and thus, by (12),

$$
(17) \qquad \left[\mathbb{E}\,V_k\big(Z^{(k)}\big) \right]^{d-1} \geq \frac{\kappa_{d-1}^k}{\kappa_k^{d-1}} \left[\mathbb{E}\,V_{d-1}\big(Z^{(d-1)}\big) \right]^k.
$$

Supppose that equality holds in (17). Then (see [30, p. 334]) equality must hold in the inequality $W_{d-2}(\Pi_{\widehat{X}})^2 \geq W_{d-3}(\Pi_{\widehat{X}})W_{d-1}(\Pi_{\widehat{X}})$. Since $\Pi_{\widehat{X}}$ is centrally symmetric and d-dimensional, this implies by [30, Th. 6.6.19] that $\Pi_{\widehat{X}}$ is a 1-tangential body, and hence a cap body, of a ball. This means that $\Pi_{\widehat{X}}$ is the convex hull of a ball B and a (possibly empty) set M of points outside B with the property that the segment joining any two points of M intersects B. If M is not empty, then $\Pi_{\widehat{X}}$ has uncountably many exposed faces which are segments of different directions. Since $\Pi_{\widehat{X}}$ is a zonoid, each of its faces is a summand of $\Pi_{\widehat{X}}$. This leads to a contradiction, hence $\Pi_{\widehat{X}}$ is a ball, and $\widehat{X}$ is isotropic.

Eliminating $V_d(\Pi_{\widehat{X}})$ from the right side of (16) by means of the inequality connecting V_1 and V_d (i.e., [34, (14.31)] for $j = 1$, $k = d$), we obtain

$$(18) \qquad \mathbb{E}\, V_k\big(Z^{(k)}\big) \geq \frac{d^k \kappa_d^k}{\kappa_{d-1}^k \kappa_k} \frac{1}{\widehat{\gamma}^k},$$

which together with (5) gives [25, Satz 3.12.3]. Equality holds precisely if $\widehat{X}$ is isotropic. Thus, *for given intensity $\widehat{\gamma}$, the isotropic mosaics yield the smallest expected volume of the typical k-face.*

We turn to the zero cell Z_0 of the mosaic X. In contrast to the case of the typical cell Z, useful explicit representations for the expectations of the functionals $Y_{r,s}(Z_0)$ are only known in the following few special cases (for proofs and references, see [34, sect. 10.4]). The first of these is

$$\mathbb{E}\, V_d(Z_0) = \frac{d!}{2^d} V_d(\Pi_{\widehat{X}}^o),$$

where $\Pi_{\widehat{X}}^o := (\Pi_{\widehat{X}})^o$ denotes the polar body of the associated zonoid $\Pi_{\widehat{X}}$. The expected total k-volume of the k-faces can be reduced to this, namely by

$$\mathbb{E}\, L_k(Z_0) = d_k^{(k)} \mathbb{E}\, V_d(Z_0) = \frac{d!}{2^d} V_{d-k}(\Pi_{\widehat{X}}) V_d(\Pi_{\widehat{X}}^o),$$

for $k = 0, \ldots, d-1$. The particular case $k = 0$ gives the expected number of vertices,

$$\mathbb{E}\, f_0(Z_0) = \frac{d!}{2^d} V_d(\Pi_{\widehat{X}}) V_d(\Pi_{\widehat{X}}^o).$$

Since the zero polytope is a.s. simple (that is, each vertex lies in precisely d facets), we have

$$\mathbb{E}\, f_1(Z_0) = \frac{d}{2} \mathbb{E}\, f_0(Z_0),$$

and for $d = 3$ we can use Euler's relation to get

$$\mathbb{E}\, f_2(Z_0) = 2 + \frac{1}{2}\mathbb{E}\, f_0(Z_0).$$

Sharp inequalities are known for the volume and the vertex number, namely

$$(19) \qquad \mathbb{E}\, V_d(Z_0) \geq d!\kappa_d \left(\frac{2\kappa_{d-1}}{d\kappa_d}\,\widehat{\gamma}\right)^{-d},$$

with equality if and only if $\widehat{X}$ is isotropic, and

$$(20) \qquad 2^d \leq \mathbb{E}\, f_0(Z_0) \leq \frac{d!\kappa_d^2}{2^d}.$$

Equality on the left side of (20) holds if and only if X is a **parallel mosaic,** which means that the hyperplanes of $\widehat{X}$ belong to d fixed translation classes. Equality on the right holds if and only if $\widehat{X}$ is **affinely isotropic.** This means that there exists a nondegenerate affine transformation α of $\mathbb{R}^d$ such that the hyperplane process $\alpha\widehat{X}$ is isotropic. For the proofs, we refer to [34, Th. 10.4.9].

The extremal property of isotropic mosaics exhibited by (19) goes much farther. *For given intensity $\widehat{\gamma}$, every moment $\mathbb{E}\, V_d(Z_0)^k$, and also every moment $\mathbb{E}\, V_d(Z)^k$, for $k \in \mathbb{N}$, attains its minimum precisely if $\widehat{X}$ is isotropic.* The proof of this result, which is due to Mecke, is also reproduced in [34, Th. 10.4.9].

3. Nonstationary Hyperplane Tessellations

We mention briefly (following [31]) how a few of the preceding results can be extended to nonstationary Poisson hyperplane tessellations. We assume again that $\widehat{X}$ is a Poisson hyperplane process in $\mathbb{R}^d$ with a locally finite intensity measure $\widehat{\Theta}$. Since $\widehat{X}$ need not be stationary now, the measure $\widehat{\Theta}$ is not necessarily translation invariant. We assume however, that it is **translation regular,** which means that it is absolutely continuous with respect to some translation invariant, locally finite measure $\widetilde{\Theta}$ on $A(d, d-1)$. For simplicity, we restrict ourselves here to the case where $\widehat{\Theta}$ has a continuous density with respect to $\widetilde{\Theta}$. In that case, the constant intensities that exist in the stationary case are replaced by measurable intensity functions, which admit intuitive interpretations. The **intensity function** $\widehat{\gamma}$ of $\widehat{X}$ can be defined by

$$\widehat{\gamma}(z) = \lim_{r \to 0} \frac{1}{V_d(rK)} \, \mathbb{E} \sum_{H \in \widehat{X}} \mathcal{H}^{d-1}\big(H \cap (rK + z)\big)$$

for $z \in \mathbb{R}^d$, where K is any convex body with $V_d(K) > 0$. It has also the representation

$$\widehat{\gamma}(z) = \lim_{r \to 0} \frac{1}{2r} \, \mathbb{E}\, \mathrm{card} \big\{ H \in \widehat{X} \; : \; H \cap (rB^d + z) \neq \emptyset \big\}.$$

For $k \in \{0, \ldots, d-1\}$, the intersection process $\widehat{X}_{d-k}$ of order $d - k$ is defined as in the stationary case; it is obtained by taking the intersections of any $d - k$ hyperplanes of $\widehat{X}$ in general position and is a.s. a simple process of k-flats. It has an intensity function given by

$$\widehat{\gamma}_{d-k}(z) = \lim_{r \to 0} \frac{1}{\kappa_{d-k} r^{d-k}} \, \mathbb{E}\, \mathrm{card} \big\{ E \in \widehat{X}_{d-k} \; : \; E \cap (rB^d + z) \neq \emptyset \big\}$$

for $z \in \mathbb{R}^d$. The inequality (13) extends to an inequality holding at every point, namely

$$(21) \qquad \widehat{\gamma}_k(z) \leq \frac{\binom{d}{k} \kappa_{d-1}^k}{d^k \kappa_{d-k} \kappa_d^{k-1}} \, \widehat{\gamma}(z)^k \qquad \text{for } z \in \mathbb{R}^d.$$

Equality for all z holds if and only if the hyperplane process $\widehat{X}$ is stationary and isotropic. Thus, here an extremal property implies the invariance of the distribution under the full group of rigid motions!

To study the tessellation induced by $\widehat{X}$, we need first a suitable notion of nondegeneracy. We say that $\widehat{X}$ is **nondegenerate** if the zero cell Z_0 is bounded with positive probability and if the following holds. Whenever $U \subset S^{d-1}$ is a measurable set and $\widehat{X}$ contains with positive probability a hyperplane with normal vector in U, then $\widehat{X}$ contains with positive probability infinitely many such hyperplanes. If $\widehat{X}$ is nondegenerate, then it can be shown that the cells induced by $\widehat{X}$ constitute a random mosaic X and that the process $X^{(k)}$ of its k-faces ($k \in \{0, \ldots, d\}$) has a locally finite intensity measure, which is also translation regular, in the sense that it is absolutely continuous with respect to some translation invariant, locally finite measure on the space of polytopes.

In the stationary case, we have defined in (3) a density $d_r^{(k)} = \gamma^{(k)}\mathbb{E}\,V_r\big(Z^{(k)}\big)$, the **specific rth intrinsic volume** of the stationary k-face process $X^{(k)}$. In the nonstationary case, where no typical k-face exists, this definition cannot be used, but it can be generalized by defining

$$d_r^{(k)}(z) := \lim_{r \to 0} \frac{1}{V_d(rB^d)}\,\mathbb{E}\sum_{K \in X^{(k)}} \Phi_r\big(K, rB^d + z\big)$$

for λ-almost all $z \in \mathbb{R}^d$. Here $\Phi_r(K, \cdot)$ is the rth curvature measure of the convex body K. It can then be shown that the relations

$$d_r^{(k)} = \binom{d - r}{d - k} d_r^{(r)}, \qquad d_r^{(r)} = \widehat{\gamma}_{d-r},$$

and hence the inequality

$$(22) \qquad d_r^{(k)} \le \binom{d - r}{d - k}\binom{d}{r}\frac{\kappa_{d-1}^{d-r}}{d^{d-r}\kappa_r\kappa_d^{d-r-1}}\,\widehat{\gamma}^{d-r},$$

hold almost everywhere. Equality in (22) holds if and only if $\widehat{X}$ is stationary and isotropic.

4. Weighted Faces

We return to the stationary Poisson hyperplane process $\widehat{X}$ and its induced tessellation X, as studied in Section 2. The results of that section reveal a clear distinction between the zero cell Z_0 and the typical cell Z, though either of them provides a natural notion of 'average cell'. Heuristically, the zero cell can also be obtained, up to translations, if in a large bounded region of space we choose a uniformly distributed random point (with respect to Lebesgue measure) and take the almost surely unique cell of X containing that point. A similar procedure makes also sense for k-faces: in a large bounded region of space we choose a random point, uniformly distributed on the k-skeleton $\mathrm{skel}_k X$ of X, with respect to the k-dimensional Hausdorff measure, and take the (almost surely unique) k-face of X containing that point. This leads (up to translations) to the notion of the weighted typical k-face. A precise formal definition using Palm theory may be sketched as follows. On $\mathsf{N}_s\big(A(d, d-1)\big)$, we can define a probability measure $\mathbb{P}_k^o$ by

$$\widehat{\gamma}_{d-k}\mathbb{P}_k^o(\mathcal{A}) = \mathbb{E}\int_{\mathrm{skel}_k X} 1_B(x)1_{\mathcal{A}}(\widehat{X} - x)\,\mathcal{H}^k(\mathrm{d}x),$$

where B is any Borel set in $\mathbb{R}^d$ with $\lambda(B) = 1$. (Details, as well as the results below, are found in [32]). There exists a hyperplane process Y_k with distribution $\mathbb{P}_k^o$, and we denote by $Z_0^{(k)}$ the (always existing and almost surely unique) k-face of the tessellation induced by Y_k that contains the origin 0. In particular, $Z_0^{(d)}$ is stochastically equivalent to the zero cell Z_0. The random polytope $Z_0^{(k)}$ is uniquely determined up to stochastic equivalence and is called the **volume-weighted typical k-face,** or briefly the **weighted typical k-face,** of X. This terminology is justified, since the distribution of $Z_0^{(k)}$ is, if translations are disregarded, the volume-weighted distribution of the typical k-face $Z^{(k)}$. In fact, for every translation invariant, nonnegative, measurable function f on $\mathcal{P}$ one has

$$\mathbb{E}\, f\big(Z_0^{(k)}\big) = \frac{1}{\mathbb{E}\, V_k\big(Z^{(k)}\big)}\, \mathbb{E}\big[f\big(Z^{(k)}\big) V_k\big(Z^{(k)}\big)\big].$$

(Recall that $V_k(K) = \mathcal{H}^k(K)$ for a k-dimensional convex body K.) Another justification of the terminology, and at the same time a precise version of the intuitive approach with which we started, is provided by the following formula. Here W can be any convex body with positive volume, and every polytope that appears is replaced by its translate with Steiner point at the origin. For $A \in \mathcal{B}(\mathcal{P})$, we have

$$\mathbb{P}\big\{Z_0^{(k)} - s\big(Z_0^{(k)}\big) \in A\big\} = \lim_{r \to \infty} \frac{\mathbb{E} \sum_{F \in X^{(k)},\, F \subset rW} \mathbf{1}_A\big(F - s(F)\big) V_k(F)}{\mathbb{E} \sum_{F \in X^{(k)},\, F \subset rW} V_k(F)}.$$

It is a useful consequence of the Poisson property of $\widehat{X}$ that the distribution of the weighted typical k-face can be determined from the distribution of the zero cell Z_0, in the following way. Recall that $\widehat{\mathbb{Q}}_{d-k}$ is the directional distribution of the intersection process $\widehat{X}_{d-k}$ of order $d-k$ and is a probability measure on the Grassmannian $G(d,k)$. Now, for $k \in \{1, \ldots, d-1\}$ and every $A \in \mathcal{B}(\mathcal{P})$ we have

$$(23) \qquad \mathbb{P}\big\{Z_0^{(k)} \in A\big\} = \int_{G(d,k)} \mathbb{P}\{Z_0 \cap L \in A\}\, \widehat{\mathbb{Q}}_{d-k}(\mathrm{d}L).$$

This can also be interpreted as follows. If we choose a random k-dimensional linear subspace $\mathcal{L}$ of $\mathbb{R}^d$ with distribution $\widehat{\mathbb{Q}}_{d-k}$ such that $\widehat{X}$ and $\mathcal{L}$ are independent, then the intersection $Z_0 \cap \mathcal{L}$ is stochastically equivalent to the weighted k-face $Z_0^{(k)}$.

The representation (23) allows us to extend the inequalities (20) to the weighted k-face. In fact, for the vertex number f_0 it yields

$$\mathbb{E}f_0\big(Z_0^{(k)}\big) = \int_{G(d,k)} \mathbb{E}f_0(Z_0 \cap L)\,\widehat{\mathbb{Q}}_{d-k}(\mathrm{d}L).$$

For fixed $L \in G(d,k)$, the polytope $Z_0 \cap L$ appearing in the integrand can also be obtained as follows. If we intersect each hyperplane of $\widehat{X}$ with L, we obtain the section process $\widehat{X} \cap L$ (see [34, sect. 4.4]), which is a stationary Poisson hyperplane process with respect to the space L. The set $Z_0 \cap L$ is the zero cell of the tessellation of L induced by $\widehat{X} \cap L$. The associated zonoid (in L) of $\widehat{X} \cap L$ is given by $\Pi_{\widehat{X}}|L$, the orthogonal projection of the associated zonoid $\Pi_{\widehat{X}}$ to L; see [34, (4.61)]. Hence, it follows from [34, Th. 10.4.9] (applied in L) that

$$(24) \qquad \mathbb{E}\,f_0\big(Z_0^{(k)}\big) = 2^{-k}k! \int_{G(d,k)} V_k(\Pi_{\widehat{X}}|L)V_k\big((\Pi_{\widehat{X}}|L)^o\big)\,\widehat{\mathbb{Q}}_{d-k}(\mathrm{d}L).$$

Here, $(\Pi_{\widehat{X}}|L)^o$ denotes the polar body of $\Pi_{\widehat{X}}|L$ in the subspace L. Since $\Pi_{\widehat{X}}|L$ is a zonoid, the inequalities

$$(25) \qquad \frac{4^k}{k!} \le V_k(\Pi_{\widehat{X}}|L)V_k\big((\Pi_{\widehat{X}}|L)^o\big) \le \kappa_k^2$$

hold. Equality on the right side (the Blaschke–Santaló inequality) holds if and only if $\Pi_{\widehat{X}}|L$ is an ellipsoid; equality on the left side (Reisner's inequality) holds if and only if $\Pi_{\widehat{X}}|L$ is a parallelepiped. From (24) and (25) we obtain the inequalities

$$(26) \qquad 2^k \le \mathbb{E}\,f_0\big(Z_0^{(k)}\big) \le 2^{-k}k!\kappa_k^2$$

for $k \in \{2,\dots,d-1\}$. Equality on the left side holds if and only if $\widehat{X}$ is a parallel mosaic. Equality on the right side holds if $\widehat{X}$ is affinely isotropic, but we don't know whether this is the only case. For the characterization of the equality case on the left side, the following geometric result is needed. Let $K \subset \mathbb{R}^d$ be a 0-symmetric zonoid with generating measure ρ, thus the support function of K is represented by

$$h(K,x) = \int_{S^{d-1}} |\langle x,v\rangle|\,\rho(\mathrm{d}v), \qquad x \in \mathbb{R}^d,$$

and ρ is a finite, even measure on S^{d-1}. Suppose that K has the property that for any $d-k$ linearly independent vectors $v_1,\dots,v_{d-k}$ in the support of the measure ρ, the orthogonal projection of K to $v_1^\perp \cap \cdots \cap v_{d-k}^\perp$ is a parallelepiped. Then K is itself a parallelepiped. This can be proved. The analogous assertion, with 'parallelepiped' replaced by 'ellipsoid', would settle the equality case on the right side of (26), but this has not been proved so far.

The inequalities (26) have been extended in [33]. Instead of weighting the typical face by the volume, we can use for weighting the total j-dimensional volume of its j-faces. Thus, for a polytope P and for $0 \le j \le \dim P$ we put

$$L_j(P) := \sum_{F \in \mathcal{F}_j(P)} V_j(F),$$

and we define the L_j-**weighted typical** k-**face** $Z_{k,j}$ of the random mosaic X as the random polytope with distribution given by

$$\mathbb{P}\{Z_{k,j} \in A\} = \frac{1}{\mathbb{E}\, L_j(Z^{(k)})} \mathbb{E}\big[\mathbf{1}_A(Z^{(k)}) L_j(Z^{(k)})\big]$$

for $A \in \mathcal{B}(\mathcal{P})$. Then the inequalities

$$2^k \le \mathbb{E} f_0(Z_{k,j}) \le 2^{j-2k} \sum_{i=0}^{k-j} 2^{2i} \binom{k-j}{i} (k-i)! \kappa_{k-i}^2$$

are valid. The equality cases are the same as for (26).

5. LARGE CELLS IN POISSON HYPERPLANE MOSAICS

As we have seen, isoperimetric problems for sizes of average cells or faces of random mosaics can be stated, and in some cases solved, if one asks for expected values of functionals measuring the size. To ask just for the shape of average cells with extremal sizes is not a meaningful question, since such extrema will not be attained. Surprisingly, however, such questions can make perfect sense if one asks for the asymptotic shape of average cells under the condition that their size (in some sense) is large. The origin of such questions is a conjecture ventured by D. G. Kendall (in the 1940s, and later popularized by him in the foreword to the first edition of [35], which appeared in 1987). Kendall considered the zero cell Z_0 of a stationary and isotropic Poisson line process in the plane and conjectured that the conditional law for its shape, under the condition of given area, converges weakly, as the area tends to infinity, to the degenerate law concentrated at the circular shape. This has been verified in various extended versions and in general dimensions, beginning with solutions and analogs in the planar case by Kovalenko [13, 14, 15] and Miles [28]. In this and the two subsequent sections we describe the main results that have been obtained on generalized versions of Kendall's problem.

The first higher-dimensional version of problems of this type was studied by Mecke and Osburg [24]. For the special case of a stationary Poisson hyperplane mosaic with spherical directional distribution concentrated (with equal masses) in $\{\pm u_1, \ldots, \pm u_d\}$, where $(u_1, \ldots, u_d)$ is an orthonormal basis of $\mathbb{R}^d$, they obtained that zero cells of large volume approximate cubical shape. This was made precise in several ways, involving monotonicity, stochastic order, and limit relations. The results were transferred to affine images of such hyperplane tessellations.

Hyperplane tessellations with more general directional distributions were investigated in [3]–[8], as we now explain. First, we consider again a stationary Poisson hyperplane process $\widehat{X}$ in $\mathbb{R}^d$, with intensity $\widehat{\gamma}$ and spherical directional distribution $\widehat{\varphi}$. We ask for the shape of its zero cell Z_0, under the condition that the zero cell is large. Here 'large' can be interpreted in terms of volume, or diameter, or many other reasonable functionals. We can put this axiomatically. By a **size functional** we understand any continuous real function $\Sigma \not\equiv 0$ on the space $\mathcal{K}_{(0)}$ of convex bodies in $\mathbb{R}^d$ containing 0 which is increasing under set inclusion and is homogeneous of some degree $k > 0$. For many such functionals, there are precise asymptotic results about the shape of Z_0 under the condition that $\Sigma(Z_0)$ is large. It turns out that such asymptotic shapes are determined by an isoperimetric inequality that connects the size functional with the **hitting functional** of the process $\widehat{X}$. This is the function Φ defined by

$$\Phi(K) := \frac{1}{2\widehat{\gamma}} \, \mathbb{E} \, \mathrm{card} \, \{H \in \widehat{X} : H \cap K \neq 0\}, \qquad K \in \mathcal{K}.$$

Explicitly, it is given by

$$\Phi(K) = \int_{S^{d-1}} h(K, u) \, \widehat{\varphi}(\mathrm{d}u).$$

Note that Φ is continuous and homogeneous of degree one. It follows from continuity and compactness that among all convex bodies $K \in \mathcal{K}_{(0)}$ with a given positive value of $\Phi(K)$, there exist convex bodies for which Σ becomes maximal. Hence, by homogeneity there is an isoperimetric-type inequality

$$(27) \qquad \Phi(K) \geq \tau\Sigma(K)^{1/k}$$

holding for all $K \in \mathcal{K}_{(0)}$, and with equality holding for some convex bodies; these are called **extremal bodies.** It turns out (in many cases) that the shapes of Σ-large zero cells approximate the shapes of extremal bodies. Since, however, these shapes can in general not be attained, we need to measure how close a shape comes to that of an extremal body. Therefore, we define, for given Σ and Φ, a **deviation functional** as any function ϑ on the space $\{K \in \mathcal{K}_{(0)} : \Sigma(K) > 0\}$ which is continuous, nonnegative, homogeneous of degree zero, and satisfies $\vartheta(K) = 0$ if and only if K is an extremal body. The existence follows by continuity, and for the same reason there exist continuous functions $f : \mathbb{R}^+ \to \mathbb{R}^+$ with $f(0) = 0$ and $f(x) > 0$ for $x > 0$ such that the following stability version of the inequality (27) holds:

$$(28) \qquad \Phi(K) \geq \bigl(1 + f(\varepsilon)\bigr)\tau\Sigma(K)^{1/k} \qquad \text{whenever} \quad \vartheta(K) \geq \varepsilon.$$

For the geometrically most interesting concrete size functionals Σ, simple explicit functions ϑ and f can be provided.

We can now estimate $\mathbb{P}\{\vartheta(Z_0) \geq \varepsilon \mid \Sigma(Z_0) \geq a\}$, the conditional probability that the zero cell deviates in shape by at least ε from an extremal body, under the condition that its Σ-size it at least a. We assume that Σ, ϑ, f with the properties listed above are given.

Theorem. *For given $\varepsilon > 0$ and $a > 0$, there exist positive constants c (depending on $\widehat{X}$, Σ, f, ε) and c_0 such that*

$$(29) \qquad \mathbb{P}\{\vartheta(Z_0) \geq \varepsilon \mid \Sigma(Z_0) \geq a\} \leq c\exp\bigl(-c_0 f(\varepsilon)\widehat{\gamma}a^{1/k}\bigr).$$

We can also condition by $\Sigma(Z_0) = a$, instead of $\Sigma(Z_0) \geq a$. Namely, the random polytope Z_0 takes its values in $\mathcal{K}$. Since this is a Polish space, the regular conditional probability distribution of Z_0 with respect to $\Sigma(Z_0)$ exists. Similarly to (29), we have

$$(30) \qquad \mathbb{P}\{\vartheta(Z_0) \geq \varepsilon \mid \Sigma(Z_0) = a\} \leq c\exp\bigl(-c_0 f(\varepsilon)\widehat{\gamma}a^{1/k}\bigr).$$

The role of the isoperimetric inequality (27) and its strengthening (28) can be explained as follows. By definition,

$$(31) \qquad \mathbb{P}\{\vartheta(Z_0) \geq \varepsilon \mid \Sigma(Z_0) \geq a\} = \frac{\mathbb{P}\{\vartheta(Z_0) \geq \varepsilon, \Sigma(Z_0) \geq a\}}{\mathbb{P}\{\Sigma(Z_0) \geq a\}},$$

and this has to be estimated from above. Let $\varepsilon > 0$ and $a > 0$ be given. Let B be an extremal body of (27). Since dilates of B are also extremal bodies and Σ is homogeneous of degree $k > 0$, we can assume that $\Sigma(B) = a$ and hence that $\Phi(B) = \tau \Sigma(B)^{1/k} = \tau a^{1/k}$. If $H \cap B = \emptyset$ for all $H \in \widehat{X}$, then $B \subset Z_0$ (since $0 \in B$), hence $\Sigma(Z_0) \geq \Sigma(B) = a$. Therefore, the denominator of (31) can be estimated by

$$\mathbb{P}\{\Sigma(Z_0) \geq a\} \geq \mathbb{P}\{\operatorname{card}\{H \in \widehat{X} : H \cap B \neq \emptyset\} = 0\}$$

$$= \exp\left(-\Phi(B)2\widehat{\gamma}\right) = \exp\left(-2\tau\widehat{\gamma}a^{1/k}\right).$$

The estimation of the numerator we explain only heuristically. Let K be a convex body satisfying

$$\vartheta(K) \geq \varepsilon, \quad \Sigma(K) \geq a.$$

Then, using (28) instead of (27),

$$\mathbb{P}\{\operatorname{card}\{H \in \widehat{X} : H \cap K \neq \emptyset\} = 0\} = \exp\left(-\Phi(K)2\widehat{\gamma}\right)$$

$$\leq \exp\left(-\left(1 + f(\varepsilon)\right)2\tau\widehat{\gamma}a^{1/k}\right).$$

An only slightly weaker inequality can be proved if the deterministic convex body K is replaced by the random polytope Z_0, namely

$$(32) \qquad \mathbb{P}\{\vartheta(Z_0) \geq \varepsilon, \, \Sigma(Z_0) \geq a\} \leq c \exp\left(-\left(1 + c_1 f(\varepsilon)\right)2\tau\widehat{\gamma}a^{1/k}\right)$$

with positive constants c, c_1. Division now gives

$$\mathbb{P}\{\vartheta(Z_0) \geq \varepsilon \mid \Sigma(Z_0) \geq a\} \leq c \exp\left(-c_1 f(\varepsilon)2\tau\widehat{\gamma}a^{1/k}\right).$$

The bulk of the work, of course, consists in the proof of the estimate (32).

The first concrete example is the case where the size functional Σ is given by the volume V_d. Denoting by

$$B = B(\widehat{X})$$

the Blaschke body of the hyperplane process $\widehat{X}$, we can express the hitting functional as a mixed volume. Since $\widehat{\varphi}$ is now the surface area measure of B, we have $\Phi(K) = dV(K, B, \ldots, B)$. Minkowski's inequality

$$(33) \qquad V(K, B, \ldots, B) \geq V_d(B)^{1-1/d} V_d(K)^{1/d}$$

is the inequality (27) in this case (with $\tau = dV_d(B)^{1-1/d}$), and it is well known that its extremal bodies are all homothetic to B. For a deviation functional we can choose

$$\vartheta(K) := \inf \left\{ \beta/\alpha \,:\, \alpha, \beta > 0, \; \alpha B \subset K + t \subset \beta B \text{ for some } t \in \mathbb{R}^d \right\}.$$

A stability version (28) of inequality (33) is known with $f(x) = \text{const} \cdot x^{d+1}$; hence we obtain

$$(34) \qquad \mathbb{P}\left\{ \vartheta(Z_0) \geq \varepsilon \mid V_d(Z_0) \geq a \right\} \leq c \exp\left(-c_0 \varepsilon^{d+1} \widehat{\gamma} a^{1/d} \right).$$

Thus, *the Blaschke body provides the shape of zero cells of large volume.* We can also deduce the existence of a (degenerate) limit distribution for the shape. By $\mathcal{S}_\mathsf{H}$ we denote the quotient space of $\mathcal{K}$ with respect to the equivalence relation given by homothety. The equivalence class of a convex body K is denoted by $s_\mathsf{H}(K)$ and is called the homothetic shape of K. We define the conditional law of the homothetic shape of Z_0, given the lower bound $a > 0$ for the volume, as the probability measure μ_a on $\mathcal{S}_\mathsf{H}$ with

$$\mu_a(A) := \mathbb{P}\left\{ s_\mathsf{H}(Z_0) \in A \mid V_d(Z_0) \geq a \right\}$$

for $A \in \mathcal{B}(\mathcal{S}_\mathsf{H})$. Then

$$(35) \qquad \lim_{a \to \infty} \mu_a = \delta_{s_\mathsf{H}(B)} \qquad \text{weakly},$$

with $\delta_{s_\mathsf{H}(B)}$ denoting the Dirac measure concentrated at $s_\mathsf{H}(B)$.

If $\widehat{X}$ is isotropic, then the Blaschke body $B(\widehat{X})$ is a ball, hence the asymptotic shape is given by the class of balls, as in Kendall's original problem.

The inequality (34) was proved in [3]. It was also remarked there that this inequality remains true if the zero cell is replaced by the typical cell. This is due to the fact that the distribution of the zero cell is, up to translations, the volume-weighted distribution of the typical cell.

The setting for extended Kendall problems was considerably broadened in [7]. Besides admitting general size functionals Σ, a wider class of Poisson hyperplane processes $\widehat{X}$ was considered, namely those with an intensity measure $\widehat{\Theta}$ of the form

$$(36) \qquad \widehat{\Theta}(A) = 2\widehat{\gamma} \int_{S^{d-1}} \int_0^\infty \mathbf{1}_A\left(H(u,t) \right) t^{r-1} \, dt \, \widehat{\varphi}(du)$$

for $A \in \mathcal{B}\big(A(d, d-1) \big)$. Here $r \geq 1$, $\widehat{\gamma} > 0$, and $\widehat{\varphi}$ is a finite (not necessarily even) measure on the unit sphere, not concentrated on a closed hemisphere. The case of a stationary hyperplane process $\widehat{X}$ is obtained if $r = 1$ and $\widehat{\varphi}$ is even. As explained in the next section, the case of $r = d$ and rotation invariant $\widehat{\varphi}$ allows one to treat the typical cell of a stationary Poisson–Voronoi tessellation in a similar way. In [7], also general versions of (30) and (35) were obtained. For stationary Poisson hyperplane tessellations, the following special size functionals were treated. The different cases may require different notions of shape, since the extremal bodies of the corresponding crucial inequality (27) may be equivalent to a fixed convex body with different meanings of 'equivalent', for example, homothetic, or equivalent by positive dilatation, or similar. If the size is measured by the diameter, then the limit shape of the zero cell is provided by the class of segments. If size is measured by thickness (minimal width), then the whole class of bodies of constant width can be considered as the asymptotic shape of the zero cell. Further size functionals studied in [7] are the inradius, the centered inradius, and the width in a given direction.

For the typical cell instead of the zero cell and for size functionals different from the volume, no such simple transfer argument as mentioned above is possible. For stationary, isotropic Poisson hyperplane tessellations, results on the asymptotic shapes of large typical cells were obtained in [8]. If the size is measured by the kth intrinsic volume, $k \in \{2, \dots, d\}$, asymptotic shapes are balls. (For the zero cell, the same was proved in [4].) For the diameter as size functional, one obtains segments as asymptotic shapes. The proof makes use of a special representation of the distribution of the typical cell with respect to the highest vertex as center function (see [34, Theorem 10.4.7]).

Extensions of some of these results to typical k-faces ($k \in \{2, \dots, d-1\}$) of stationary Poisson hyperplane tessellations were investigated by Hug and Schneider [9]. This requires the additional condition that the direction of the face lies in a sufficiently small neighbourhood of a given direction. This was continued in [11], where the regular conditional distributions of, possibly weighted, typical faces under the hypothesis of given direction were studied, with a particular view to the shape of large faces.

6. LARGE CELLS IN POISSON–VORONOI MOSAICS

A much studied class of tessellations are the Voronoi or Dirichlet mosaics. If $S \subset \mathbb{R}$ is a nonempty, locally finite set and $x \in S$, then the **Voronoi cell** $C(x, S)$ of x (with respect to S) is defined as the set of all points in $\mathbb{R}^d$ for which x is a nearest point in S, thus

$$C(x, S) = \big\{ y \in \mathbb{R}^d \; : \; \|y - x\| \leq \|y - s\| \text{ for all } s \in S \big\}.$$

The point x is called the **nucleus** of the Voronoi cell $C(x, S)$.

Now let $\widetilde{X}$ be a stationary Poisson point process in $\mathbb{R}^d$ with intensity $\widetilde{\gamma} > 0$ (and, as always assumed, locally finite intensity measure). Then the collection $\big\{ C(x, \widetilde{X}) \; : \; x \in \widetilde{X} \big\}$ is a stationary random mosaic. It is called the **Poisson–Voronoi mosaic** induced by $\widetilde{X}$. We denote it by X. Since the intensity measure of $\widetilde{X}$ is locally finite and translation invariant, it is a constant multiple of Lebesgue measure and hence is invariant under rotations. Therefore, the random mosaic X is isotropic.

We are interested in the asymptotic shape of the typical cell Z of X, under the condition that it is large. It follows from Slivnyak's theorem on Poisson processes that the typical cell Z is stochastically equivalent to $C\big(0, \widetilde{X} \cup \{0\}\big)$, the Voronoi cell of 0 for the point process $\widetilde{X} \cup \{0\}$, which is obtained from $\widetilde{X}$ by adding a point at 0. For $x \in \mathbb{R}^d$, let $H(x)$ be the mid-hyperplane of 0 and x, that is, the set of all points having equal distance from 0 and x. For $x \neq 0$, let $H^-(x)$ be the closed halfspace bounded by $H(x)$ that contains 0. By the definition of Voronoi cells, we have

$$C\big(0, \widetilde{X} \cup \{0\}\big) = \bigcap_{x \in \widetilde{X}} H^-(x);$$

hence $C\big(0, \widetilde{X} \cup \{0\}\big)$ is the zero cell, Z_0, of the mosaic induced by the hyperplane process $\widehat{X} := \big\{ H(x) \; : \; x \in \widetilde{X} \big\}$. This is a (non-stationary) Poisson process, and its intensity measure $\widehat{\Theta}$ can be represented by

$$(37) \qquad \widehat{\Theta}(A) - 2^d \widetilde{\gamma} \int_{S^{d-1}} \int_0^\infty \mathbf{1}_A\big(H(u, t) \big) \, t^{d-1} \, \mathrm{d}t \, \sigma(\mathrm{d}u)$$

for $A \in \mathcal{B}\big(A(d, d{-}1) \big)$, where σ denotes spherical Lebesgue measure. This is of type (36), with $\widehat{\gamma} = 2^d d \kappa_d \widetilde{\gamma}$. Therefore, the methods described in [7] can be used to obtain results on asymptotic shapes for Z_0 (which is stochastically equivalent to the typical cell Z of X), under the condition that it has large size. In [4], the following results were obtained.

For $K \in \mathcal{K}_{(0)}$ with $K \neq \{0\}$, we measure the deviation from a centered ball by the function

$$\vartheta(K) := \frac{R_0 - r_0}{R_0 + r_0},$$

where R_0 is the radius of the smallest ball with center 0 containing K and r_0 is the radius of the largest ball with center 0 contained in K. Let $k \in \{1, \dots, d\}$, let V_k denote the kth intrinsic volume. If $\varepsilon \in (0, 1)$ and if $a > 0$ is sufficiently large, then

$$\mathbb{P}\{\vartheta(Z_0) \geq \varepsilon \mid V_k(Z_0) \geq a\} \leq c \exp\left(-c_0 \varepsilon^{(d+3)/2} \widetilde{\gamma} a^{d/k}\right),$$

where the constant c depends on d and ε, while c_0 depends only on d. A similar result was obtained in [4], with the size of Z_0 measured by the centered inradius.

The general methods of [7] allow also the treatment of size functionals where the resulting asymptotic shapes are of lower dimension. For example, if the size of Z_0 is measured by the largest distance of a vertex from the nucleus, then the limit shape is given by the class of all segments with one endpoint at the origin.

Lower-dimensional typical faces of stationary Poisson–Voronoi tessellations were studied by Hug and Schneider [10]. Under the condition of large inradius, the relative boundary of such a typical face lies, with high probability, in a narrow spherical annulus.

7. Large Cells in Poisson–Delaunay Mosaics

As in the previous section, let $\widetilde{X}$ be a stationary Poisson point process in $\mathbb{R}^d$ with intensity $\widetilde{\gamma} > 0$. Together with the Voronoi mosaic induced by $\widetilde{X}$ comes a certain dual of it, the Delaunay mosaic. We recall here its definition without recourse to the Voronoi mosaic. With probability one, any $d + 1$ points of $\widetilde{X}$ lie on a unique sphere. If the open ball bounded by this sphere does not contain a point of $\widetilde{X}$, then the convex hull of the $d + 1$ points is called a **cell**. The collection of all cells obtained in this way is a tessellation of $\mathbb{R}^d$ into simplices. In this way, a stationary random mosaic Y is defined, which is called the **Poisson–Delaunay mosaic** induced by $\widetilde{X}$.

For a d-dimensional simplex S, there is a unique sphere through its vertices, and we denote by $z(S)$ the center of this sphere, also called the **circumcenter** of S, and by $R(S)$ the radius of the sphere. Let Z be the typical cell of Y with respect to the center function z. Then Z is a d-dimensional random simplex with circumcenter 0. For its distribution, there is an explicit integral representation due to Miles; see [34, Theorem 10.4.4]. This was used in [5], [6] to obtain results on asymptotic shapes of large typical cells. We describe briefly a general result obtained in [6]. Let Δ_0 be the subspace of $\mathcal{P}$ consisting of all d-dimensional simplices with circumcenter 0. By a **size functional** we understand now a positive, continuous function Σ on Δ_0 which is homogeneous of some degree $k > 0$ and which, if restricted to the simplices S with $R(S) = 1$, has the property that Σ attains a maximum (denoted by τ) and that $V_d/\Sigma^{1/k}$ is bounded. By homogeneity, we then have

$$(38) \qquad \Sigma(S) \leq \tau R(S)^{1/k}$$

for all $S \in \Delta_0$. Every simplex S for which (38) holds with equality is called an **extremal simplex**. For given Σ, a **deviation functional** is defined as a nonnegative, continuous function ϑ on Δ_0 which is homogeneous of degree zero and satisfies $\vartheta(S) = 0$ if and only if S is an extremal simplex. For given Σ and ϑ, a **stability function** is a continuous function $f : [0,1] \to [0,1]$ with the properties that $f(0) = 0$, $f(x) > 0$ for $x > 0$ and

$$(39) \qquad \Sigma(S) \leq \left(1 - f(\varepsilon)\right) \tau R(S)^{1/k} \qquad \text{whenever} \quad \vartheta(S) \geq \varepsilon.$$

Now suppose that Σ, ϑ, f with these properties are given. If $\varepsilon \in (0,1)$ and if $a > 0$ is sufficiently large, then

$$(40) \qquad \mathbb{P}\left\{ \vartheta(Z) \geq \varepsilon \mid \Sigma(Z) \geq a \right\} \leq c \exp\left(- c_0 f(\varepsilon) \tilde{\gamma} a^{d/k} \right),$$

with constants c, c_0 independent of a.

For concrete size functionals Σ, this yields results on asymptotic shapes if the extremal simplices of the isoperimetric inequality (38) can be determined. This can be surprisingly difficult; for example, it is still not known whether the simplices of extremal mean width inscribed to the unit sphere are regular. Other cases are simpler. For $\Sigma = V_d$, the volume, it is easy to see that all maximal simplices are regular, and in [5] a stability result of type (39), with a stability function of optimal order, was obtained, for the following deviation functional. For a simplex $S \in \Delta_0$, let $\vartheta(S)$ be the smallest number η for which there exists a regular simplex $T \in \Delta_0$ with

$R(S) = 1$ such that for each vertex p of $R(S)^{-1}S$ there is a vertex q of T with $\|p - q\| \leq \eta$, and conversely. The version of (40) proved in [5] reads

$$\mathbb{P}\{\vartheta(Z) \geq \varepsilon \mid V_d(Z) \geq a\} \leq c \exp\left(-c_0 \varepsilon^2 \tilde{\gamma} a\right),$$

where c depends only on d and ε and c_0 depends on d.

Further, in [6] the following concrete cases of the general result (40) were treated. For each of the size functionals: surface area, inradius, minimal width, the asymptotic shape of the typical cell is that of the regular simplices. For the case of the surface area, it follows from a more general result of Tanner [36] that the extremal simplices of (38) are the regular ones. The case of the inradius is easier, and for the minimal width, a result of Alexander [1] was used. If the diameter is chosen as size functional, then the asymptotic shapes of large typical cells are provided by the diametral simplices. A simplex S is called diametral if a longest edge of S is a diameter of the circumsphere of S. In the plane, these are the right-angled triangles.

8. General Mosaics

For random mosaics that are not of the special types described in the previous sections, only very few extremal results have been obtained. In the plane, we can mention inequalities due to Mecke [17], which are parallel to results on deterministic mosaics by L. Fejes Tóth. Let X be a stationary random mosaic in $\mathbb{R}^2$. Let Λ denote the mean total length of the edges of X per unit area; with the notation of [34, sect. 10.1], this is $\Lambda = d_1^{(1)} = \overline{V}_1(X^{(1)})$. Further, let N ($= n_{20} = n_{21}$ in the notation of [34]) be the mean number of edges of the typical cell of X. Mecke has proved the following two theorems.

Suppose that all cells of X have the same area F. Then

$$\Lambda^2 \geq \frac{N}{F} \tan \frac{\pi}{N}.$$

Equality holds if and only if $N \in \{3, 4, 6\}$ and all cells of X are regular N-gons.

Suppose that all cells of X have the same perimeter U. Then

$$\Lambda \geq \frac{2N}{U} \tan \frac{\pi}{N}.$$

Equality holds if and only if $N \in \{3, 4, 6\}$ and all cells of X are regular N-gons.

For general mosaics in higher dimensions, we know only of one result exhibiting an extremal property. Let X be a stationary random mosaic in $\mathbb{R}^d$. An inequality of Wieacker [37] connects the volume of the Blaschke body $B(X)$ with the expected volume of the zero cell Z_0, namely

$$(41) \qquad V_d\big(B(X)\big)^{d-1} \mathbb{E} V_d(Z_0) \geq 1.$$

Since this result is not mentioned in [34], we present here the proof in the style of [34]. First we show that

$$(42) \quad \int_{S^{d-1}} f(u)\, S_{d-1}\big(B(X), \mathrm{d}u\big) = \mathbb{E}\left(V_d(Z_0)^{-1} \int_{S^{d-1}} f(u)\, S_{d-1}(Z_0, \mathrm{d}u) \right)$$

for any nonnegative measurable function f on S^{d-1}. In fact, writing

$$g(K) := \int_{S^{d-1}} f(u)\, S_{d-1}(K, \mathrm{d}u) \qquad \text{for } K \in \mathcal{K},$$

we get from [34, Th. 10.4.1] and Fubini's theorem for kernels (with $\mathbb{Q}^{(d)}$, $\gamma^{(d)}$, Z as in Section 2)

$$\mathbb{E}\left(V_d(Z_0)^{-1} \int_{S^{d-1}} f(u)\, S_{d-1}(Z_0, \mathrm{d}u) \right) = \mathbb{E}\big(V_d(Z_0)^{-1} g(Z_0) \big) = \gamma^{(d)} \mathbb{E}\, g(Z)$$

$$= \gamma^{(d)} \int_{\mathcal{K}_0} g(K)\, \mathbb{Q}^{(d)}(\mathrm{d}K) = \gamma^{(d)} \int_{\mathcal{K}_0} \int_{S^{d-1}} f(u)\, S_{d-1}(K, \mathrm{d}u)\, \mathbb{Q}^{(d)}(\mathrm{d}K)$$

$$= \int_{S^{d-1}} f(u)\, S_{d-1}\big(B(X), \mathrm{d}u\big),$$

which proves (42). Now we use (24), together with Minkowski's inequality for mixed volumes and Jensen's inequality for concave functions, to obtain

$$V_d\big(B(X)\big) = \frac{1}{d} \int_{S^{d-1}} h\big(B(X), u\big)\, S_{d-1}\big(B(X), \mathrm{d}u\big)$$

$$= \frac{1}{d} \mathbb{E}\left(V_d(Z_0)^{-1} \int_{S^{d-1}} h\big(B(X), u\big)\, S_{d-1}(Z_0, \mathrm{d}u) \right)$$

$$= \mathbb{E}\big(V_d(Z_0)^{-1} V\big(B(X), Z_0, \ldots, Z_0\big) \big)$$

$$\geq V_d\big(B(X)\big)^{\frac{1}{d}}\mathbb{E}\big(V_d(Z_0)^{-\frac{1}{d}}\big)$$

$$\geq V_d\big(B(X)\big)^{\frac{1}{d}}\big(\mathbb{E}V_d(Z_0)\big)^{-\frac{1}{d}}.$$

This gives (41). Equality holds if and only if there exists a convex body K such that a.s. every realization of X consists of translates of K (so the randomness affects only the translations).

References

[1] Alexander, R., The width and diameter of a simplex, *Geom. Dedicata,* **6** (1977), 87–94.

[2] Fejes Tóth, L., *Reguläre Figuren,* Akadémiai Kiadó, Budapest and Teubner, Leipzig (1965).

[3] Hug, D., Reitzner, M. and Schneider, R., The limit shape of the zero cell in a stationary Poisson hyperplane tessellation, *Ann. Probab.,* **32** (2004), 1140–1167.

[4] Hug, D., Reitzner, M. and Schneider, R., Large Poisson–Voronoi cells and Crofton cells, *Adv. Appl. Prob.,* **36** (2004), 667–690.

[5] Hug, D. and Schneider, R., Large cells in Poisson–Delaunay tessellations, *Discrete Comput. Geom.,* **31** (2004), 503–514.

[6] Hug, D. and Schneider, R., Large typical cells in Poisson–Delaunay mosaics, *Rev. Roumaine Math. Pures Appl.,* **50** (2005), 657–670.

[7] Hug, D. and Schneider, R., Asymptotic shapes of large cells in random tessellations, *Geom. Funct. Anal.,* **17** (2007), 156–191.

[8] Hug, D. and Schneider, R., Typical cells in Poisson hyperplane tessellations, *Discrete Comput. Geom.,* **38** (2007), 305–319.

[9] Hug, D. and Schneider, R., Large faces in Poisson hyperplane mosaics, *Ann. Probab.,* **38** (2010), 1320–1344.

[10] Hug, D. and Schneider, R., Faces of Poisson–Voronoi mosaics, *Probab. Theory Related Fields,* **151** (2011), 125–151.

[11] Hug, D. and Schneider, R., Faces with given directions in anisotropic Poisson hyperplane mosaics, *Adv. Appl. Prob.,* **43** (2011), 308–321.

[12] Hug, D. and Schneider, R., Reverse inequalities for zonoids and their application, *Adv. Math.,* **228** (2011), 2634–2646.

[13] Kovalenko, I. N., A proof of a conjecture of David Kendall on the shape of random polygons of large area, (Russian) *Kibernet. Sistem. Anal.* (1997), 3–10, 187; Engl. transl., *Cybernet. Systems Anal.*, **33** (1997), 461–467.

[14] Kovalenko, I. N., An extension of a conjecture of D. G. Kendall concerning shapes of random polygons to Poisson Voronoï cells, in: Engel, P. *et al.* (eds.) *Voronoï's impact on modern science,* Book I. Transl. from the Ukrainian, Kyiv: Institute of Mathematics. Proc. Inst. Math. Natl. Acad. Sci. Ukr., *Math. Appl.,* **212** (1998), 266–274.

[15] Kovalenko, I. N., A simplified proof of a conjecture of D. G. Kendall concerning shapes of random polygons, *J. Appl. Math. Stochastic Anal.,* **12** (1999), 301–310.

[16] Mecke, J., Inequalities for intersection densities of superpositions of stationary Poisson hyperplane processes, in: Jensen, E. B., Gundersen, H. J. G. (eds.) *Proc. Second Int. Workshop Stereology, Stochastic Geometry,* pp. 115–124, Aarhus (1983).

[17] Mecke, J., Isoperimetric properties of stationary random mosaics, *Math. Nachr.,* **117** (1984), 75–82.

[18] Mecke, J., On some inequalities for Poisson networks, *Math. Nachr.,* **128** (1986), 81–86.

[19] Mecke, J., An extremal property of random flats, *J. Microsc.,* **151** (1988), 205–209.

[20] Mecke, J., On the intersection density of flat processes, *Math. Nachr.,* **151** (1991), 69–74.

[21] Mecke, J., Inequalities for the anisotropic Poisson polytope, *Adv. Appl. Prob.,* **27** (1995), 56–62.

[22] Mecke, J., Inequalities for mixed stationary Poisson hyperplane tessellations, *Adv. Appl. Prob.,* **30** (1998), 921–928.

[23] Mecke, J., On the relationship between the 0-cell and the typical cell of a stationary random tessellation, *Pattern Recognition,* **32** (1999), 1645–1648.

[24] Mecke, J. and Osburg, I., On the shape of large Crofton parallelotopes, *Math. Notae,* **41** (2001/02) (2003), 149–157.

[25] Mecke, J., Schneider, R., Stoyan, D. and Weil, W., *Stochastische Geometrie,* DMV-Seminar **16**, Birkhäuser, Basel (1990).

[26] Mecke, J. and Thomas, C., On an extreme value problem for flat processes, *Commun. Stat., Stochastic Models (2),* **2** (1986), 273–280.

[27] Miles, R. E., Random polytopes: the generalisation to n dimensions of the intervals of a Poisson process, Ph.D. Thesis, Cambridge University (1961).

[28] Miles, R. E., A heuristic proof of a long-standing conjecture of D. G. Kendall concerning the shapes of certain large random polygons, *Adv. Appl. Prob.,* **27** (1995), 397–417.

[29] Okabe, A., Boots, B., Sugihara, K. and Chiu, S. N., *Spatial Tessellations; Concepts and Applications of Voronoi Diagrams,* 2nd ed., Wiley, Chichester (2000).

[30] Schneider, R., *Convex Bodies: the Brunn–Minkowski Theory,* Cambridge University Press, Cambridge (1993).

[31] Schneider, R., Nonstationary Poisson hyperplanes and their induced tessellations, *Adv. Appl. Prob.,* **35** (2003), 139–158.

[32] Schneider, R., Weighted faces of Poisson hyperplane tessellations, *Adv. Appl. Prob.,* **41** (2009), 682–694.

[33] Schneider, R., Vertex numbers of weighted faces in Poisson hyperplane mosaics, *Discrete Comput. Geom.,* **44**, 599–607 (2010).

[34] Schneider, R. and Weil, W., *Stochastic and Integral Geometry,* Springer, Berlin, Heidelberg (2008).

[35] Stoyan, D., Kendall, W. S. and Mecke, J., *Stochastic Geometry and Its Applications,* 2nd ed., Wiley, Chichester (1995).

[36] Tanner, R. M., Some content maximizing properties of the regular simplex, *Pacific J. Math.,* **52** (1974), 611–616.

[37] Wieacker, J. A., Geometric inequalities for random surfaces, *Math. Nachr.,* **142** (1989), 73–106.

Rolf Schneider

Mathematisches Institut,
Albert-Ludwigs-Universität,
Eckerstr. 1,
79104 Freiburg i.Br.,
Germany

e-mail: rolf.schneider@math.uni-freiburg.de

BOLYAI SOCIETY
MATHEMATICAL STUDIES, 24

Geometry –
Intuitive, Discrete, and Convex
pp. 331–389.

Conflict-Free Coloring and its Applications

SHAKHAR SMORODINSKY

Let $H = (V, E)$ be a hypergraph. A *conflict-free* coloring of H is an assignment of colors to V such that, in each hyperedge $e \in E$, there is at least one uniquely-colored vertex. This notion is an extension of the classical graph coloring. Such colorings arise in the context of frequency assignment to cellular antennae, in battery consumption aspects of sensor networks, in RFID protocols, and several other fields. Conflict-free coloring has been the focus of many recent research papers. In this paper, we survey this notion and its combinatorial and algorithmic aspects.

1. Introduction

1.1. Notations and Definitions

In order to introduce the main notion of this paper, we start with several basic definitions: Unless otherwise stated, the term log denotes the base 2 logarithm.

A *hypergraph* is a pair $(V, \mathcal{E})$ where V is a set and $\mathcal{E}$ is a collection of subsets of V. The elements of V are called *vertices* and the elements of $\mathcal{E}$ are called *hyperedges*. When all hyperedges in $\mathcal{E}$ contain exactly two elements of V then the pair $(V, \mathcal{E})$ is a *simple graph*. For a subset $V' \subset V$ refer to the hypergraph $H(V') = (V', \{S \cap V' | S \in \mathcal{E}\})$ as the *sub-hypergraph* induced by V'. A k-coloring, for some $k \in \mathbb{N}$, of (the vertices of) H is a function $\varphi : V \to \{1, \ldots, k\}$. Let $H = (V, \mathcal{E})$ be a hypergraph. A k-coloring φ of H is called *proper* or *non-monochromatic* if every hyperedge $e \in \mathcal{E}$ with $|e| \geq 2$ is non-monochromatic. That is, there exists at least two vertices $x, y \in e$

such that $\varphi(x) \neq \varphi(y)$. Let $\chi(H)$ denote the least integer k for which H admits a proper coloring with k colors.

In this paper, we focus on the following colorings which are more restrictive than proper coloring:

Definition 1.1 (Conflict-Free and Unique-Maximum Colorings). Let $H = (V, \mathcal{E})$ be a hypergraph and let $C : V \to \{1, \ldots, k\}$ be some coloring of H. We say that C is a *conflict-free* coloring (*CF-coloring* for short) if every hyperedge $e \in \mathcal{E}$ contains at least one uniquely colored vertex. More formally, for every hyperedge $e \in \mathcal{E}$ there is a vertex $x \in e$ such that $\forall y \in e, y \neq x \Rightarrow C(y) \neq C(x)$. We say that C is a *unique-maximum* coloring (*UM-coloring* for short) if the maximum color in every hyperedge is unique. That is, for every hyperedge $e \in \mathcal{E}$, $\left| e \cap C^{-1}(\max_{v \in e} C(v)) \right| = 1$.

Let $\chi_{\mathrm{cf}}(H)$ (respectively, $\chi_{\mathrm{um}}(H)$) denote the least integer k for which H admits a CF-coloring (respectively, a UM-coloring) with k colors. Obviously, every UM-coloring of a hypergraph H is also a CF-coloring of H, and every CF-coloring of H is also a proper coloring of H. Hence, we have the followng inequalities:

$$\chi(H) \leq \chi_{\mathrm{cf}}(H) \leq \chi_{\mathrm{um}}(H).$$

Notice that for simple graphs, the three notions of coloring (non-monochromatic, CF and UM) coincide. Also, for 3-uniform hypergraphs (i.e., every hyperedge has cardinality 3), the two first notions (non-monochromatic and CF) coincide. However, already for 3-uniform hypergraphs there can be an arbitrarily large gap between $\chi_{\mathrm{cf}}(H)$ and $\chi_{\mathrm{um}}(H)$. Consider, for example, two sets A and B each of cardinality $n > 1$. Let $H = (A \cup B, \mathcal{E})$ where $\mathcal{E}$ consists of all triples of elements e such that $e \cap A \neq \emptyset$ and $e \cap B \neq \emptyset$. In other words $\mathcal{E}$ consists of all triples containing two elements from one of the sets A or B and one element from the other set. It is easily seen that $\chi_{\mathrm{cf}}(H) = 2$ by simply coloring all elements of A with 1 and all elements of B with 2. It is also not hard to verify that $\chi_{\mathrm{um}}(H) \geq n$ (in fact $\chi_{\mathrm{um}}(H) = n+1$). Indeed, let C be a UM-coloring of H. If all elements of A are colored with distinct colors we are done. Otherwise, there exist two elements u, v in A with the same color, say i. We claim that all elements of B are colored with colors greater than i. Assume to the contrary that there is an element $w \in B$ with color $C(w) = j \leq i$. However, in that case the hyperedge $\{u, v, w\}$ does not have the unique-maximum property. Hence all colors of B are distinct for otherwise if there are two vertices w_1, w_2 with the same color, again the hyperedge $\{w_1, w_2, u\}$ does not have the unique-maximum property.

Let us describe a simple yet an important example of a hypergraph H and analyze its chromatic number $\chi(H)$ and its CF-chromatic number $\chi_{cf}(H)$. The vertices of the hypergraph consist of the first n integers $[n] = \{1, \ldots, n\}$. The hyperedge-set is the set of all (non-empty) subsets of $[n]$ consisting of consecutive elements of $[n]$, e.g., $\{2, 3, 4\}$, $\{2\}$, the set $[n]$, etc. We refer to such hypergraphs as *hypergraphs induced by points on the line with respect to intervals* or as the *discrete intervals hypergraph*. Trivially, we have $\chi(H) = 2$. We will prove the following proposition:

Proposition 1.2. $\chi_{cf}(H) = \chi_{um}(H) = \lfloor \log n \rfloor + 1$.

Proof. First we prove that $\chi_{um}(H) \leq \lfloor \log n \rfloor + 1$. Assume without loss of generality that n is of the form $n = 2^k - 1$ for some integer k. If $n < 2^k - 1$ then we can add the vertices $n+1, n+2, \ldots, 2^k - 1$ and this can only increase the UM-chromatic number. In this case we will see that $\chi_{um}(H) \leq k$ and that for $n \geq 2^k$ $\chi_{cf}(H) \geq k + 1$. The proof is by induction on k. For $k = 1$ the claim holds trivially. Assume that the claim holds for some integer k and let $n = 2^{k+1} - 1$. Consider the median vertex 2^k and color it with a unique (maximum color), say $k + 1$, not to be used again. By the induction hypothesis, the set of elements to the right of 2^k, namely the set $\{2^k + 1, 2^k + 2, \ldots, 2^{k+1} - 1\}$ can be colored with k colors, say '1', '2', ..., 'k', so that any of its subsets of consecutive elements has unique maximum color. The same holds for the set of elements to the left of 2^k. We will use the same set of k colors for the right set and the left set (and color the median with the unique color '$k+1$'). It is easily verified that this coloring is indeed a UM-coloring for H. Thus we use a total of $k + 1$ colors and this completes the induction step.

Next, we need to show that for $n \geq 2^k$ we have $\chi_{cf}(H) \geq k + 1$. Again, the proof is by induction on k. The base case $k = 0$ is trivial. For the induction step, let $k > 0$ and put $n = 2^k$. Let C be some CF-coloring of the underlying discrete intervals hypergraph. Consider the hyperedge $[n]$. There must be a uniquely colored vertex in $[n]$. Let x be this vertex. Either to the right of x or to its left we have at least 2^{k-1} vertices. That is, there is a hyperedge $S \subset [n]$ that does not contain x such that $|S| \geq 2^{k-1}$, so, by the induction hypothesis, any CF-coloring for S uses at least k colors. Thus, together with the color of x, C uses at least $k+1$ colors in total. This completes the induction step. ∎

The notion of CF-coloring was first introduced and studied in [47] and [25]. This notion attracted many researchers and has been the focus of many research papers both in the computer science and mathematics communities. Recently, it has been studied also in the infinite settings of the so-called *almost disjoint set systems* by Hajnal et al. [27]. In this survey, we mostly consider hypergraphs that naturally arise in geometry. These come in two types:

- **Hypergraphs induced by regions:** Let $\mathcal{R}$ be a finite collection of regions (i.e., subsets) in $\mathbb{R}^d$, $d \geq 1$. For a point $p \in \mathbb{R}^d$, define $r(p) = \{R \in \mathcal{R} : p \in R\}$. The hypergraph $(\mathcal{R}, \{r(p)\}_{p \in \mathbb{R}^d})$, denoted $H(\mathcal{R})$, is called the *hypergraph* induced *by* $\mathcal{R}$. Since $\mathcal{R}$ is finite, so is the power set $2^{\mathcal{R}}$. This implies that the hypergraph $H(\mathcal{R})$ is finite as well.

- **Hypergraphs induced by points with respect to regions:** Let $P \subset \mathbb{R}^d$ and let $\mathcal{R}$ be a family of regions in $\mathbb{R}^d$. We refer to the hypergraph $H_{\mathcal{R}}(P) = (P, \{P \cap S \mid S \in \mathcal{R}\})$ as the *hypergraph induced by P with respect to* $\mathcal{R}$. When $\mathcal{R}$ is clear from the context we sometimes refer to it as *the hypergraph induced by P*. In the literature, hypergraphs that are induced by points with respect to geometric regions of some specific kind are sometimes referred to as *range spaces*.

Definition 1.3 (Delaunay-Graph). For a hypergraph $H = (V, \mathcal{E})$, denote by $G(H)$ the *Delaunay-graph* of H which is the graph $(V, \{S \in \mathcal{E} \mid |S| = 2\})$.

In most of the coloring solutions presented in this paper we will see that, in fact, we get the stronger UM-coloring. It is also interesting to study hypergraphs for which $\chi_{cf}(H) < \chi_{um}(H)$. This line of research has been pursued in [15, 17]

1.2. Motivation

We start with several motivations for studying CF-colorings and UM-colorings.

1.2.1. Wireless Networks. Wireless communication is used in many different situations such as mobile telephony, radio and TV broadcasting, satellite communication, etc. In each of these situations a frequency assignment problem arises with application-specific characteristics. Researchers have developed different modeling approaches for each of the features of the problem, such as the handling of interference among radio signals, the availability of frequencies, and the optimization criterion.

The work of Even et al. [25] and of Smorodinsky [47] proposed to model frequency assignment to cellular antennas as CF-coloring. In this new model, one can use a very "small" number of distinct frequencies in total, to assign to a large number of antennas in a wireless network. Cellular networks are heterogeneous networks with two different types of nodes: *base-stations* (that act as servers) and *clients*. The base-stations are interconnected by an external fixed backbone network. Clients are connected only to base stations; connections between clients and base-stations are implemented by radio links. Fixed frequencies are assigned to base-stations to enable links to clients. Clients, on the other hand, continuously scan frequencies in search of a base-station with good reception. This scanning takes place automatically and enables smooth transitions between base-stations when a client is mobile. Consider a client that is within the reception range of two base stations. If these two base stations are assigned the same frequency, then mutual interference occurs, and the links between the client and each of these conflicting base stations are rendered too noisy to be used. A base station may serve a client provided that the reception is strong enough and interference from other base stations is weak enough. The fundamental problem of frequency assignment in cellular network is to assign frequencies to base stations so that every client is served by some base station. The goal is to minimize the number of assigned frequencies since the available spectrum is limited and costly.

The problem of frequency assignment was traditionally treated as a graph coloring problem, where the vertices of the graph are the given set of antennas and the edges are those pairs of antennas that overlap in their reception range. Thus, if we color the vertices of the graph such that no two vertices that are connected by an edge have the same color, we guarantee that there will be no conflicting base stations. However, this model is too restrictive. In this model, if a client lies within the reception range of say, k antennas, then every pair of these antennas are conflicting and therefore they must be assigned k distinct colors (i.e., frequencies). But note that if one of these antennas is assigned a color (say 1) that no other antenna is assigned (even if all other antennas are assigned the same color, say 2) then we use a total of two colors and this client can still be served. See Figure 1 for an illustration with three antennas.

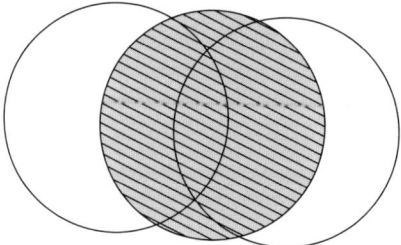

Fig. 1. An example of three antennas presented as discs in the plane. In the classical model three distinct colors are needed where as in the new model two colors are enough as depicted here

A natural question thus arises: Suppose we are given a set of n antennas. The location of each antenna (base station) and its radius of transmission is fixed and is known (and is modeled as a disc in the plane). We seek the least number of colors that always suffice such that each of the discs is assigned one of the colors and such that every covered point p is also covered by some disc D whose assigned color is distinct from all the colors of the other discs that cover p. This is a special case of CF-coloring where the underlying hypergraph is induced by a finite family of discs in the plane.

1.2.2. RFID networks. Radio frequency identification (RFID) is a technology where a reader device can "sense" the presence of a close by object by reading a tag device attached to the object. To improve coverage, multiple RFID readers can be deployed in the given region. RFID systems consist of readers and tags. A tag has an ID stored in its memory. The reader is able to read the IDs of the tags in the vicinity by using wireless protocol. In a typical RFID application, tags are attached to objects of interest, and the reader detects the presence of an object by using an available mapping of IDs to objects. We focus on passive tags i.e., tags that do not carry a battery. The power needed for passive tags to transmit their IDs to the reader is supplied by the reader itself. Assume that we are given a set D of readers where each reader is modeled by some disc in the plane. Let P be a set of tags (modeled as points) that lie in the union of the discs in D. Suppose that all readers in D use the same wireless frequency. For the sake of simplicity, suppose also that each reader is only allowed to be activated once. The goal is to schedule for each reader $d \in D$ a time slot $t(d)$ for which the reader d will be active. That is, at time $t(d)$ reader d would initiate a 'read' action. We further assume that a given tag $p \in P$ can be read by reader $d \in D$ at time t if $p \in d$ and d is initiating a 'read' action at time t (namely,

$t = t(d))$ and no other reader d' for which $p \in d'$ is active at time t. We say that P is read by our schedule, if for every $p \in P$ there is at least one $d \in D$ and a time t such that p is read by d at time t. Obviously, we would like to minimize the total time slots used in the schedule. Thus our goal is to find a function $t : D \to \{1, \ldots, k\}$ which is conflict-free for the hypergraph $H(D)$. Since we want to minimize the total time slots used, again the question of what is the minimum number of colors that always suffice to CF-color any hypergraph induced by a finite set of n discs is of interest.

1.2.3. Vertex ranking. Let $G = (V, E)$ be a simple graph. An *ordered coloring* (also a *vertex ranking*) of G is a coloring of the vertices $\chi : V \to \{1, \ldots, k\}$ such that whenever two vertices u and v have the same color i then every simple path between u and v contains a vertex with color greater than i. Such a coloring has been studied before and has several applications. It was studied in the context of VLSI design [46] and in the context of parallel Cholesky factorization of matrices [37]. The vertex ranking problem is also interesting for the Operations Research community. It has applications in planning efficient assembly of products in manufacturing systems [31]. In general, it seems that the vertex ranking problem can model situations where inter-related tasks have to be accomplished fast in parallel, with some constrains (assembly from parts, parallel query optimization in databases, etc.). See also [32, 45]

The vertex ranking coloring is yet another special form of UM-coloring. Given a graph G, consider the hypergraph $H = (V, E')$ where a subset $V' \subseteq V$ is a hyperedge in E' if and only if V' is the set of vertices in some simple path of G. It is easily observed that an ordered coloring of G is equivalent to a UM-coloring of H.

1.3. A General Conflict-Free coloring Framework

Let P be a set of n points in $\mathbb{R}^2$ and let $\mathcal{D}$ be the set of all planar discs. In [25, 47] it was proved that $\chi_{\mathrm{um}}(H_{\mathcal{D}}(P)) = O(\log n)$ and that this bound is asymptotically tight since for any $n \subset \mathbb{N}$ there exist hypergraphs induced by sets of n points in the plane (w.r.t discs) which require $\Omega(\log n)$ in any CF-coloring. In fact, Pach and Tóth [43] proved a stronger lower-bound by showing that for any set P of n points it holds that $\chi_{\mathrm{cf}}(H_{\mathcal{D}}(P)) = \Omega(\log n)$. The proofs of [25, 47] are algorithmic and rely on two crucial properties: The first property is that the Delaunay graph $G(H_{\mathcal{D}}(P))$ always contains a "large" independent set. The second is the following shrinkability property of discs: For every disc d containing a set of $i \geq 2$ points of P there is another disc d' such that $d' \cap P \subseteq d \cap P$ and $|d' \cap P| = 2$.

In [25, 47] it was also proved that, if D is a set of n discs in the plane, then $\chi_{\text{um}}(H(D)) = O(\log n)$. This bound was obtained by a reduction to a three-dimensional problem of UM-coloring a set of n points in $\mathbb{R}^3$ with respect to lower half-spaces. Later, Har-Peled and Smorodinsky [28] generalized this result to pseudo-discs using a probabilistic argument. Pach and Tardos [40] provided several non-trivial upper-bounds on the CF-chromatic number of arbitrary hypergraphs. In particular they showed that for every hypergraph H with m hyperedges

$$\chi_{\text{cf}}(H) \leq 1/2 + \sqrt{2m + 1/4}.$$

Smorodinsky [48] introduced the following general framework for UM-coloring any hypergraph. This framework holds for arbitrary hypergraphs and the number of colors used is related to the chromatic number of the underlying hypergraph. Informally, the idea is to find a proper coloring with very 'few' colors and assign to all vertices of the largest color class the final color '1', discard all the colored elements and recursively continue on the remaining sub-hypergraph. See Algorithm 1 below.

Algorithm 1 UMcolor(H): *UM-coloring of a hypergraph $H = (V, \mathcal{E})$.*

1: $i \leftarrow 0$: *i denotes an unused color*
2: **while** $V \neq \emptyset$ **do**
3:      **Increment:** $i \leftarrow i + 1$
4:      **Auxiliary coloring:** find a proper coloring χ of the induced sub-hypergraph $H(V)$ with "few" colors
5:      $V' \leftarrow$ **Largest color class of** χ
6:      **Color:** $f(x) \leftarrow i$, $\forall x \in V'$
7:      **Prune:** $V \leftarrow V \setminus V'$
8: **end while**

Theorem 1.4 ([48]). *Algorithm 1 outputs a valid UM-coloring of H.*

Proof. Formally, Algorithm 1 is not well defined as its output depends on the auxiliary coloring of step 4 of the algorithm. Nevertheless, we regard step 4 as given to us by some 'black' box and we treat this aspect of the algorithm later on. For a hyperedge $e \in \mathcal{E}$, let i be the maximal index (color) for which there is a vertex $v \in e$ colored with i. We claim that there is exactly one such vertex. Indeed, assume to the contrary that there is another such vertex $v' \in e$. Consider the ith iteration and let V' denote the set of vertices of V that are colored with color greater or equal to i. Namely, V' is the set of vertices that 'survived' all the prune steps up to iteration i and reached iteration i. Let χ denote the auxiliary proper coloring for the hypergraph $H(V')$ in iteration i. Since $e' = e \cap V'$ is a hyperedge of $H(V')$ and v and v' belong to the same color class of χ and $v, v' \in e'$ and since χ is a non-monochromatic coloring, there must exist a third vertex $v'' \in e'$ such that $\chi(v'') \neq \chi(v)$. This means that the final color of v'' is greater than i, a contradiction to the maximality of i in e. This completes the proof of the theorem. ∎

The number of colors used by Algorithm 1 is the number of iterations that are performed (i.e., the number of prune steps). This number depends on the 'black-box' auxiliary coloring provided in step 4 of the algorithm. If the auxiliary coloring χ uses a total of C_i colors on $|V_i|$ vertices, where V_i is the set of input vertices at iteration i, then by the pigeon-hole principle one of the colors is assigned to at least $\frac{|V_i|}{C_i}$ vertices so in the prune step of the same iteration at least $\frac{|V_i|}{C_i}$ vertices are discarded. Thus, after l iterations of the algorithm we are left with at most $|V| \cdot \Pi_{i=1}^{l}\left(1 - \frac{1}{C_i}\right)$ vertices. If this number is less than 1, then the number of colors used by the algorithm is at most l. If for example $C_i = 2$ for every iteration, then the algorithm discards at least $\frac{|V_i|}{2}$ vertices in each iteration so the number of vertices left after l iterations is at most $|V|\left(1 - \frac{1}{2}\right)^l$ so for $l = \lfloor \log n \rfloor + 1$ this number is less than 1. Thus the number of iterations is bounded by $\lfloor \log n \rfloor + 1$ where n is the number of vertices of the input hypergraph. In the next section we analyze the chromatic number $\chi(H)$ for several geometrically induced hypergraphs and use Algorithm 1 to obtain bounds on $\chi_{um}(H)$.

We note that, as observed above, for a hypergraph H that admits a proper coloring with "few" colors hereditarily (that is, every induced sub-hypergraph admits a proper coloring with "few" colors), H also admits a UM-coloring with few colors. The following theorem summarizes this fact:

Theorem 1.5 ([48]). *Let $H = (V, \mathcal{E})$ be a hypergraph with n vertices, and let $k \in \mathbb{N}$ be a fixed integer, $k \geq 2$. If every induced sub-hypergraph $H' \subseteq H$ satisfies $\chi(H') \leq k$, then $\chi_{um}(H) \leq \log_{1+\frac{1}{k-1}} n = O(k \log n)$.*

Remark 1.6. We note that the parameter k in Theorem 1.5 can be replaced with a non-constant function $k = k(H')$. For example, if $k(H') = (n')^\alpha$ where $0 < \alpha \le 1$ is a fixed real and n' is the number of vertices of H', an easy calculation shows that $\chi_{\text{um}}(H) = O(n^\alpha)$ where n is the number of vertices of H.

As we will see, for many of the hypergraphs that are mentioned in this survey, the two numbers $\chi(H), \chi_{\text{um}}(H)$ are only a polylogarithmic (in $|V|$) factor apart. For the proof to work, the requirement that a hypergraph H admits a proper coloring with few colors hereditarily is necessary. One example is the 3-uniform hypergraph H with $2n$ vertices given above. We have $\chi(H) = 2$ and $\chi_{\text{um}}(H) = n+1$. Obviously H does not admit a proper 2-coloring hereditarily.

2. CONFLICT-FREE COLORING OF GEOMETRIC HYPERGRAPHS

2.1. Discs and Pseudo-Discs in the Plane

2.1.1. Discs in $\mathbb{R}^2$. In [48] it was shown that the chromatic number of a hypergraph induced by a family of n discs in the plane is bounded by four. That is, for a finite family D of n discs in the plane we have:

Theorem 2.1 ([48]). $\chi(H(D)) \le 4$

Combining Theorem 1.5 and Theorem 2.1 we obtain the following:

Theorem 2.2 ([48]). *Let $\mathcal{D}$ be a set of n discs in the plane. Then* $\chi_{\text{um}}(H(\mathcal{D})) \le \log_{4/3} n$.

Proof. We use Algorithm 1 and the auxiliary proper four coloring provided by Theorem 2.1 in each prune step. Thus in each step i we discard at least $|V_i|/4$ discs so the total number of iterations is bounded by $\log_{4/3} n$. ∎

Remark. The existence of a four coloring provided in Theorem 2.1 is algorithmic and uses the algorithm provided in the Four-Color Theorem [8, 9] which runs in linear time. It is easy to see that the total running time used by algorithm 1 for this case is therefore $O(n \log n)$. The bound in Theorem 2.2 holds also for the case of hypergraphs induced by points in the plane with respect to discs. This follows from the fact that such a hypergraph H satisfies $\chi(H) \leq 4$. Indeed, the Delaunay graph $G(H)$ is planar (and hence four colorable) and any disc containing at least 2 points also contains an edge of $G(H)$ [25].

Smorodinsky [48] proved that there exists an absolute constant C such that for any family $\mathcal{P}$ of pseudo-discs in the plane $\chi(H(\mathcal{P})) \leq C$. Hence, by Theorem 1.5 we have $\chi_{\text{um}}(H(\mathcal{P})) = O(\log n)$. It is not known what is the exact constant and it might be possible that it is still 4. By taking 4 pairwise (openly-disjoint) touching discs, one can verify that it is impossible to find a proper coloring of the discs with less than 4 colors.

There are natural geometric hypergraphs which require n distinct colors even in any proper coloring. For example, one can place a set P of n points in general position in the plane (i.e., no three points lie on a common line) and consider those ranges that are defined by rectangles. In any proper coloring of P (w.r.t rectangles) every two such points need distinct colors since for any two points p, q there is a rectangle containing only p and q.

One might wonder what makes discs more special than other shapes? Below, we show that a key property that allows CF-coloring discs with a "small" number of colors unlike rectangles is the so called "low" *union-complexity* of discs.

Definition 2.3. Let $\mathcal{R}$ be a family of n simple Jordan regions in the plane. The *union complexity* of $\mathcal{R}$ is the number of vertices (i.e., intersection of boundaries of pairs of regions in $\mathcal{R}$) that lie on the boundary $\partial \bigcup_{r \in \mathcal{R}} r$.

As mentioned already, families of discs or pseudo-discs in the plane induce hypergraphs with chromatic number bounded by some absolute constant. The proof of [48] uses the fact that pseudo-discs have "linear union complexity" [33].

The following theorem bounds the chromatic number of a hypergraph induced by a finite family of regions $\mathcal{R}$ in the plane as a function of the union complexity of $\mathcal{R}$:

Theorem 2.4 ([48]). *Let $\mathcal{R}$ be a set of n simple Jordan regions and let $\mathcal{U} : \mathbb{N} \to \mathbb{N}$ be a function such that $U(m)$ is the maximum union complexity of any k regions in $\mathcal{R}$ over all $k \leq m$, for $1 \leq m \leq n$. We assume that $\frac{U(m)}{m}$ is a non-decreasing function. Then, $\chi(H(\mathcal{R})) = O\left(\frac{U(n)}{n}\right)$. Furthermore, such a coloring can be computed in polynomial time under a proper and reasonable model of computation.*

As a corollary of Theorem 2.4, for any family $\mathcal{R}$ of n planar Jordan regions for which the union-complexity function $\mathcal{U}(n)$ is linear, we have that $\chi(H(\mathcal{R})) = O(1)$. Hence, combining Theorem 2.4 with Theorem 1.5 we have:

Theorem 2.5 ([48]). *Let $\mathcal{R}$ be a set of n simple Jordan regions and let $\mathcal{U} : \mathbb{N} \rightarrow \mathbb{N}$ be a function such that $U(m)$ is the maximum complexity of any k regions in $\mathcal{R}$ over all $k \leq m$, for $1 \leq m \leq n$. If $\mathcal{R}$ has linear union complexity in the sense that $\mathcal{U}(n) \leq Cn$ for some constant C, then $\chi_{\mathrm{um}}(H(\mathcal{R})) = O(\log n)$.*

2.2. Axis-Parallel rectangles

2.2.1. hypergraphs induced by axis-parallel rectangles. As mentioned already, a hypergraph induced by n rectangles in the plane might need n colors in any proper coloring. However, in the special case of axis-parallel rectangles, one can obtain non-trivial upper bounds. Notice that axis-parallel rectangles might have quadratic union complexity so using the above framework yields only the trivial upper bound of n. Nevertheless, in [48] it was shown that any hypergraph that is induced by a family of n axis-parallel rectangles, admits an $O(\log n)$ proper coloring. This bound is asymptotically tight as was shown recently by Pach and Tardos [41].

Theorem 2.6 ([48]). *Let $\mathcal{R}$ be a set of n axis-parallel rectangles in the plane. Then $\chi(H(\mathcal{R})) \leq 8 \log n$.*

Plugging this fact into Algorithm 1 yields:

Theorem 2.7 ([48]). *Let $\mathcal{R}$ be a set of n axis-parallel rectangles in the plane. Then $\chi_{\mathrm{um}}(H(\mathcal{R})) = O(\log^2 n)$.*

Remark. Notice that in particular there exists a family $\mathcal{R}$ of n axis-parallel rectangles for which $\chi_{\mathrm{cf}}(H(\mathcal{R})) = \Omega(\log n)$. Another example of a hypergraph H induced by n axis-parallel squares with $\chi(H) = 2$ and $\chi_{\mathrm{cf}}(H) = \Omega(\log n)$ is given in Figure 2. This hypergraph is, in fact, isomorphic to the discrete interval hypergraph with n vertices.

Problem 1. Close the asymptotic gap between the best known upper bound $O(\log^2 n)$ and the lower bound $\Omega(\log n)$ on the CF-chromatic number of hypergraphs induced by n axis-parallel rectangles in the plane.

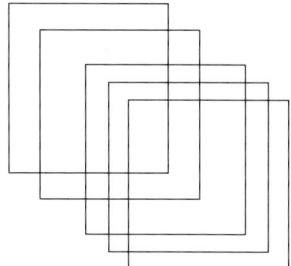

Fig. 2. An example of n axis-parallel squares inducing the hypergraph H with $\chi(H) = 2$ and $\chi_{\mathrm{cf}}(H) = \Omega(\log n)$

2.2.2. Points with respect to axis-parallel rectangles. Let $\mathcal{R}$ be the family of all axis-parallel rectangles in the plane. For a finite set P in the plane, let $H(P)$ denote the hypergraph $H_{\mathcal{R}}(P)$. Let $D(P)$ denote the Delaunay graph of $H(P)$. It is easily seen that $\chi(D(P)) = \chi(H(P))$ since every axis-parallel rectangle containing at least two points, also contains an edge of $D(P)$.

The following problem seems to be rather elusive:

Problem 2. Let $\mathcal{R}$ be the family of all axis-parallel rectangles in the plane. Let $d = d(n)$ be the least integer such that for any set P of n points in the plane $\chi(D(P)) \le d(n)$. Provide sharp asymptotic bounds on $d(n)$.

It was first observed in [28] that $d(n) = O(\sqrt{n})$ by a simple application of the classical Erdős-Szekeres theorem for a sequence of reals. This theorem states that in a sequence of $k^2 + 1$ reals there is always a monotone subsequence of length at least $k + 1$ (see, e.g., [52]).

One can show that for any set P of n points in the plane there is a subset $P' \subset P$ of size $\Omega(\sqrt{n})$ which is independent in the graph $D(P)$. To see this, sort the points $P = \{p_1, \ldots, p_n\}$ according to their x-coordinate. Write the sequence of y-coordinates of the points in P $y_1, \ldots, y_n$. By the Erdős-Szekeres theorem, there is a subsequence $y_{i_1}, \ldots, y_{i_k}$ with $k = \Omega(\sqrt{n})$ which is monotone. We refer to the corresponding subset of P as a *monotone chain*. Notice that by taking every other point in the monotone chain, the set $p_{i_1}, p_{i_3}, p_{i_5}, \ldots$ is a subset of size $k/2 = \Omega(\sqrt{n})$ which is independent in $D(P)$. See Figure 3 for an illustration. In order to complete the coloring it is enough to observe that one can iteratively partition P into $O(\sqrt{n})$ independent sets of $D(P)$.

The bounds on $d(n)$ were recently improved and the best known bounds are stated below:

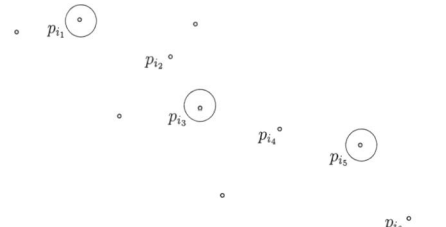

Fig. 3. The circled points form an independent set in the Delaunay graph $D(P)$

Upper bound: [14] $d(n) = \tilde{O}(n^{0.368})$

Lower bound: [20] $d(n) = \Omega\left(\frac{\log n}{\log^2 \log n}\right)$.

We give a short sketch of the ideas presented in [3] in order to obtain the upper bound $d(n) = \tilde{O}(n^{0.382})$ where $\tilde{O}$ denotes the fact that a factor of polylog is hiding in the big-O notation. Our presentation of the ideas is slightly different from [23, 3] since our aim is to bound $d(n)$ which corresponds to coloring the Delaunay graph of n points rather than CF-coloring the points themselves. However, as mentioned above, such a bound implies also a similar bound on the CF-chromatic number of the underlying hypergraph. Assume that $d(n) \geq c \log n$ for some fixed constant c. We will show that $d(n) = O(n^\alpha)$ for all $\alpha > \alpha_0 = \frac{3-\sqrt{5}}{2}$. The proof relies on the following key ingredient, first proved in [23]. For a point set P in the plane, let G_r be an $r \times r$ grid such that each row of G_r and each column of G_r contains at most $\lceil n/r \rceil$ points of P. Such a grid is easily seen to exists. A coloring of P is called a *quasi-coloring* with respect to G_r if every rectangle that is fully contained in a row of G_r or fully contained in a column of G_r is non-monochromatic. In other words, when coloring P, we do not care about rectangles that are not fully contained in a row or fully contained in a column (or contain only one point).

Lemma 2.8 ([23, 3]). *Let P be a set of n points in the plane. If $\Omega(\log n) = d(n) = O(n^\alpha)$ then for every r, P admits a quasi-coloring with respéct to G_r with $\tilde{O}\left(\left(\frac{n}{r}\right)^{2\alpha - \alpha^2}\right)$ colors.*

The proof of the lemma uses a probabilistic argument. We first color each column in G_r independently with $d(n/r)$ colors. Then for each column we permute the colors randomly and then re-color all points in a given row that were assigned the same initial color. We omit the details of the proof and its probabilistic analysis.

Next, we choose an appropriate subset $P' \subset P$ which consists of $O(r)$ monotone chains and with the following key property: If a rectangle S contains points from at least two rows of G_r and at least two columns of G_r, then S also contains a point of P'. Note that a chain can be colored with 2 colors so altogether one can color P' with $O(r)$ colors, not to be used for $P \setminus P'$. Thus a rectangle that is not fully contained in a row or a column of G_r is served by the coloring. Hence, it is enough to quasi-color the points of $P \setminus P'$ with respect to G_r. By the above lemma, the total number of colors required for such a coloring is $\tilde{O}\left(\left(\frac{n}{r}\right)^{2\alpha - \alpha^2} + r\right)$. Choosing $r = n^{\frac{2\alpha - \alpha^2}{1 + 2\alpha - \alpha^2}}$ we obtain the bound $\tilde{O}\left(n^{\frac{2\alpha - \alpha^2}{1 + 2\alpha - \alpha^2}}\right)$. Thus, taking α_0 to satisfy the equality

$$\alpha_0 = \frac{2\alpha_0 - \alpha_0^2}{1 + 2\alpha_0 - \alpha_0^2}.$$

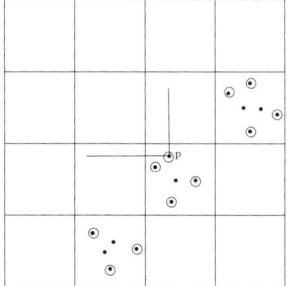

Fig. 4. The grid G_r (for $r = 4$) and one of its positive diagonals. The circled points are taken to be in P' and the square points are in $P \setminus P'$. The point p is an extreme point of type 2 in that diagonal and is also an extreme point of type 1 in the negative diagonal that contains the grid cell of p

To complete the proof, we need to construct the set P'. Consider the diagonals of the grid G_r. See Figure 4 for an illustration. In each positive diagonal we take the subset of (extreme) points of type 2 or 4, where a point p is said to be of type 2 (respectively, 4) if the 2'nd quadrant (respectively, the 4'th quadrant) with respect to p (i.e., the subset of all points above and to the left of p) does not contain any other point from the diagonal. Similarly, for diagonals with negative slope we take the points of type 1 and 3. If a point belongs to more than one type (in the two diagonals that contain the point) then we arbitrarily choose one of the colors it gets from one of the diagonals. It is easy to see that the set P' admits a proper coloring with $O(r)$ colors, as there are only $2r - 1$ positive diagonals and

$2r-1$ negative diagonals, and in each diagonal the extreme points of a fixed type form a monotone chain.

As mentioned, reducing the gap between the best known asymptotic upper and lower bounds mentioned above is a very interesting open problem.

2.3. Shallow Regions

As mentioned already, for every n there are sets D of n discs in the plane such that $\chi_{\mathrm{cf}}(H(D)) = \Omega(\log n)$. For example, one can place n unit discs whose centers all lie on a line, say the x-axis, such that the distance between any two consecutive centers is less than $1/n$. It was shown in [25] that, for such a family D, $\chi_{\mathrm{cf}}(H(D)) = \Omega(\log n)$ since $H(D)$ is isomorphic to the discrete interval hypergraph with n vertices. However, in this case there are points that are covered by many of the discs of D (in fact, by all of them). This leads to the following fascinating problem: What happens if we have a family of n discs D with the property that every point is covered by at most k discs of D, for some parameter k. It is not hard to see that in such a case, one can color D with $O(k)$ colors such that any two intersecting discs have distinct colors. However, we are interested only in CF-coloring of D. Let us call a family of regions, with the property that no point is covered by more than k of the regions, a k-*shallow* family.

Problem 3. What is the minimum integer $f = f(k)$ such that for any finite family of k-shallow discs D, we have: $\chi_{\mathrm{cf}}(H(D)) \le f(k)$?

As mentioned already, it is easy to see that $f(k) = O(k)$. However, it is conjectured that the true upper bound should be polylogarithmic in k.

In the further restricted case that any disc in D intersects at most k other discs, Alon and Smorodinsky [5] proved that $\chi_{\mathrm{cf}}(H(D)) = O(\log^3 k)$ and this was recently improved by Smorodinsky [49] to $\chi_{\mathrm{cf}}(H(D)) = O(\log^2 k)$. Both bounds also hold for families of pseudo-discs. We sketch the proof of the following theorem:

Theorem 2.9 ([49])**.** *Let D be a family of n discs in the plane such that any disc in D intersects at most k other discs in D. Then $\chi_{\mathrm{cf}}(H(D)) = O(\log^2 k)$.*

The proof of Theorem 2.9 is probabilistic and uses the Lovász Local Lemma [6]. We start with a few technical lemmas:

Denote by $E_{\le \ell}(D)$ the subset of hyperedges of $H(D)$ of cardinality less than or equal to ℓ.

Lemma 2.10. *Let D be a finite set of n planar discs. Then $|E_{\leq k}(D)| = O(kn)$.*

Proof. This easily follows from the fact that discs have linear union-complexity [33] and the Clarkson–Shor probabilistic technique [21]. We omit the details of the proof. ∎

Lemma 2.11. *Let D be a set of n planar discs, and let $\ell > 1$ be an integer. Then the hypergraph $(D, E_{\leq \ell}(D))$ can be CF-colored with $O(\ell)$ colors.*

Remark. In fact, the proof of Lemma 2.11 which can be found in [7] provides a stronger coloring. The coloring has the property that every hyperedge in $E_{\leq \ell}(D)$ is colorful (i.e., all vertices have distinct colors). Such a coloring is referred to as ℓ-colorful coloring and is discussed in more details in Subsection 3.2.

Lemma 2.12. *Let D be a set of discs such that every disc intersects at most k others. Then there is a constant C such that D can be colored with two colors (red and blue) and such that for every face $f \in \mathcal{A}(D)$ with depth at least $C \ln k$, there are at least $\frac{|d(f)|}{3}$ red discs containing f and at least $\frac{|d(f)|}{3}$ blue discs containing f, where $d(f)$ is the set of all discs containing the face f.*

Proof. Consider a random coloring of the discs in D, where each disc $d \in D$ is colored independently red or blue with probability $\frac{1}{2}$. For a face f of the arrangement $\mathcal{A}(D)$ with $|d(f)| \geq C \ln k$ (for some constant C to be determined later), let A_f denote the "bad" event that either less than $\frac{|d(f)|}{3}$ of the discs in $d(f)$ or more than $\frac{2|d(f)|}{3}$ of them are colored blue. By the Chernoff inequality (see, e.g., [6]) we have:

$$Pr[A_f] \leq 2e^{-\frac{|d(f)|}{72}} \leq 2e^{-\frac{C \ln k}{72}}.$$

We claim that for every face f, the event A_f is mutually independent of all but at most $O(k^3)$ other events. Indeed A_f is independent of all events A_s for which $d(s) \cap d(f) = \emptyset$. By assumption, $|d(f)| \leq k + 1$. Observe also that a disc that contains f, can contain at most $O(k^2)$ other faces, simply because the arrangement of k discs consists of at most $O(k^2)$ faces. Hence, the claim follows.

Let C be a constant such that:

$$e \cdot 2e^{-\frac{C \ln k}{72}} \cdot 2k^3 < 1.$$

By the Lovász Local Lemma, (see, e.g., [6]) we have:

$$Pr[\bigwedge_{|d(f)| \geq C \ln k} \bar{A}_f] > 0.$$

In particular, this means that there exists a coloring for which every face f with $|d(f)| \geq C \ln k$ has at least $\frac{|d(f)|}{3}$ red discs containing f and at least $\frac{|d(f)|}{3}$ blue discs containing it, as asserted. This completes the proof of the lemma. ■

Proof of Theorem 2.9. Consider a coloring of D by two colors as in Lemma 2.12. Let B_1 denote the set of discs in D colored blue. We will color the discs of B_1 with $O(\ln k)$ colors such that $E_{\leq 2C \ln k}(B_1)$ is conflict-free, as guaranteed by Lemma 2.11, and recursively color the discs in $D \setminus B_1$ with colors disjoint from those used to color B_1. This is done, again, by splitting the discs in $D \setminus B_1$ into a set of red discs and a set B_2 of blue discs with the properties guaranteed by Lemma 2.12. We repeat this process until every face of the arrangement $\mathcal{A}(D')$ (of the set D' of all remaining discs) has depth at most $C \ln k$. At that time, we color D' with $O(\ln k)$ colors as described in Lemma 2.11. To see that this coloring scheme is a valid conflict-free coloring, consider a point $p \in \bigcup_{d \in D} d$. Let $d(p) \subset D$ denote the subset of all discs in D that contain p. Let i be the largest index for which $d(p) \cap B_i \neq \emptyset$. If i does not exist (namely, $d(p) \cap B_i = \emptyset \ \forall i$) then by Lemma 2.12 $|d(p)| \leq C \ln k$. However, this means that $d(p) \in E_{\leq C \ln k}(D)$ and thus $d(p)$ is conflict-free by the coloring of the last step. If $|d(p) \cap B_i| \leq 2C \ln k$ then $d(p)$ is conflict free since one of the colors in $d(p) \cap B_i$ is unique according to the coloring of $E_{\leq c \ln k}(B_i)$. Assume then, that $|d(p) \cap B_i| > 2C \ln k$. Let x denote the number of discs containing p at step i. By the property of the coloring of step i, we have that $x \geq 3C \ln k$. This means that after removing B_i, the face containing p is also contained in at least $C \ln k$ other discs. Hence, p must also belong to a disc of B_{i+1}, a contradiction to the maximality of i. To argue about the number of colors used by the above procedure, note that in each prune step, the depth of every face with depth $i \geq C \ln k$ is reduced with a factor of at least $\frac{1}{3}$. We started with a set of discs such that the maximal depth is $k + 1$. After the first step, the maximal depth is $\frac{2}{3}k$ and for each step we used $O(\ln k)$ colors so, in total, we have that the maximum number of colors $f(k, r)$, needed for CF-coloring a family of discs with maximum depth r such that each disc intersects at most k others satisfies the recursion:

$$f(k, r) \le O(\ln k) + f\left(k, \frac{2}{3}r\right).$$

This gives $f(k, r) = O(\ln k \log r)$. Since, in our case $r \le k + 1$, we obtain the asserted upper bound. This completes the proof of the theorem. ∎

Remark. Theorem 2.9 works almost verbatim for any family of regions (not necessarily convex) with linear union complexity. Thus, for example, the result applies to families of homothetics or more generally to any family of pseudo-discs, since pseudo-discs have linear union complexity ([33]). We also note that, as in other cases mentioned so far, it is easily seen that the proof of the bound of Theorem 2.9 holds for UM-coloring.

The proof of Theorem 2.9 is non-constructive since it uses the Lovász Local Lemma. However, we can use the recently discovered algorithmic version of the Local Lemma of Moser and Tardos [39] to obtain a constructive proof of Theorem 2.9.

Problem 4. As mentioned, the only lower bound that is known for this problem is $\Omega(\log k)$ which is obvious from taking the lower bound construction of [25] with k discs. It would be interesting to close the gap between this lower bound and the upper bound $O(\log^2 k)$.

The following is a rather challenging open problem:

Problem 5. Obtain a CF-coloring of discs with maximum depth $k+1$ (i.e., no point is covered by more than $k + 1$ discs) with only polylogarithmic (in k) many colors. Obviously, the assumption of this subsection that a disc can intersect at most k others is much stronger and implies maximum depth $k + 1$. However, the converse is not true. Assuming only bounded depth does not imply the former. In bounded depth, we still might have discs intersecting many (possibly all) other discs.

3. EXTENSIONS OF CF-COLORING

3.1. k-CF coloring

We generalize the notion of CF-coloring of a hypergraph to k-*CF-coloring*. Informally, we think of a hyperedge as being 'served' if there is a color that appears in the hyperedge (at least once and) at most k times, for some fix prescribed parameter k. For example, we will see that when the underlying hypergraph is induced by n points in $\mathbb{R}^3$ with respect to the family of all balls, there are n points for which any CF-coloring needs n colors but there exists a 2-CF-coloring with $O(\sqrt{n})$ colors (and a k-CF-coloring with $O(n^{1/k})$ colors for any fixed $k \geq 2$). We also show that any hypergraph $(V, \mathcal{E})$ with a finite VC-dimension c, can be k-CF-colored with $O(\log |P|)$ colors, for a reasonably large k. This relaxation of the model is applicable in the wireless scenario since the real interference between conflicting antennas (i.e., antennas that are assigned the same frequency and overlap in their coverage area) is a function of the number of such antennas. This suggests that if for any given point, there is some frequency that is assigned to at most a "small" number of antennas that cover this point, then this point can still be served using that frequency because the interference between a small number of antennas is low. This feature is captured by the following notion of k-CF-coloring.

Definition 3.1. k-CF-coloring of a hypergraph: Let $H = (V, \mathcal{E})$ be a hypergraph. A function $\chi : V \to \{1, \dots, i\}$ is a k-*CF-coloring* of H if for every $S \in \mathcal{E}$ there exists a color j such that $1 \leq |\{v \in S | \chi(v) = j\}| \leq k$; that is, for every hyperedge $S \in \mathcal{E}$ there exists at least one color j such that j appears (at least once and) at most k times among the colors assigned to vertices of S.

Let $\chi_{kCF}(H)$ denote the minimum number of colors needed for a k-CF-coloring of H.

Note that a 1-CF-coloring of a hypergraph H is simply a CF-coloring.

Here we modify Algorithm 1 to obtain a k-CF coloring of any hypergraph. We need yet another definition of the following relaxed version of a proper coloring:

Definition 3.2. Let $H = (V, \mathcal{E})$ be a hypergraph. A coloring φ of H is called k-*weak* if every hyperedge $e \in \mathcal{E}$ with $|e| \geq k$ is non-monochromatic.

That is, for every hyperedge $e \in \mathcal{E}$ with $|e| \geq k$ there exists at least two vertices $x, y \in e$ such that $\varphi(x) \neq \varphi(y)$.

Notice that a k-weak coloring (for $k \geq 2$) of a hypergraph $H = (V, \mathcal{E})$ is simply a proper coloring for the hypergraph $(V, \mathcal{E}_{\geq k})$ where $\mathcal{E}_{\geq k}$ is the subset of hyperedges in $\mathcal{E}$ with cardinality at least k. This notion was used implicitly in [28, 48] and then was explicitly defined and studied in the Ph.D. of Keszegh [35, 34]. It is also related to the notion of cover-decomposability and polychromatic colorings (see, e.g., [26, 42, 44]).

We are ready to generalize Algorithm 1. See Algorithm 2 below.

Algorithm 2 k-CFcolor(H): *k-Conflict-Free-color a hypergraph $H = (V, \mathcal{E})$.*

1: $i \leftarrow 0$: *i denotes an unused color*
2: **while** $V \neq \emptyset$ **do**
3:    **Increment:** $i \leftarrow i + 1$
4:    **Auxiliary coloring:** find a weak $k+1$-coloring χ of $H(V)$ with "few" colors
5:    $V' \leftarrow$ **Largest color class of** χ
6:    **Color:** $f(x) \leftarrow i$, $\forall x \in V'$
7:    **Prune:** $V \leftarrow V \setminus V'$, $H \leftarrow H(V)$
8: **end while**

Theorem 3.3 ([28]). *Algorithm 2 outputs a valid k-CF-coloring of H.*

Proof. The proof is similar to the proof provided in Section 1.3 for the validity of Algorithm 1. In fact, again, the coloring provided by Algorithm 2 has the stronger property that for any hyperedge $S \in \mathcal{E}$ the maximal color appears at most k times. ∎

As a corollary similar to the one mentioned in Theorem 1.5, for a hypergraph H that admit a $k+1$-weak coloring with "few" colors hereditarily, H also admits a k-CF-coloring with few colors. The following theorem summarizes this fact:

Theorem 3.4 ([28]). *Let $H = (V, \mathcal{E})$ be a hypergraph with n vertices, and let $l, k \in \mathbb{N}$ be two fixed integers, $k \geq 2$. Assume that every induced sub-hypergraph $H' \subseteq H$ admits a $k+1$-weak coloring with at most l colors. Then H admits a k-CF-coloring with at most $\log_{1+\frac{1}{l-1}} n = O(l \log n)$ colors.*

Proof. The proof is similar to the proof of Theorem 1.5 ∎

3.1.1. CF-Coloring of Balls in Three Dimensions.

Lemma 3.5. *Let $\mathcal{B}$ be the set of balls in three dimensions. There exists a hypergraph H induced by a finite set P of n points in $\mathbb{R}^3$ with respect to $\mathcal{B}$ such that $\chi_{1CF}(H) = n$. The same holds for the set $\mathcal{H}$ of halfspaces in $\mathbb{R}^d$, for $d > 3$.*

Proof. Take P to be a set of n points on the positive portion of the moment curve $\gamma = \{(t, t^2, t^3) \mid t \geq 0\}$ in $\mathbb{R}^3$. It is easy to verify that any pair of points $p, q \in P$ are connected in the Delaunay triangulation of P implying that there exists a ball whose intersection with P is $\{p, q\}$. Thus, all points must be colored using different colors.

The second claim follows by taking P to be n distinct points on the moment curve $\{(t, t^2, \ldots, t^d)\}$ in $\mathbb{R}^d$ (i.e, P is the set of vertices of a so-called *cyclic-polytope* $C(n, d)$. See, e.g., [50]). ∎

Theorem 3.6 ([28, 47]). *Let P be a set of n points in $\mathbb{R}^3$. Put $H = H_{\mathcal{B}}(P)$. Then $\chi_{kCF}(H) = O(n^{1/k})$, for any fixed constant $k \geq 1$.*

Proof. As is easily seen by Algorithm 2, it is enough to prove that H admits a $k + 1$-weak coloring with $O(n^{1/k})$ colors. If so, then in every iteration we discard at least $\Omega(|P_i|^{1-\frac{1}{k}})$ elements so the total number of iterations (colors) used is $O(n^{1/k})$. The proof that H admits a $k+1$-weak coloring with $O(n^{1/k})$ colors uses the probabilistic method. We provide only a brief sketch of the proof. It is enough to consider all balls containing exactly $k + 1$ points since if a ball contains more than $k+1$ points then by perturbation and shrinking arguments it will also contain a subset of $k+1$ points that can be cut-off by a ball. So we may assume that in the underlying hypergraph $H = (P, \mathcal{E})$, all hyperedges have cardinality $k + 1$ (such a hypergraph is also called a $k + 1$-uniform hypergraph). So we want to color the set P with $O(n^{1/k})$ colors such that any hyperedge in $\mathcal{E}$ is non-monochromatic. By the Clarkson–Shor technique, it is easy to see that the number of hyperedges in $\mathcal{E}$ is $O(k^2 n^2)$. Thus the average degree of a vertex in H is $O(n)$ where the constant of proportionality depends on k. It is well known that such a hypergraph has chromatic number $O(n^{1/k})$. This is proved via the probabilistic method. The main ingredient is the Lovász Local Lemma (see, e.g., [6]). ∎

In a similar way we have:

Theorem 3.7 ([28, 47]). *Let* $\mathcal{R}$ *be a set of* n *balls in* $\mathbb{R}^3$. *Then*

$$\chi_{kCF}(H(\mathcal{R})) = O(n^{1/k}).$$

3.1.2. VC-dimension and k-CF coloring.

Definition 3.8. Let $H = (V, \mathcal{E})$ be a hypergraph. The *Vapnik-Chervonenkis* dimension (or *VC-dimension*) of H, denoted by $VC(H)$, is the maximal cardinality of a subset $V' \subset V$ such that $\{V' \cap r \mid r \in \mathcal{E}\} = 2^{V'}$ (such a subset is said to be *shattered*). If there are arbitrarily large shattered subsets in V then $VC(H)$ is defined to be ∞. See [38] for discussion of VC-dimension and its applications.

There are many hypergraphs with finite VC-dimension that arise naturally in combinatorial and computational geometry. One such example is the hypergraph $H = (\mathbb{R}^d, \mathcal{H}_d)$, where $\mathcal{H}_d$ is the family of all (open) halfspaces in $\mathbb{R}^d$. Any set of $d + 1$ affinely independent points is shattered in this space, and, by Radon's theorem, no set of $d + 2$ points is shattered. Therefore $VC(H) = d + 1$.

Definition 3.9. Let $(V, \mathcal{E})$ be a hypergraph with $|V| = n$ and let $0 < \epsilon \leq 1$. A subset $N \subset V$ is called an ϵ-*net* for $(V, \mathcal{E})$ if for every hyperedge $S \in \mathcal{E}$ with $|S| \geq \epsilon n$ we have $S \cap N \neq \emptyset$.

Thus, an ϵ-net is a *hitting set* of all 'heavy' hyperedges, namely, those containing at least ϵn vertices.

An important consequence of the finiteness of the VC-dimension is the existence of small ϵ-nets, as shown by Haussler and Welzl in [29], where the notion of VC-dimension of a hypergraph was introduced to computational geometry.

Theorem 3.10 ([29]). *For any hypergraph* $H = (V, \mathcal{E})$ *with finite VC-dimension* d *and for any* $\epsilon > 0$, *there exists an* ϵ-*net* $N \subset V$ *of size* $O\left(\frac{d}{\epsilon} \log \frac{d}{\epsilon}\right)$.

Remark. In fact, Theorem 3.10 is valid also in the case where H is equipped with an arbitrary probability measure μ. An ϵ-net in this case is a subset $N \subset V$ that meets all hyperedges with measure at least ϵ.

Since all hypergraphs mentioned so far have finite VC-dimension, and since some of them sometimes must be CF-colored with n colors, there is no direct relationship between a finite VC-dimension of a hypergraph and the existence of a CF-coloring of that hypergraph with a small number of colors. In this subsection we show that such a relationship does exist, if we are interested in k-CF-coloring with a reasonably large k.

We first introduce a variant of the general framework for k-CF-coloring of a hypergraph $H = (V, \mathcal{E})$. In this framework we modify lines 4 and 5 in Algorithm 2. In Algorithm 2 we first find a $k + 1$-weak coloring of the underlying hypergraph (line 4) which is a partition of the vertices into sets such that each set has the following property: Every set in the partition cannot fully contain a hyperedge with cardinality at least $k+1$. Equivalently, every color class $V' \subset V$ has the property that every hyperedge containing at least $k + 1$ vertices of V' also contain vertices of $V \setminus V'$. We modify that framework by directly finding a "large" such subset in the hypergraph.

Definition 3.11. Let $H = (V, \mathcal{E})$ be a hypergraph. A subset $V' \subset V$ is k-*admissible* if for any hyperedge $S \in \mathcal{E}$ with $|S \cap V'| > k$ we have $S \cap (V \setminus V') \neq \emptyset$.

Assume that we are given an algorithm **A** that computes, for any hypergraph $H = (V, \mathcal{E})$, a non-empty k-admissible set $V' = \mathbf{A}(H)$. We can now use algorithm **A** to k-CF-color the given hypergraph (i) Compute a $k + 1$-admissible set $V' = \mathbf{A}(H)$, and assign to all the elements in V' the color 1. (ii) Color the remaining elements in $V \setminus V'$ recursively, where in the ith stage we assign the color i to the vertices in the resulting $k + 1$-admissible set. We denote the resulting coloring by $C_A(H)$.

The proof of the following theorem is, yet, again, similar to that of Theorem 1.4.

Theorem 3.12 ([28, 47]). *Given a hypergraph* $H = (V, \mathcal{E})$, *the coloring* $C_A(H)$ *is a valid* k-*CF coloring of* S.

Lemma 3.13. *Let* $H = (V, \mathcal{E})$ *with* $|V| = n$ *be a hypergraph with VC-dimension* d. *For any* $k \geq d$ *there exists a* k-*admissible set* $V' \subset V$ *with respect to* H *of size* $\Omega\big(n^{1-(d-1)/k}\big)$.

Proof. Any coloring of V is valid as far as the small hyperedges of $\mathcal{E}$ are concerned; namely, those are the hyperedges that contain at most k vertices.

Thus, let $\mathcal{E}'$ be the subset of hyperedges of $\mathcal{E}$ of size larger than k. By Sauer's Lemma (see, e.g., [6]) we have that $|\mathcal{E}'| \leq |\mathcal{E}| \leq n^d$.

Next, we randomly color V by black and white, where an element is being colored in black with probability p, where p would be specified shortly. Let I be the set of points of V colored in black. If a hyperedge $r \in \mathcal{E}'$ is colored only in black, we remove one of the vertices of r from I. Let I' be the resulting set. Clearly, I' is a k-admissible set for H.

Furthermore, by linearity of expectation, the expected size of I' is at least

$$pn - \sum_{r \in \mathcal{E}'} p^{|r|} \geq pn - \sum_{r \in \mathcal{E}'} p^{k+1} \geq pn - p^{k+1}n^d.$$

Setting $p = \big((k+1)n^{d-1}\big)^{-1/k}$, we have that the expected size of I' is at least $pn - p^{k+1}n^d = pn(1 - 1/(k+1)) = \Omega\big(n^{1-(d-1)/k}\big)$, as required. ∎

As was already seen, for geometric hypergraphs one might be able to get better bounds than the one guaranteed by Lemma 3.13.

Theorem 3.14 ([28, 47]). *Let $H = (V, \mathcal{E})$ with $|V| = n$ be a finite hypergraph with VC-dimension d. Then for $k \geq d \log n$ there exists a k-CFcoloring of H with $O(\log n)$ colors.*

Proof. By Lemma 3.13 the hypergraph H contains a k-admissible set of size at least $n/2$. Plugging this fact to the algorithm suggested by Theorem 3.12 completes the proof of the theorem. ∎

As remarked above, Theorem 3.14 applies to all hypergraphs mentioned in this paper. Note also, that Lemma 3.13 gives us a trade off between the number of colors and the threshold size of the coloring. As such, the bound of Theorem 3.14 is just one of a family of such bounds implied by Lemma 3.13.

3.2. k-Strong CF-Coloring

Here, we focus on the notion of *k-strong-conflict free* (abbreviated, *kSCF*) which is yet another extension of the notion of CF-coloring of hypergraphs.

Definition 3.15 (*k*-strong conflict-free coloring:). Let $H = (V, \mathcal{E})$ be a hypergraph and let $k \in \mathbb{N}$ be some fixed integer. A coloring of V is called *k strong-conflict-free* for H (*k*SCF for short) if for every hyperedge $e \in \mathcal{E}$ with $|e| \geq k$ there exists at least k vertices in e, whose colors are unique among the colors assigned to the vertices of e and for each hyperedge $e \in \mathcal{E}$ with $|e| < k$ all vertices in e get distinct colors. Let $f_H(k)$ denote the least integer l such that H admits a *k*SCF-coloring with l colors.

Abellanas et al. [2] were the first to study *k*SCF-coloring[1]. They focused on the special case of hypergraphs induced by n points in $\mathbb{R}^2$ with respect to discs. They showed that in this case the hypergraph admits a *k*SCF-coloring with $O\left(\frac{\log n}{\log \frac{ck}{ck-1}}\right)$ $(= O(k \log n))$ colors, for some absolute constant c.

The following notion was recently introduced and studied by Aloupis et al. [7] for the special case of hypergraphs induced by discs:

Definition 3.16 (*k*-colorful coloring). Let $H = (V, \mathcal{E})$ be a hypergraph, and let φ be a coloring of H. A hyperedge $e \in \mathcal{E}$ is said to be *k-colorful* with respect to φ if there exist k vertices in e that are colored distinctively under φ. The coloring φ is called *k-colorful* if every hyperedge $e \in \mathcal{E}$ is $\min\{|e|, k\}$-colorful. Let $c_H(k)$ denote the least integer l such that H admits a *k*-colorful coloring with l colors.

Aloupis et al. [7] introduced this notion explicitly and were motivated by a problem related to battery lifetime in sensor networks. This notion is also related to the notion of polychromatic colorings. In polychromatic colorings, the general question is to estimate the minimum number $f = f(k)$ such that one can k-color the hypergraph with the property that all hyperedges of cardinality at least $f(k)$ are colorful in the sense that they contain a representative color of each color class. (see, e.g., [26, 13, 44] for additional details on the motivation and related problems).

Remark. Every *k*SCF-coloring of a hypergraph H is a *k*-colorful coloring of H. However, the opposite claim is not necessarily true.

The following connection between *k*-colorful coloring and strong-conflict-free coloring of hypergraphs was proved by Horev et al. in [30]. If a hypergraph H admits a *k*-colorful coloring with a "small" number of colors (hereditarily) then it also admits a $(k - 1)$SCF-coloring with a "small" number of colors. This connection is analogous to the connection between non-monochromatic coloring and CF-coloring as appear in Theorem 1.5 and

---

[1]They referred to such a coloring as *k*-conflict-free coloring.

the connection between $k + 1$-weak coloring and k-CF-coloring as appear in Theorem 3.4. We start by introducing the general framework of [30] for kSCF-coloring a given hypergraph.

A Framework For Strong-Conflict-Free Coloring. Let H be a hypergraph with n vertices and let k and l be some fixed integers such that H admits the hereditary property that every vertex-induced sub-hypergraph H' of H admits a k-colorful coloring with at most l colors. Then H admits a $(k - 1)SCF$-coloring with $O(l \log n)$ colors. For the case when l is replaced with the function $kn(H')^{\alpha}$ we get a better bound without the $\log n$ factor. The proof is constructive. The following framework (denoted as Algorithm 3) produces a valid $(k - 1)$SCF coloring for a hypergraph H.

Algorithm 3 (k-1)SCF-color(H): $(k - 1)$-*Strong Conflict-Free-color a hypergraph* $H = (V, \mathcal{E})$.

1: $i \leftarrow 1$ *i denotes an unused color*
2: **while** $V \neq \emptyset$ **do**
3:    **Increment:** $i \leftarrow i + 1$
4:    **Auxiliary Coloring:** find a k-colorful coloring φ of $H(V)$ with "few" colors
5:    $V' \leftarrow$ **Largest color class of** φ
6:    **Color:** $\chi(x) \leftarrow i$, $\forall x \in V'$
7:    **Prune:** $V \leftarrow V \setminus V'$.
8:    **Increment:** $i \leftarrow i + 1$.
9: **end while**
10: **Return** χ.

Note that Algorithm 3 is a generalization of Algorithm 1. Indeed for $k = 2$ the two algorithms become identical since a 2-colorful coloring is equivalent to a proper coloring. Arguing about the number of colors used by the algorithm is identical to the arguments as in the coloring produced by Algorithm 1. The proof or correctness is slightly more subtle.

For a hypergraph $H = (V, \mathcal{E})$, we write $n(H)$ to denote the number of vertices of H. As a corollary of the framework described in Algorithm 3 we obtain the following theorems:

Theorem 3.17 ([30]). *Let $H = (V, \mathcal{E})$ be a hypergraph with n vertices, and let $k, \ell \in \mathbb{N}$ be fixed integers, $k \geq 2$. If every induced sub-hypergraph $H' \subseteq H$ satisfies $c_{H'}(k) \leq \ell$, then $f_H(k - 1) \leq \log_{1+\frac{1}{\ell-1}} n = O(l \log n)$.*

Theorem 3.18 ([30]). *Let $H = (V, \mathcal{E})$ be a hypergraph with n vertices, and let $k \geq 2$ be a fixed integer. let $0 < \alpha \leq 1$ be a fixed real. If every induced sub-hypergraph $H' \subseteq H$ satisfies $c_{H'}(k) = O(kn(H')^{\alpha})$, then $f_H(k-1) = O(kn^{\alpha})$.*

As a corollary of Theorem 3.17 and a result of Aloupis et al. [7] on k-colorful coloring of discs or points with respect to discs we obtain the following:

Theorem 3.19 ([30]). *If H is a hypergraph induced by n discs in the plane or a hypergraph induced by n points in the plane with respect to discs then $f_H(k) = O(k \log n)$.*

Proof. The proof follows by combining the fact that $c_H(k) = O(k)$ [7] with Theorem 3.17 ∎

Theorem 3.21 below provides an upper bound on the number of colors required by $kSCF$-coloring of geometrically induced hypergraphs as a function of the union-complexity of the regions that induce the hypergraphs.

Recall that, for a set $\mathcal{R}$ of n simple closed planar Jordan regions, $\mathcal{U}_{\mathcal{R}}$: $\mathbb{N} \to \mathbb{N}$ is the function defined in Theorem 2.4.

Theorem 3.20 ([30]). *Let $k \geq 2$, let $0 \leq \alpha \leq 1$, and let c be a fixed constant. Let $\mathcal{R}$ be a set of n simple closed Jordan regions such that $\mathcal{U}_{\mathcal{R}}(m) \leq cm^{1+\alpha}$, for $1 \leq m \leq n$, and let $H = H(\mathcal{R})$. Then $c_H(k) = O(kn^{\alpha})$.*

Combining Theorem 3.17 with Theorem 3.20 (for $\alpha = 0$) and Theorem 3.18 with Theorem 3.20 (for $0 < \alpha < 1$) yields the following result:

Theorem 3.21 ([30]). *Let $k \geq 2$, let $0 \leq \alpha \leq 1$, and let c be a constant. Let $\mathcal{R}$ be a set of n simple closed Jordan regions such that $\mathcal{U}_{\mathcal{R}}(m) = cm^{1+\alpha}$, for $1 \leq m \leq n$. Let $H = H(\mathcal{R})$. Then:*

$$f_H(k-1) = \begin{cases} O(k \log n), & \alpha = 0, \\ O(kn^{\alpha}), & 0 < \alpha \leq 1. \end{cases}$$

Axis-parallel rectangles: Consider $kSCF$-colorings of hypergraphs induced by axis-parallel rectangles in the plane. As mentioned before, axis-parallel rectangles might have quadratic union-complexity. For a hypergraph H induced by axis-parallel rectangles, Theorem 3.21 states that

$f_H(k-1) = O(kn)$. This bound is meaningless, since the bound $f_H(k-1) \leq n$ is trivial. Nevertheless, the following theorem provides a better upper bound for this case:

Theorem 3.22 ([30]). *Let $k \geq 2$. Let $\mathcal{R}$ be a set of n axis-parallel rectangles, and let $H = H(\mathcal{R})$. Then $f_H(k-1) = O(k \log^2 n)$.*

In order to obtain Theorem 3.22 we need the following theorem:

Theorem 3.23 ([30]). *Let $H = H(\mathcal{R})$, be the hypergraph induced by a family $\mathcal{R}$ of n axis-parallel rectangles in the plane, and let $k \in \mathbb{N}$ be an integer, $k \geq 2$. For every induced sub-hypergraph $H' \subseteq H$ we have: $c_{H'}(k) \leq k \log n$.*

The proof of Theorem 3.22 is therefore an easy consequence of Theorem 3.23 combined with Theorem 3.17.

Har-Peled and Smorodinsky [28] proved that any family $\mathcal{R}$ of n axis-parallel rectangles admit a CF-coloring with $O(\log^2 n)$ colors. Their proof uses the probabilistic method. They also provide a randomized algorithm for obtaining CF-coloring with at most $O(\log^2 n)$ colors. Later, Smorodinsky [48] provided a deterministic polynomial-time algorithm that produces a CF-coloring for n axis-parallel rectangles with $O(\log^2 n)$ colors. Theorem 3.22 thus generalizes the results of [28] and [48]. The upper bound provided in Theorem 3.21 for $\alpha = 0$ is optimal. Specifically, there exist matching lower bounds on the number of colors required by any $kSCF$-coloring of hypergraphs induced by (unit) discs in the plane.

Theorem 3.24 ([1]).

(i) *There exist families $\mathcal{R}$ of n (unit) discs for which $f_{H(\mathcal{R})}(k) = \Omega(k \log n)$*

(ii) *There exist families $\mathcal{R}$ of n axis-parallel squares for which $f_{H(\mathcal{R})}(k) = \Omega(k \log n)$.*

Notice that for axis-parallel rectangles there is a logarithmic gap between the best known upper and lower bounds.

Theorems 3.17 and 3.18 asserts that in order to attain upper bounds on $f_H(k)$, for a hypergraph H, one may concentrate on attaining a bound on $c_H(k)$. Given a k-colorful coloring of H, Algorithm 3 obtains a strong-conflict-free coloring of H in a constructive manner. Here computational efficiency is not of main interest. However, it can be seen that for certain families of geometrically induced hypergraphs, Algorithm 3 is efficient. In particular, for hypergraphs induced by discs or axis-parallel rectangles, Algorithm 3 has a low degree polynomial running time. Colorful-colorings of such hypergraphs can be computed once the arrangement of the discs is computed together with the depth of every face.

3.3. List Colorings

In view of the motivation for CF-coloring in the context of wireless antennae, it is natural to assume that each antenna is restricted to use some subset of the spectrum of frequencies and that different antennae might have different such subsets associated with them (depending, for example, on the physical location of the antenna). Thus, it makes sense to study the following more restrictive notion of coloring:

Let $H = (V, \mathcal{E})$ with $V = \{v_1, \ldots, v_n\}$ be a hypergraph and let $\mathcal{L} = \{L_1, \ldots, L_n\}$ be a family of subsets of the integers. We say that H admits a proper coloring from $\mathcal{L}$ (respectively, a CF-coloring from $\mathcal{L}$, a UM-coloring from $\mathcal{L}$) if there exists a proper coloring (respectively a CF-coloring, a UM-coloring) $C \colon V \to \mathbb{N}$ such that $C(v_i) \in L_i$ for $i = 1, \ldots, n$.

Definition 3.25. We say that a hypergraph $H = (V, \mathcal{E})$ is *k-choosable* (respectively, *k-CF-choosable, k-UM-choosable*) if for every family $\mathcal{L} = \{L_1, \ldots, L_n\}$ such that $|L_i| \geq k$ for $i = 1, \ldots, n$, H admits a proper-coloring (respectively a CF-coloring, a UM-coloring) from $\mathcal{L}$.

We are interested in the minimum number k for which a given hypergraph is k-choosable (respectively, k-CF-choosable, k-UM-choosable). We refer to this number as the choice-number (respectively the *CF-choice-number, UM-choice-number*) of H and denote it by $ch(H)$ (respectively $ch_{\mathrm{cf}}(H)$, $ch_{\mathrm{um}}(H)$). Obviously, if the choice-number (respectively, the CF-choice-number, UM-choice-number) of H is k then it can be properly colored (respectively CF-colored, UM-colored) with at most k colors, as one can proper color (respectively, CF-color, UM-color) H from $\mathcal{L} = \{L_1, \ldots, L_n\}$ where for every i we have $L_i = \{1, \ldots, k\}$. Thus,

$$ch(H) \geq \chi(H).$$

$$ch_{\mathrm{cf}}(H) \geq \chi_{\mathrm{cf}}(H).$$

$$ch_{\mathrm{um}}(H) \geq \chi_{\mathrm{um}}(H).$$

Hence, any lower bound on the number of colors required by a proper coloring of H (respectively, a CF-coloring, a UM-coloring of H) is also a lower bound on the choice number (respectively, the CF-choice-number, the UM-choice-number) of H.

The study of choice numbers in the special case of graphs was initiated by Vizing [51] and by Erdős Rubin and Taylor [24]. The study of the CF-choice number and the UM-choice number of hypergraphs was initiated very recently by Cheilaris, Smorodinsky and Sulovský [16].

Let us return to the discrete interval hypergraph H_n with n vertices, which was described in the introduction. As was shown already, we have $\chi_{cf}(H_n) = \chi_{um}(H_n) = \lfloor \log_2 n \rfloor + 1$. In particular we have the lower bound $ch_{cf}(H_n) \geq \lfloor \log_2 n \rfloor + 1$. Hence, the following upper-bound is tight:

Proposition 3.26. *For* $n \geq 1$, $ch_{cf}(H_n) \leq \lfloor \log_2 n \rfloor + 1$.

Proof. Assume, without loss of generality, that $n = 2^{k+1} - 1$. We will show that H_n is $k + 1$ CF-choosable. The proof is by induction on k. Let $\mathcal{L} = \{L_i\}_{i \in [n]}$, such that $|L_i| = k + 1$, for every i. Consider the median vertex $p = 2^k$. Choose a color $x \in L_p$ and assign it to p. Remove x from all other lists (for lists containing x), i.e., consider $\mathcal{L}' = \{L'_i\}_{i \in [n] \setminus p}$ where $L'_i = L_i \setminus \{x\}$. Note that all lists in $\mathcal{L}'$ have size at least k. The induction hypothesis is that we can CF-color any set of points of size $2^k - 1$ from lists of size k. Indeed, the number of vertices smaller (respectively, larger) than p is exactly $2^k - 1$. Thus, we CF-color vertices smaller than p and independently vertices larger than p, both using colors from the lists of $\mathcal{L}'$. Intervals that contain the median vertex p also have the conflict-free property, because color x is used only in p. This completes the induction step and hence the proof of the proposition. ∎

Note that, even in the discrete interval hypergraph, it is a more difficult problem to obtain any non-trivial upper bound on the UM-choice number. A divide and conquer approach, along the lines of the proof of Proposition 3.26 is doomed to fail. In such an approach, some vertex close to the median must be found, a color must be assigned to it from its list, and this color must be deleted from all other lists. However, vertices close to the median might have only "low" colors in their lists. Thus, while we are guaranteed that a vertex close to the median is uniquely colored for intervals containing it, such a unique color is not necessarily the maximal color for such intervals.

Instead, Cheilaris et al. used a different approach. This approach provides a general framework for UM-coloring hypergraphs from lists. Moreover, when applied to many geometric hypergraphs, it provides asymptotically tight bounds for the UM-choice number.

Below, we give an informal description of that approach, which is then summarized in Algorithm 4. It is similar in spirit to Algorithm 1.

Start by sorting the colors in the union of all lists in increasing order. Let c denote the minimum color. Let $V^c \subseteq V$ denote the subset of vertices containing c in their lists. Note that V^c might contain very few vertices, in fact, it might be that $|V^c| = 1$. We simultaneously color a suitable subset $U \subseteq V^c$ of vertices in V^c with c. We make sure that U is independent in the sub-hypergraph $H(V^c)$. The exact way in which we choose U is crucial to the performance of the algorithm and is discussed below. Next, for the uncolored vertices in $V^c \setminus U$, we remove the color c from their lists. This is repeated for every color in the union $\bigcup_{v \in V} L_v$ in increasing order of the colors. The algorithm stops when all vertices are colored. Notice that such an algorithm might run into a problem, when all colors in the list of some vertex are removed before this vertex is colored. Later, we show that if we choose the subset $U \subseteq V^c$ in a clever way and the lists are sufficiently large, then we avoid such a problem.

Algorithm 4 UMColorGeneric(H, $\mathcal{L}$): Unique-maximum color hypergraph $H = (V, \mathcal{E})$ from lists of family $\mathcal{L}$

1: **while** $V \neq \emptyset$ **do**
2:     $c \leftarrow \min \bigcup_{v \in V} L_v$ {c is the minimum color in the union of the lists}
3:     $V^c \leftarrow \{v \in V \mid c \in L_v\}$ {V^c is the subset of remaining vertices containing c in their lists}
4:     $U \leftarrow$ a "good" independent subset of the induced sub-hypergraph $H(V^c)$
5:     **for** $x \in U$ **do** {for every vertex in the independent set,}
6:        $f(x) \leftarrow c$ {color it with color c}
7:     **end for**
8:     **for** $v \in V^c \setminus U$ **do** {for every uncolored vertex in V^c,}
9:        $L_v \leftarrow L_v \setminus \{c\}$ {remove c from its list}
10:     **end for**
11:     $V \leftarrow V \setminus U$ {remove the colored vertices}
12: **end while**
13: **Return** f.

As mentioned, Algorithm 4 might cause some lists to run out of colors before coloring all vertices. However, if this does not happen, it is proved that the algorithm produces a UM-coloring.

Lemma 3.27 ([16]). *Provided that the lists associated with the vertices do not run out of colors during the execution of Algorithm 4, then the algorithm produces a UM-coloring from $\mathcal{L}$.*

Proof. The proof is similar to the validity proof of Algorithm 1 and we omit the details. ■

The key ingredient, which will determine the necessary size of the lists of $\mathcal{L}$, is the particular choice of the independent set in the above algorithm. We assume that the hypergraph $H = (V, \mathcal{E})$ is hereditarily k-colorable for some fixed positive integer k. Recall that, as shown before, this is the case in many geometric hypergraphs. We must also put some condition on the size of the lists in the family $\mathcal{L} = \{L_v\}_{v \in V}$. With some hindsight, we require

$$\sum_{v \in V} \lambda^{-|L_v|} < 1,$$

where $\lambda := \frac{k}{k-1}$.

Theorem 3.28 ([16]). *Let $H = (V, \mathcal{E})$ be a hypergraph which is hereditarily k-colorable and set $\lambda := \frac{k}{k-1}$. Let $\mathcal{L} = \{L_v\}_{v \in V}$, such that $\sum_{v \in V} \lambda^{-|L_v|} < 1$. Then, H admits a UM-coloring from $\mathcal{L}$.*

Notice, that in particular for a hypergraph H which is hereditarily k-colorable we have:

$$ch_{\mathrm{um}}(H) \leq \log_\lambda n + 1 = O(k \log n)$$

Thus, Theorem 3.28 subsumes all the theorems (derived from Algorithm 1) that are mentioned in Section 2.

Proof. The proof of Theorem 3.28 is constructive and uses a potential method: This method gives priority to coloring vertices that have fewer remaining colors in their lists, when choosing the independent sets. Towards that goal, we define a potential function on subsets of uncolored vertices and we choose the independent set with the highest potential (the potential quantifies how dangerous it is that some vertex in the set will run out of colors in its list).

For an uncolored vertex $v \in V$, let $r_t(v)$ denote the number of colors remaining in the list of v in the beginning of iteration t of the algorithm. Obviously, the value of $r_t(v)$ depends on the particular run of the algorithm. For a subset of uncolored vertices $X \subseteq V$ in the beginning of iteration t, let $P_t(X) := \sum_{v \in X} \lambda^{-r_t(v)}$. We define the potential in the beginning of iteration t to be $P_t := P_t(V_t)$, where V_t denotes the subset of all uncolored vertices in the beginning of iteration t. Notice that the value of the potential in the beginning of the algorithm (i.e., in the first iteration) is

$$P_1 = \sum_{v \in V} \lambda^{-|L_v|} < 1.$$

Our goal is to show that, with the right choice of the independent set in each iteration, we can make sure that for any iteration t and every vertex $v \in V_t$ the inequality $r_t(v) > 0$ holds. In order to achieve this, we will show that, with the right choice of the subset of vertices colored in each iteration, the potential function P_t is non-increasing in t. This will imply that for any iteration t and every uncolored vertex $v \in V_t$ we have:

$$\lambda^{-r_t(v)} \leq P_t \leq P_1 < 1$$

and hence $r_t(v) > 0$, as required.

Assume that the potential function is non-increasing up to iteration t. Let P_t be the value of the potential function in the beginning of iteration t and let c be the color associated with iteration t. Recall that V_t denotes the set of uncolored vertices that are considered in iteration t, and $V^c \subseteq V_t$ denotes the subset of uncolored vertices that contain the color c in their lists. Put $P' = P_t(V_t \setminus V^c)$ and $P'' = P_t(V^c)$. Note that $P_t = P' + P''$. Let us describe how we find the independent set of vertices to be colored at iteration t. First, we find an auxiliary proper coloring of the hypergraph $H[V^c]$ with k colors (here we use the hereditary k-colorability property of the hypergraph). Consider the color class U which has the largest potential $P_t(U)$. Since the vertices in V^c are partitioned into at most k independent subsets $U_1, \ldots, U_k$ and $P'' = \sum_{i=1}^{k} P_t(U_i)$, then by the pigeon-hole principle there is an index j for which $P_t(U_j) \geq P''/k$. We choose $U = U_j$ as the independent set to be colored at iteration t. Notice that, in this case, the value $r_{t+1}(v) = r_t(v) - 1$ for every vertex $v \in V^c \setminus U$, and all vertices in U are colored. For vertices in $V_t \setminus V^c$, there is no change in the size of their lists. Thus, the value P_{t+1} of the potential function at the end of iteration t (and in the beginning of iteration $t+1$) is $P_{t+1} \leq P' + \lambda(1 - \frac{1}{k})P''$. Since $\lambda = \frac{k}{k-1}$, we have that $P_{t+1} \leq P' + P'' = P_t$, as required. ∎

3.3.1. A relation between chromatic and choice number in general hypergraphs.

Using a probabilistic argument, Cheilaris et al. [16] proved the following general theorem for arbitrary hypergraphs and arbitrary colorings with the so-called refinement property:

Definition 3.29. We call C' a *refinement* of a coloring C if $C(x) \neq C(y)$ implies $C'(x) \neq C'(y)$. A class $\mathcal{C}$ of colorings is said to have *the refinement property* if every refinement of a coloring in the class is also in the class.

The class of conflict-free colorings and the class of proper colorings are examples of classes which have the refinement property. On the other hand, the class of unique-maximum colorings does not have this property.

For a class $\mathcal{C}$ of colorings, one can naturally extend the notions of chromatic number $\chi_{\mathcal{C}}$ and choice number $ch_{\mathcal{C}}$ to $\mathcal{C}$.

Theorem 3.30 ([16]). *For every class of colorings $\mathcal{C}$ that has the refinement property and every hypergraph H with n vertices, $ch_{\mathcal{C}}(H) \leq \chi_{\mathcal{C}}(H) \cdot \ln n + 1$.*

Proof. If $k = \chi_{\mathcal{C}}(H)$, then there exists a $\mathcal{C}$-coloring C of H with colors $\{1, \ldots, k\}$, which induces a partition of V into k classes: $V_1 \cup V_2 \cup \cdots \cup V_k$. Consider a family $\mathcal{L} = \{L_v\}_{v \in V}$, such that for every v, $|L_v| = k^* > k \cdot \ln n$. We wish to find a family $\mathcal{L}' = \{L'_v\}_{v \in V}$ with the following properties:

1. For every $v \in V$, $L'_v \subseteq L_v$.

2. For every $v \in V$, $L'_v \neq \emptyset$.

3. For every $i \neq j$, if $v \in V_i$ and $u \in V_j$, then $L'_v \cap L'_u = \emptyset$.

Obviously, if such a family $\mathcal{L}'$ exists, then there exists a $\mathcal{C}$-coloring from $\mathcal{L}'$: For each $v \in V$, pick a color $x \in L'_v$ and assign it to v.

We create the family $\mathcal{L}'$ randomly as follows: For each element in $\cup \mathcal{L}$, assign it uniformly at random to one of the k classes of the partition $V_1 \cup \cdots \cup V_k$. For every vertex $v \in V$, say with $v \in V_i$, we create L'_v, by keeping only elements of L_v that were assigned through the above random process to v's class, V_i.

The family $\mathcal{L}'$ obviously has properties 1 and 3. We will prove that with positive probability it also has property 2.

For a fixed v, the probability that $L'_v = \emptyset$ is at most

$$\left(1 - \frac{1}{k}\right)^{k^*} \leq e^{-k^*/k} < e^{-\ln n} = \frac{1}{n}$$

and therefore, using the union bound, the probability that for at least one vertex v, $L'_v = \emptyset$, is at most

$$n\left(1 - \frac{1}{k}\right)^{k^*} < 1.$$

Thus, there is at least one family $\mathcal{L}'$ where property 2 also holds, as claimed. ∎

Corollary 3.31. *For every hypergraph H,*

$$ch_{cf}(H) \leq \chi_{cf}(H) \cdot \ln n + 1.$$

Corollary 3.32. *For every hypergraph H,*

$$ch(H) \leq \chi(H) \cdot \ln n + 1.$$

The argument in the proof of Theorem 3.30 is a generalization of an argument first given in [24], proving that any bipartite graph with n vertices is $O(\log n)$-choosable (see also [4]).

4. Non-Geometric Hypergraphs

Pach and Tardos [40] investigated the CF-chromatic number of arbitrary hypergraphs and proved that the inequality:

$$\chi_{cf}(H) \leq 1/2 + \sqrt{2m + 1/4}$$

holds for every hypergraph H with m edges, and that this bound is tight. Cheilaris et al. [16] strengthened this bound in two ways by proving that:

$$ch_{um}(H) \leq 1/2 + \sqrt{2m + 1/4}.$$

If, in addition, every hyperedge contains at least $2t - 1$ vertices (for $t \geq 3$) then Pach and Tardos showed that:

$$\chi_{cf}(H) = O(m^{\frac{1}{t}} \log m).$$

Using the Lovász Local Lemma, they show that the same result holds for hypergraphs, in which the size of every edge is at least $2t-1$ and every edge intersects at most m other edges.

Hypergraphs induced by neighborhoods in graphs. A particular interest arises when dealing with hypergraphs induced by neighborhoods of vertices of a given graph. Given a graph $G = (V, E)$ and a vertex $v \in V$, denote by $N_G(v) = N(v)$ the set of all neighbors of v in G together with v and refer to it as the *neighborhood of v*. Call the set $\dot{N}(G) = N_G(v) \setminus \{v\}$ the *pointed neighborhood of v*. The hypergraph H associated with the neighborhoods of G has its vertex set $V(H) = V$ and its edge set $E(H) = \{N_G(v)|v \in V\}$ and the hypergraph $\dot{H}$ associated with the pointed neighborhoods of G has $V(\dot{H}) = V$ and $E(\dot{H}) = \{\dot{N}_G(v)|v \in V\}$. The *conflict-free chromatic parameter* $\kappa_{CF}(G)$ is defined simply as $\chi_{\mathrm{cf}}(H)$ and the *pointed* version of this parameter $\dot{\kappa}_{CF}(G)$ is defined analogously as $\chi_{\mathrm{cf}}(\dot{H})$.

We start with an example taken from [40] in order to provide some basic insights into the relation between these two parameters. Let K'_s be the graph obtained from the complete graph K_s on s vertices by subdividing each edge with a new vertex. Each pair of the s original vertices form the pointed neighborhood of one of the new vertices, so all original vertices must receive different colors in any conflict-free coloring of the corresponding hypergraph $\dot{H}$. Thus, we have $\dot{\kappa}_{CF}(K'_s) \geq s$ and it is easy to see that equality holds here. On the other hand, K'_s is bipartite and any proper coloring of a graph is also a conflict-free coloring of the hypergraph formed by the neighborhoods of its vertices. This shows that $\kappa_{CF}(K'_s) = 2$, for any $s \geq 2$. The example illustrates that the pointed conflict-free chromatic parameter of a graph cannot be bounded from above by any function of its non-pointed variant. For many other graphs, the non-pointed parameter can be larger than the pointed parameter. For instance, let G denote the graph obtained from the complete graph K_4 by subdividing a single edge with a vertex. It is easy to check that $\kappa_{CF}(G) = 3$, while $\dot{\kappa}_{CF}(G) = 2$. However, it is not difficult to verify that

$$\kappa_{CF}(G) \leq 2\dot{\kappa}_{CF}(G)$$

for any graph G. This inequality holds, because in a conflict-free coloring of the pointed neighborhoods, each neighborhood $N(x)$ also has a vertex whose color is not repeated in $N(x)$, unless x has degree one in the subgraph spanned by one of the color classes. One can fix this by carefully splitting each color class into two. The following theorems were proved in [40]:

Theorem 4.1 ([40]). *The conflict-free chromatic parameter of any graph G with n vertices satisfies $\kappa_{CF}(G) = O(\log^2 n)$. The corresponding coloring can be found by a deterministic polynomial time algorithm.*

Theorem 4.2 ([40]). *There exist graphs of n vertices with conflict-free chromatic parameter* $\Omega(\log n)$.

Problem 6. Close the gap between the last two bounds.

For graphs with maximum degree Δ, a slightly better upper-bound is known:

Theorem 4.3 ([40]). *The conflict-free chromatic parameter of any graph G with maximum degree* Δ *satisfies* $\kappa_{CF}(G) = O(\log^{2+\epsilon} \Delta)$ *for any* $\epsilon > 0$. *The corresponding coloring can be found by a deterministic polynomial time algorithm.*

Hypergraphs induced by simple paths in graphs. As mentioned in the introduction, a particular interest is in hypergraphs induced by simple paths in a given graph: Recall the that given a graph G, we consider the hypergraph $H = (V, E')$ where a subset $V' \subset V$ is a hyperedge in E' if and only if V' is the set of vertices in some simple path of G. As mentioned before, the parameter $\chi_{\text{um}}(H)$ is known as the vertex ranking number of G and was studied in other context in the literature (see, e.g., [32, 45]). An interesting question arises when trying to understand the relation between the two parameters $\chi_{\text{cf}}(H)$ and $\chi_{\text{um}}(H)$. This line of research was pursued in [15] and [17]. Cheilaris and Tóth proved the following:

Theorem 4.4 ([17]). *(i) Let G be a simple graph and let H be the hypergraph induced by paths in G as above: Then* $\chi_{\text{um}}(H) \leq 2^{\chi_{\text{cf}}(H)} - 1$.

(ii) There is is a sequence of such hypergraphs $\{H_i\}_{i=1}^{\infty}$ *induced by paths such that*

$$\lim_{n \to \infty} \frac{\chi_{\text{um}}(H_n)}{\chi_{\text{cf}}(H_n)} = 2.$$

Narrowing the gaps between the two parameters for such hypergraphs is an interesting open problem:

Problem 7. Let $f(k)$ denote the function of the least integer such that for every hypergraph H induced by path in a graph G we have that $\chi_{\text{um}}(H) \leq f(\chi_{\text{cf}}(H))$. Find the asymptotic behavior of f.

5. ALGORITHMS

Until now we were mainly concerned with the combinatorial problem of obtaining bounds on the CF-chromatic number of various hypergraphs. We now turn our attention to the computational aspect of the corresponding optimization problem. Even et al. [25] proved that given a finite set D of discs in the plane, it is NP-hard to compute an optimal CF-coloring for $H(D)$; namely, a CF-coloring of $H(D)$ using a minimum number of colors. This hardness result holds even if all discs have the same radius. However, as mentioned in the introduction, any set D of n discs admits a CF-coloring that uses $O(\log n)$ colors and such a coloring can be found in deterministic polynomial time (in fact in $O(n \log n)$ time). This trivially implies that such an algorithm serves as an $O(\log n)$ approximation algorithm for the corresponding optimization problem.

5.1. Approximation Algorithms

Given a finite set D of discs in the plane, the *size ratio* of D denoted by $\rho = \rho(D)$ is the ratio between the maximum and minimum radii of discs in D. For simplicity, we may assume that the smallest radius is 1. For each $i \geq 1$, let D^i denote the subset of discs in D whose radius is in the range $[2^{i-1}, 2^i)$. Let $\phi_{2^i}(D^i)$ denote the maximum number of centers of discs in D^i that are contained in a $2^i \times 2^i$ square. Refer to $\phi_{2^i}(D^i)$ as the *local density* of D^i (with respect to $2^i \times 2^i$ square). For a set of points X in $\mathbb{R}^2$ let $D_r(X)$ denote the set of $|X|$ discs with radius r centered at the points of X. The following algorithmic results were provided in [25].

Theorem 5.1 ([25]).

1. Given a finite set D of discs with size-ratio ρ, there exists a polynomial-time algorithm that compute a CF-coloring of D using

$$O\left(\min\{(\log \rho) \cdot \max_i\{\log \phi_{2^i}(D^i)\}, \log |D|\}\right)$$

colors.

2. Given a finite set of centers $X \subset \mathbb{R}^2$, there exists a polynomial-time algorithm that computes a UM-coloring χ of the hypergraph

induced X with respect to all discs using $O(\log |X|)$ colors. This is equivalent to the following: If we color $D_r(X)$ by assigning each disc $d \in D_r(X)$ the color of its center then this is a valid UM-coloring of the hypergraph $H(D_r(X))$ for every radius r.

The tightness of Theorem 5.1 follows from the fact that for any integer n, there exists a set D of n unit discs with $\phi_1(D) = n$ for which $\Omega(\log n)$ colors are necessary in every CF-coloring of D.

In the first part of Theorem 5.1 the discs are not necessarily congruent. That is, the size-ratio ρ may be bigger than 1. In the second part of Theorem 5.1, the discs are congruent (i.e., the size-ratio equals 1). However, the common radius is not determined in advance. Namely, the order of quantifiers in the second part of the theorem is as follows: Given the locations of the disk centers, the algorithm computes a coloring of the centers (of the discs) such that this coloring is conflict-free *for every* radius r.

Building on Theorem 5.1, Even et al. [25] also obtain two bi-criteria CF-coloring algorithms for discs having the same (unit) radius. In both cases the algorithm uses only few colors. In the first case this comes at a cost of not serving a small area that is covered by the discs (i.e., an area close to the boundary of the union of the discs). In the second case, all the area covered by the discs is served, but the discs are assumed to have a slightly larger radius. A formal statement of these bi-criteria results is as follows:

Theorem 5.2 ([25]). *For every $0 < \varepsilon < 1$ and every finite set of centers $X \subset \mathbb{R}^2$, there exist polynomial-time algorithms that compute colorings as follows:*

1. *A coloring χ of $D_1(X)$ using $O\left(\log \frac{1}{\varepsilon}\right)$ colors for which the following holds: The area of the set of points in $\bigcup D_1(X)$ that are not served with respect to χ is at most an ε-fraction of the total area of $D_1(X)$.*

2. *A coloring of $D_{1+\varepsilon}(X)$ that uses $O\left(\log \frac{1}{\varepsilon}\right)$ colors such that every point in $\bigcup D_1(X)$ is served.*

In other words, in the first case, the portion of the total area that is not served is an exponentially small fraction as a function of the number of colors. In the second case, the increase in the radius of the discs is exponentially small as a function of the number of colors.

The following problem seems like a non-trivial challenge.

Problem 8. Is there a constant factor approximation algorithm for finding an optimal CF-coloring for a finite set of discs in the plane?

Remark. In the special case that all discs are congruent (i.e., have the same radius) Lev-Tov and Peleg [36] have recently provided a constant-factor approximation algorithm.

5.1.1. An $O(1)$-Approximation for CF-Coloring of Rectangles and Regular Hexagons.

Recall that Theorem 2.7 states that every set of n axis-parallel rectangles can be CF-colored with $O(\log^2 n)$ colors and such a coloring can be found in polynomial time.

Let $\mathcal{R}$ denote a set of axis-parallel rectangles. Given a rectangle $R \in \mathcal{R}$, let $w(R)$ ($h(R)$, respectively) denote the width (height, respectively) of R. The *size-ratio* of $\mathcal{R}$ is defined by $\max \left\{ \frac{w(R_1)}{w(R_2)}, \frac{h(R_1)}{h(R_2)} \right\}_{R_1, R_2 \in \mathcal{R}}$.

The size ratio of a collection of regular hexagons is simply the ratio of the longest side length and the shortest side length.

Theorem 5.3 ([25]). *Let $\mathcal{R}$ denote either a set of axis-parallel rectangles or a set of homothets of a regular hexagons. Let ρ denote the size-ratio of $\mathcal{R}$ and let $\chi_{\mathrm{opt}}(\mathcal{R})$ denote an optimal CF-coloring of $\mathcal{R}$.*

1. *If $\mathcal{R}$ is a set of rectangles, then there exists a polynomial-time algorithm that computes a CF-coloring χ of $\mathcal{R}$ such that $|\chi(\mathcal{R})| = O((\log \rho + 1)^2 \cdot |\chi_{\mathrm{opt}}(\mathcal{R})|)$.*

2. *If $\mathcal{R}$ is a set of hexagons, then there exists a polynomial-time algorithm that computes a CF-coloring χ of $\mathcal{R}$ such that $|\chi(\mathcal{R})| = O((\log \rho + 1) \cdot |\chi_{\mathrm{opt}}(\mathcal{R})|)$.*

For a constant size-ratio ρ, Theorem 5.3 implies a constant approximation algorithm.

5.2. Online CF-Coloring

Recall the motivation to study CF-coloring in the context of cellular antanae. To capture a dynamic scenario where antennae can be added to the network, Chen et al. [18] introduced an online version of the CF coloring problem. As we shall soon see, the online version of the problem is considerably harder, even in the one-dimensional case, where the static version (i.e., CF-coloring the discrete intervals hypergraph) is trivial and fully understood.

5.2.1. Points with respect to intervals. Let us start with the simplest possible example where things become highly non-trivial in an online setting. We start with the dynamic extension of the discrete interval hypergraph case. Thats is, we deal with coloring of points on the line, with respect to interval ranges. We maintain a finite set $P \subset \mathbb{R}$. Initially, P is empty, and an adversary repeatedly insert points into P, one point at a time. We denote by $P(t)$ the set P after the tth point has been inserted. Each time a new point p is inserted, we need to assign a color $c(p)$ to it, which is a positive integer. Once the color has been assigned to p, it cannot be changed in the future. The coloring should remain a valid CF-coloring at all times. That is, as in the static case, for any interval I that contains points of $P(t)$, there is a color that appears exactly once in I.

We begin by examining a natural, simple, and obvious coloring algorithm (referred to as the UniMax greedy algorithm) which might be inefficient in the worst case. Chen et al. [18] presented an efficient 2-stage variant of the UniMax greedy algorithm and showed that the maximum number of colors that it uses is $\Theta(\log^2 n)$.

As in the case in most CF-coloring of hypergraphs that were tackled so far, we wish to maintain the unique maximum invariant. At any given step t the coloring of $P(t)$ is a UM-coloring.

The following simple-minded algorithm for coloring an inserted point p into the current set $P(t)$ is used. We say that the newly inserted point p *sees* a point x if all the colors of the points between p and x (exclusive) are smaller than $c(x)$. In this case we also say that p sees the color $c(x)$. Then p gets the smallest color that it does not see. (Note that a color can be seen from p either to the left or to the right, but not in both directions; see below.) Refer to this algorithm as the *Unique Maximum Greedy* algorithm, or the UniMax greedy algorithm, for short.

Below is an illustration of the coloring rule of the UniMax greedy algorithm. The left column gives the colors (integers in the range $1, 2, \ldots, 6$) assigned to the points in the current set P and the location of the next point to be inserted (indicated by a period). The right column gives the colors "seen" by the new point. The colors seen to the left precede the $\cdot$, and those seen to the right succeed the ·.

1·	[1·]
1 · 2	[1 · 2]
1 · 32	[1 · 3]
12 · 32	[2 · 3]
121 · 32	[21 · 3]
121 · 432	[21 · 4]
121 · 3432	[21 · 34]
1215 · 3432	[5 · 34]
1215 · 13432	[5 · 134]
12152 · 13432	[52 · 134]
121526 · 13432	[6 · 134]

Correctness. The correctness of the algorithm is established by induction on the insertion order. First, note that no color can be seen twice from p: This is obvious for two points that lie both to the left or both to the right of p. If p sees the same color at a point u to its left and at a point v to its right, then the interval $[u, v]$, before p is inserted, does not have a unique maximum color; thus this case is impossible, too. Next, if p is assigned color c, any interval that contains p still has a unique maximum color: This follows by induction when the maximum color is greater than c. If the maximum color is c, then it cannot be shared by another point u in the interval, because then p would have seen the nearest such point and thus would not be assigned color c. It is also easy to see that the algorithm assigns to each newly inserted point the smallest possible color that maintains the invariant of a unique maximum color in each interval. This makes the algorithm *greedy* with respect to the unique maximum condition.

Special insertion orders. Denote by $C(P(t))$ the sequence of colors assigned to the points of $P(t)$, in left-to-right order along the line.

The *complete binary tree sequence* S_k of order k is defined recursively as $S_1 = (1)$ and $S_k = S_{k-1} \| (k) \| S_{k-1}$, for $k > 1$, where $\|$ denotes concatenation. Clearly, $|S_k| = 2^k - 1$.

For each pair of integers $a < b$, denote by $C_0(a, b)$ the following special sequence. Let k be the integer satisfying $2^{k-1} \le b < 2^k$. Then $C_0(a, b)$ is the subsequence of S_k from the ath place to the bth place (inclusive). For example, $C_0(5, 12)$ is the subsequence $(1, 2, 1, 4, 1, 2, 1, 3)$ of $(1, 2, 1, 3, 1, 2, 1, 4, 1, 2, 1, 3, 1, 2, 1)$.

Lemma 5.4. (a) *If each point is inserted into P to the right of all preceding points, then $C(P(t)) = C_0(1,t)$.*

(b) *If each point is inserted into P to the left of all preceding points, then $C(P(t)) = C_0(2^k - t, 2^k - 1)$, where k satisfies $2^{k-1} \le t < 2^k$.*

Proof. The proof is easy and is left as an exercise to the reader. ■

Unfortunately, the UniMax greedy algorithm might be very inefficient as was shown in [18]:

Theorem 5.5 ([18]). *The UniMax greedy algorithm may require $\Omega(\sqrt{n})$ colors in the worst case for a set of n points.*

Problem 9. Obtain an upper bound for the maximum number of colors that the algorithm uses for n inserted points. It is conjectured that the bound is close to the $\Omega(\sqrt{n})$ lower bound. At the moment, there is no known sub-linear upper bound.

Related algorithms. *The First-Fit algorithm—another greedy strategy.* The UniMax greedy algorithm is greedy for maintaining the unique maximum invariant. Perhaps it is more natural to consider a greedy approach in which we want only to enforce the standard CF property. That is, we want to assign to each newly inserted point the *smallest* color for which the CF property continues to hold. There are cases where this *First-Fit* greedy algorithm uses fewer colors than the UniMax greedy algorithm: Consider an insertion of five points in the order (1 3 2 4 5). The UniMax greedy algorithm produces the color sequence (1 3 2 1 4), whereas the First-Fit algorithm produces the coloring (1 3 2 1 2). Unfortunately, Bar-Noy et al. [11] have shown that there are sequences with $2i + 3$ elements that force the algorithm to use $i + 3$ colors, and this bound is tight.

CF coloring for unit intervals. Consider the special case where we want the CF property to hold only for *unit intervals*. In this case, $O(\log n)$ colors suffice: Partition the line into the unit intervals $J_i = [i, i+1)$ for $i \in \mathbb{Z}$. Color the intervals J_i with even i as white, and those with odd i as black. Note that any unit interval meets only one white and one black interval. We color the points in each J_i independently, using the same set of "light colors" for each white interval and the same set of "dark colors" for each black interval. For each J_i, we color the points that it contains using the UniMax greedy algorithm, except that new points inserted into J_i between two previously inserted points get a special color, color 0. It is easily checked that the resulting coloring is CF with respect to unit intervals. Since we effectively insert points into any J_i only to the left or to the right of the previously inserted points, Lemma 5.4(c) implies that the algorithm uses only $O(\log n)$ (light and dark) colors. We remark that this algorithm satisfies the unique maximum color property for unit-length intervals.

We note that, in contrast to the static case (which can always be solved with $O(1)$ colors), $\Omega(\log n)$ colors may be needed in the worst case. Indeed, consider a left-to-right insertion of n points into a sufficiently small interval. Each contiguous subsequence σ of the points will be a suffix of the whole sequence at the time the rightmost element of σ is inserted. Since such a suffix can be cut off the current set by a unit interval, it must have a unique color. Hence, at the end of insertion, *every* subsequence must have a unique color, which implies (see [25, 47]) that $\Omega(\log n)$ colors are needed.

An efficient online deterministic algorithm for points with respect to intervals. We describe an efficient online algorithm for coloring points with respect to intervals that was obtained in [18]. This is done by modifying the UniMax greedy algorithm into a deterministic 2-stage coloring scheme. It is then shown that it uses only $O(\log^2 n)$ colors. The algorithm is referred to as the *leveled UniMax greedy algorithm*.

Let x be the point which we currently insert. We assign a color to x in two steps. First we assign x to a *level*, denoted by $\ell(x)$. Once x is assigned to level $\ell(x)$ we give it an actual color among the set of colors dedicated to $\ell(x)$. We maintain the invariant that each color is used by at most one level. Formally, the colors that we use are pairs $(\ell(x), c(x)) \in \mathbb{Z}^2$, where $\ell(x)$ is the level of x and $c(x)$ is its integer color within that level.

Modifying the definition from the UniMax greedy algorithm, we say that point x *sees* point y (or that point y is *visible* to x) if and only if for every point z between x and y, $\ell(z) < \ell(y)$. When x is inserted, we set $\ell(x)$ to be the smallest level ℓ such that either to the left of x or to the right of x (or in both directions) there is no point y visible to x at level ℓ.

To give x a color, we now consider only the points of level $\ell(x)$ that x can see. That is, we discard every point y such that $\ell(y) \neq \ell(x)$, and every point y such that $\ell(y) = \ell(x)$ and there is a point z between x and y such that $\ell(z) > \ell(y)$. We apply the UniMax greedy algorithm so as to color x with respect to the sequence P_x of the remaining points, using the colors of level $\ell(x)$ only. That is, we give x the color $(\ell(x), c(x))$, where $c(x)$ is the smallest color that ensures that the coloring of P_x maintains the unique maximum color condition. This completes the description of the algorithm. See Figure 5 for an illustration.

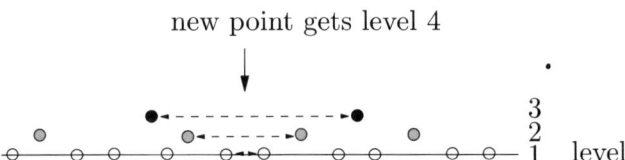

Fig. 5. Illustrating the 2-stage deterministic algorithm. An insertion order that realizes the depicted assignment of levels to points is to first insert all level-1 points from left to right, then insert the level-2 points from left to right, and then the level-3 points

We begin the analysis of the algorithm by making a few observations on its performance.

(a) Suppose that a point x is inserted and is assigned to level $i > 1$. Since x was not assigned to any level $j < i$, it must see a point ℓ_j at level j that lies to its left, and another such point r_j that lies to its right. Let $E_j(x)$ denote the interval $[\ell_j, r_j]$. Note that, by definition, these intervals are *nested*, that is, $E_j(x) \subset E_k(x)$ for $j < k < i$. See Figure 5.

(b) We define a *run* at level i to be a maximal sequence of points $x_1 < x_2 < \cdots < x_k$ at level i, such that all points between x_1 and x_k that are distinct from $x_2, x_3, \ldots, x_{k-1}$ are assigned to levels smaller than i. Whenever a new point x is assigned to level i and is inserted into a run of that level, it is always inserted either to the left or to the right of all points in the run. Moreover, the actual color that x gets is determined solely from the colors of the points already in the run. See Figure 5.

(c) The runs keep evolving as new points are inserted. A run may either grow when a new point of the same level is inserted at its left or right end (note that other points at smaller levels may separate the new point from the former end of the run) or split into two runs when a point of a higher level is inserted somewhere between its ends.

(d) As in observation (a), the points at level i define *intervals*, called *i-intervals*. Any such interval E is a contiguous subsequence $[x, y]$ of P, so that x and y are both at level i and all the points between x and y have smaller levels. E is formed when the second of its endpoints, say x, is inserted. We say that x *closes* the interval E and refer to it as a *closing point*. Note that, by construction, x cannot close another interval.

(e) Continuing observation (a), when x is inserted, it *destroys* the intervals $E_j(x)$, for $j < i$, into which it is inserted, and only these intervals. That is, each of these intervals now contains a point with a level greater than that of its endpoints, so it is no longer a valid interval. We charge x to the set of the closing endpoints of all these intervals. Clearly, none of these points will ever be charged again by another insertion (since it is the closing endpoint of only one interval, which is now destroyed). We maintain a forest F, whose nodes are all the points of P. The leaves of F are all the points at level 1. When a new point x is inserted, we make it a new root of F, and the parent of all the closing points that it charges. Since these points have smaller levels than x, and since none of these points becomes a child of another parent, it follows that F is indeed a forest.

Note that the nonclosing points can only be roots of trees of F. Note also that a node at level i has exactly $i - 1$ children, exactly one at each level $j < i$. Hence, each tree of F is a *binomial tree* (see [22]); if its root has level i, then it has 2^i nodes.

This implies that if m is the maximal level assigned after n points have been inserted, then we must have $2^m \leq n$, or $m \leq \log n$. That is, the algorithm uses at most $\log n$ levels.

We next prove that the algorithm uses only $O(\log n)$ colors at each level. We recall the way runs evolve: They grow by adding points at their right or left ends, and split into prefix and suffix subruns, when a point with a larger level is inserted in their middle.

Lemma 5.6. *At any time during the insertion process, the colors assigned to the points in a run form a sequence of the form $C_0(a, b)$. Moreover, when the jth smallest color of level i is given to a point x, the run to which x is appended has at least $2^{j-2} + 1$ elements (including x).*

Proof. The proof proceeds by induction through the sequence of insertion steps and is based on the following observation. Let σ be a contiguous subsequence of the complete binary tree sequence S_{k-1}, and let x be a point added, say, to the left of σ. If we assign to x color $c(x)$, using the UniMax greedy algorithm, then $(c(x)) \parallel \sigma$ is a contiguous subsequence of either S_{k-1} or S_k. The latter happens only if σ contains $S_{k-2} \parallel (k-1)$ as a prefix. Symmetric properties hold when x is inserted to the right of σ. We omit the straightforward proof of this observation. ∎

As a consequence we have.

Theorem 5.7 ([18]). (a) *The algorithm uses at most* $(2+\log n)\log n$ *colors.*

(b) *At any time, the coloring is a valid CF-coloring.*

(c) *In the worst case the algorithm may be forced to use* $\Omega(\log^2 n)$ *colors after* n *points are inserted.*

Proof. (a) We have already argued that the number of levels is at most $\log n$. Within a level i, the kth smallest color is assigned when a run contains at least 2^{k-2} points. Hence $2^{k-2} \le n$, or $k \le 2 + \log n$, and (a) follows.

To show (b), consider an arbitrary interval I. Let ℓ be the highest level of a point in I. Let $\sigma = (y_1, y_2, \ldots, y_j)$ be the sequence of the points in I of level ℓ. Since ℓ is the highest level in I, σ is a contiguous subsequence of some run, and, by Lemma 5.6, the sequence of the colors of its points is also of the form $C_0(a', b')$. Hence, there is a point $y_i \in \sigma$ which is uniquely colored among $y_1, y_2, \ldots, y_j$ by a color of level ℓ.

To show (c), we construct a sequence P so as to force its coloring to proceed level by level. We first insert 2^{k-1} points from left to right, thereby making them all be assigned to level 1 and colored with k different colors of that level. Let P_1 denote the set of these points. We next insert a second batch of 2^{k-2} points from left to right. The first point is inserted between the first and second points of P_1, the second point between the third and fourth points of P_1, and so on, where the jth new point is inserted between the $(2j-1)$th and $(2j)$th points of P_1. By construction, all points in the second batch are assigned to level 2, and they are colored with $k-1$ different colors of that level. Let P_2 denote the set of all points inserted so far. P_2 is the concatenation of 2^{k-2} triples, where the levels in each triple are $(1, 2, 1)$. We now insert a third batch of 2^{k-3} points from left to right. The first point is inserted between the first and second triples of P_2, the second point between the third and fourth triples of P_2, and so on, where the jth new point is inserted between the $(2j-1)$th and $(2j)$th triples of P_2. By construction, all points in the third batch are assigned to level 3, and they are colored with $k-2$ different colors of that level.

The construction is continued in this manner. Just before inserting the ith batch of 2^{k-i} points, we have a set P_{i-1} of $2^{k-1} + \cdots + 2^{k-i+1}$ points, which is the concatenation of 2^{k-i+1} tuples, where the sequences of levels in each of these tuples are all identical and equal to the "complete binary tree sequence" $C_0(1, 2^{i-1} - 1)$, as defined above (whose elements now encode levels rather than colors). The points of the ith batch are inserted from left to right, where the jth point is inserted between the $(2j-1)$th and $(2j)$th tuples of P_{i-1}. By construction, all points in the ith batch are assigned to level i and are colored with $k - i + 1$ different colors of that level. Proceeding in this manner, we end the construction by inserting the $(k - 1)$th batch, which consists of a single point that is assigned to level k. Altogether we have inserted $n = 2^k - 1$ points and forced the algorithm to use $k + (k - 1) + \cdots + 1 = k(k + 1)/2 = \Omega(\log^2 n)$ different colors. ∎

Given that the only known lower bound for this online CF-coloring problem is $\Omega(\log n)$ which holds also in the static problem, its a major open problem to close the gap with the $O(\log^2 n)$ upper bound provided by the algorithm above.

Problem 10. Find a deterministic online CF-coloring for coloring points with respect to intervals which uses $o(\log^2 n)$ colors in the worst case or improve the $\Omega(\log n)$ lower bound.

Other Online models. For the case of online CF-coloring points with respect to intervals, other models of a weaker adversary were studied in [12]. For example, a natural assumption is that the adversary reveals, for a newly inserted point, its final position among the set of all points in the end of the online input. This is referred to as the *online absolute positions model.* In this model an online CF-coloring algorithm that uses at most $O(\log n)$ colors is presented in [12].

5.2.2. Points with respect to halfplanes or unit discs. In [18] it was shown that the two-dimensional variant of online CF-coloring a given sequence of inserted points with respect to arbitrary discs is hopeless as there exists sequences of n points for which every CF-coloring requires n distinct colors. However if we require a CF-coloring with respect to congruent discs or with respect to half-planes, there is some hope. Even though no efficient deterministic online algorithms are known for such cases, some efficient randomized algorithms that uses expected $O(\log n)$ colors are provided in [19, 10] under the assumption that the adversary is oblivious to the random bits used by the algorithm.

Chen Kaplan and Sharir [19] introduced an $O(log^3n)$ deterministic algorithm for online CF-coloring any n nearly-equal axis-parallel rectangles in the plane.

5.2.3. Degenerate hypergraphs. Next, we describe the general framework of [10] for online CF-coloring any hypergraph. This framework is used to obtain efficient randomized online algorithms for hypergraphs provided that a special parameter referred to as the *degeneracy* of the underlying hypergraph is small. This notion extends the notion of a degenerate graph to that of a hypergraph:

Definition 5.8. Let $k > 0$ be a fixed integer and let $H = (V, E)$ be a hypergraph on the n vertices $v_1, \ldots, v_n$. For a permutation $\pi \colon \{1, \ldots, n\} \to \{1, \ldots, n\}$ define the n partial sums, indexed by $t = 1, \ldots, n$,

$$S_t^\pi = \sum_{j=1}^t d(v_{\pi(j)}),$$

where

$$d(v_{\pi(j)}) = \left| \left\{ i < j \mid \{v_{\pi(i)}, v_{\pi(j)}\} \in G(H(\{v_{\pi(1)}, \ldots, v_{\pi(j)}\})) \right\} \right|,$$

that is, $d(v_{\pi(j)})$ is the number of neighbors of $v_{\pi(j)}$ in the Delaunay graph of the hypergraph induced by $\{v_{\pi(1)}, \ldots, v_{\pi(j)}\}$. Assume that for all permutations π and for every $t \in \{1, \ldots, n\}$ we have

(1) $S_t^\pi \le kt.$

Then, we say that H is *k-degenerate*.

Let $H = (V, E)$ be any hypergraph. We define a framework that colors the vertices of V in an online fashion, i.e., when the vertices of V are revealed by an adversary one at a time. At each time step t, the algorithm must assign a color to the newly revealed vertex v_t. This color cannot be changed in future times $t' > t$. The coloring has to be conflict-free for all the induced hypergraphs $H(V_t)$ with $t = 1, \ldots, n$, where $V_t \subseteq V$ is the set of vertices revealed by time t.

For a fixed positive integer h, let $A = \{a_1, \ldots, a_h\}$ be a set of h *auxiliary* colors. This auxiliary colors set should not be confused with the set of *main* colors used for the conflict-free coloring: $\{1, 2, \ldots\}$. Let $f \colon \mathbb{N}^+ \to A$ be some fixed function. In the following, we define the framework that depends on the choice of the function f and the parameter h.

A table (to be updated online) is maintained with row entries indexed by the variable i with range in $\mathbb{N}^+$. Each row entry i at time t is associated with a subset $V_t^i \subseteq V_t$ in addition to an auxiliary proper non-monochromatic coloring of $H(V_t^i)$ with at most h colors. The subsets V_t^i are nested. Namely, $V_t^{i+1} \subset V_t^i$ for every i. Informally, we think of a newly inserted vertex as trying to reach its final entry by some decision process. It starts with entry 1 and continue "climbing" to higher levels as long as it does not succeed to get its final color. We say that $f(i)$ is the auxiliary color that *represents* entry i in the table. At the beginning all entries of the table are empty. Suppose all entries of the table are updated until time $t-1$ and let v_t be the vertex revealed by the adversary at time t. The framework first checks if an auxiliary color can be assigned to v_t such that the auxiliary coloring of V_{t-1}^1 together with the color of v_t is a proper non-monochromatic coloring of $H(V_{t-1}^1 \cup \{v_t\})$. Any (proper non-monochromatic) coloring procedure can be used by the framework. For example a first-fit greedy method in which all colors in the order $a_1, \ldots, a_h$ are checked until one is found. If such a color cannot be found for v_t, then entry 1 is left with no changes and the process continues to the next entry. If however, such a color can be assigned, then v_t is added to the set V_{t-1}^1. Let c denote such an auxiliary color assigned to v_t. If this color is the same as $f(1)$ (the auxiliary color that represents entry 1), then the final color in the online conflict-free coloring of v_t is 1 and the updating process for the t-th vertex stops. Otherwise, if an auxiliary color cannot be found or if the assigned auxiliary color is not the same as $f(1)$, then the updating process continues to the next entry. The updating process stops at the first entry i for which v_t is both added to V_t^i and the auxiliary color assigned to v_t is the same as $f(i)$. Then, the main color of v_t in the final conflict-free coloring is set to i. See Figure 6 for an illustration.

It is possible that v_t never gets a final color. In this case we say that the framework does not halt. However, termination can be guaranteed by imposing some restrictions on the auxiliary coloring method and the choice of the function f. For example, if first-fit is used for the auxiliary colorings at any entry and if f is the constant function $f(i) = a_1$, for all i, then the framework is guaranteed to halt for any time t. Later, a randomized online algorithm based on this framework is derived under the oblivious adversary model. This algorithm always halts, or to be more precise halts with probability 1, and moreover it halts after a "small" number of entries with high probability. We prove that the above framework produces a valid conflict-free coloring in case it halts.

Lemma 5.9. *If the above framework halts for any vertex v_t then it produces a valid online conflict-free coloring of H.*

$f : N \to \{a,b,c\}$	v_1	v_3	v_4	v_2	
$f(1) = a$	a	c	a	b	
$f(2) = b$		b		a	
$f(3) = a$				a	
$\bullet$					
$\bullet$					
$\bullet$					
$\bullet$					

Fig. 6. An example of the updating process of the table for the hypergraph induced by points with respect to intervals. 3 auxiliary colors denoted $\{a,b,c\}$ are used. In each line i the auxiliary coloring is given. It serves as a proper coloring for the hypergraphs $H(V_t^i)$ induced by the subset V_t^i of all points revealed up to time t that reached line i. The first point v_1 is inserted to the left. The second point v_2 to the right and the third point v_3 in the middle, etc. For instance, at the first entry (i.e., line) of the table, the auxiliary color of v_2 is b. In the second line it is a and in the third line it is a. Since $f(3) = a$, the final color of v_2 is 3. Similarly, the final color of v_1 is 1, of v_3 is 2, and of v_4 is 1

Proof. Let $H(V_t)$ be the hypergraph induced by the vertices already revealed at time t. Let S be a hyperedge in this hypergraph and let j be the maximum integer for which there is a vertex v of S colored with j. We claim that exactly one such vertex in S exists. Assume to the contrary that there is another vertex v' in S colored with j. This means that at time t both vertices v and v' were present at entry j of the table (i.e., $v, v' \in V_t^j$) and that they both got an auxiliary color (in the auxiliary coloring of the set V_t^j) which equals $f(j)$. However, since the auxiliary coloring is a proper non-monochromatic coloring of the induced hypergraph at entry j, $S \cap V_t^j$ is not monochromatic so there must exist a third vertex $v'' \in S \cap V_t^j$ that was present at entry j and was assigned an auxiliary color different from $f(j)$. Thus, v'' got its final color in an entry greater than j, a contradiction to the maximality of j in the hyperedge S. This completes the proof of the lemma. ∎

The above algorithmic framework can also describe some well-known deterministic algorithms. For example, if first-fit is used for auxiliary colorings and f is the constant function, $f(i) = a_1$, for all i, then, for the hypergraph induced by points on a line with respect to intervals, the algorithm derived from the framework becomes identical to the UniMax greedy algorithm described above.

An online randomized conflict-free coloring algorithm. We devise a randomized online conflict-free coloring algorithm in the oblivious adversary model. In this model, the adversary has to commit to a permutation according to the order of which the vertices of the hypergraph are revealed to the algorithm. Namely, the adversary does not have access to the random bits that are used by the algorithm. The algorithm always produces a valid coloring and the number of colors used is related to the degeneracy of the underlying hypergraph in a manner described in the following theorem.

Theorem 5.10 ([10]). *Let $H = (V, E)$ be a k-degenerate hypergraph on n vertices. Then, there exists a randomized online conflict-free coloring algorithm for H which uses at most $O\left(\log_{1+\frac{1}{4k+1}} n\right) = O(k \log n)$ colors with high probability against an oblivious adversary.*

The algorithm is based on the framework presented above. In order to define the algorithm, we need to state what is (a) the set of auxiliary colors of each entry, (b) the function f, and (c) the algorithm we use for the auxiliary coloring at each entry. We use the set of auxiliary colors $A = \{a_1, \ldots, a_{2k+1}\}$. For each entry i, the representing color $f(i)$ is chosen uniformly at random from A. We use a first-fit algorithm for the auxiliary coloring.

Our assumption on the hypergraph H (being k-degenerate) implies that at least half of the vertices up to time t that *reached* entry i (but not necessarily added to entry i), denoted by $X_{\geq i}^t$, have been actually given some auxiliary color at entry i (that is, $|V_t^i| \geq \frac{1}{2}|X_{\geq i}^t|$). This is due to the fact that at least half of those vertices v_t have at most $2k$ neighbors in the Delaunay graph of the hypergraph induced by $X_{\geq i}^{t-1}$ (since the sum of these quantities is at most $k|X_{\geq i}^t|$ and since $V_t^i \subseteq X_{\geq i}^t$). Therefore, since we have $2k + 1$ colors available, there is always an available color to assign to such a vertex. The following lemma shows that if we use one of these available colors then the updated coloring is indeed a proper non-monochromatic coloring of the corresponding induced hypergraph as well.

Lemma 5.11. *Let $H = (V, E)$ be a k-degenerate hypergraph and let V_t^j be the subset of V at time t and at level j as produced by the above algorithm. Then, for any j and t if v_t is assigned a color distinct from all its neighbors* in the Delaunay graph $G(H(V_t^j))$ then this color together with the colors assigned to the vertices V_{t-1}^j is also a proper non-monochromatic coloring of the hypergraph $II(V_t^j)$.

Proof. Follows from Lemma 5.9 ∎

We also prove that for every vertex v_t, the algorithm always halts, or more precisely halts with probability 1.

Proposition 5.12. *For every vertex v_t, the algorithm halts with probability 1.*

Proof.

$$\text{Pr[algorithm does not halt for } v_t] =$$

$$\text{Pr[algorithm does not assign a main color to } v_t \text{ in any entry]} \le$$

$$\text{Pr[algorithm does not assign a main color to } v_t \text{ in any empty entry]} =$$

$$\text{Pr}[\bigcap_{i:\text{ empty entry}} (\text{algorithm does not assign a main color to } v_t \text{ in entry } i)] =$$

$$\prod_{i:\text{ empty entry}} \text{Pr[algorithm does not assign a main color to } v_t \text{ in entry } i] =$$

$$\prod_{i:\text{ empty entry}} (1 - h^{-1}) = \lim_{j \to \infty} (1 - h^{-1})^j = 0$$

and therefore Pr[algorithm halts for v_t] $= 1$. ∎

We proceed to the analysis of the number of colors used by the algorithm, proving theorem 5.10.

Lemma 5.13. *Let $H = (V, E)$ be a hypergraph and let C be a coloring produced by the above algorithm on an online input $V = \{v_t\}$ for $t = 1, \ldots, n$. Let X_i (respectively $X_{\ge i}$) denote the random variable counting the number of points of V that were assigned a final color at entry i (respectively a final color at some entry $\ge i$). Let $\mathbf{E}_i = \mathbf{E}[X_i]$ and $\mathbf{E}_{\ge i} = \mathbf{E}[X_{\ge i}]$ (note that $X_{\ge i+1} = X_{\ge i} - X_i$). Then:*

$$\mathbf{E}_{\ge i} \le \left(\frac{4k+1}{4k+2}\right)^{i-1} n.$$

Proof. By induction on i. The case $i = 1$ is trivial. Assume that the statement holds for i. To complete the induction step, we need to prove that $\mathbf{E}_{\geq i+1} \leq \left(\frac{4k+1}{4k+2}\right)^i n$. By the conditional expectation formula, we have for any two random variables X, Y that $\mathbf{E}[X] = \mathbf{E}[\mathbf{E}[X \mid Y]]$. Thus,

$$\mathbf{E}_{\geq i+1} = \mathbf{E}[\mathbf{E}[X_{\geq i+1} \mid X_{\geq i}]] = \mathbf{E}[\mathbf{E}[X_{\geq i} - X_i \mid X_{\geq i}]] = \mathbf{E}[X_{\geq i} - \mathbf{E}[X_i \mid X_{\geq i}]].$$

It is easily seen that $\mathbf{E}[X_i \mid X_{\geq i}] \geq \frac{1}{2}\frac{X_{\geq i}}{2k+1}$ since at least half of the vertices of $X_{\geq i}$ got an auxiliary color by the above algorithm. Moreover each of those elements that got an auxiliary color had probability $\frac{1}{2k+1}$ to get the final color i. This is the only place where we need to assume that the adversary is oblivious and does not have access to the random bits. Thus,

$$\mathbf{E}[X_{\geq i} - \mathbf{E}[X_i \mid X_{\geq i}]] \leq \mathbf{E}\left[X_{\geq i} - \frac{1}{2(2k+1)}X_{\geq i}\right]$$

$$= \frac{4k+1}{4k+2}\mathbf{E}[X_{\geq i}] \leq \left(\frac{4k+1}{4k+2}\right)^i n,$$

by linearity of expectation and by the induction hypotheses. This completes the proof of the lemma. ∎

Lemma 5.14. *The expected number of colors used by the above algorithm is at most $\log_{\frac{4k+2}{4k+1}} n + 1$.*

Proof. Let I_i be the indicator random variable for the following event: some points are colored with a main color in entry i. We are interested in the number of colors used, that is $Y := \sum_{i=1}^{\infty} I_i$. Let $b(k,n) = \log_{\frac{4k+2}{4k+1}} n$. Then,

$$\mathbf{E}[Y] = \mathbf{E}\left[\sum_{1 \leq i} I_i\right] \leq \mathbf{E}\left[\sum_{1 \leq i \leq b(k,n)} I_i\right] + \mathbf{E}[X_{\geq b(k,n)+1}] \leq b(k,n) + 1,$$

by Markov's inequality and lemma 5.13. ∎

We notice that:

$$b(k,n) = \frac{\ln n}{\ln \frac{4k+2}{4k+1}} \leq (4k+2)\ln n = O(k \log n).$$

We also have the following concentration result:

$\Pr[\text{more than } c \cdot b(k, n) \text{ colors are used}]$

$$= \Pr[X_{\geq c \cdot b(k,n)+1} \geq 1] \leq \mathbf{E}_{\geq c \cdot b(k,n)+1} \leq \frac{1}{n^{c-1}},$$

by Markov's inequality and by lemma 5.13.

This completes the performance analysis of the algorithm.

Remark. In the above description of the algorithm, all the random bits are chosen in advance (by deciding the values of the function f in advance). However, one can be more efficient and calculate the entry $f(i)$ only at the first time we need to update entry i, for any i. Since at each entry we need to use $O(\log k)$ random bits and we showed that the number of entries used is $O(k \log n)$ with high probability then the total number of random bits used by the algorithm is $O(k \log k \log n)$ with high probability.

Acknowledgments. I would like to thank the two anonymous referees for providing very valuable comments and suggestions.

REFERENCES

[1] M. A. Abam, M. de Berg, and S.-H. Poon, Fault-tolerant conflict-free coloring. In *Proceedings of the 20th Annual Canadian Conference on Computational Geometry (CCCG), Montreal, Canada, August 13–15*, 2008.

[2] M. Abellanas, P. Bose, J. Garcia, F. Hurtado, M. Nicolas, and P. A. Ramos, On properties of higher order delaunay graphs with applications. In *EWCG*, 2005.

[3] D. Ajwani, K. Elbassioni, S. Govindarajan, and S. Ray, Conflict-free coloring for rectangle ranges using $\tilde{O}(n^{.381+\epsilon})$ colors. In *SPAA '07: Proc. 19th ACM Symp. on Parallelism in Algorithms and Architectures*, pages 181–187, 2007.

[4] N. Alon, Choice numbers of graphs: a probabilistic approach. 1:107–114, 1992.

[5] N. Alon and S. Smorodinsky, Conflict-free colorings of shallow discs, *Int. J. Comput. Geometry Appl.*, 18(6):599–604, 2008.

[6] N. Alon and J. H. Spencer, *The Probabilistic Method*. Wiely Inter-Science, 2000.

[7] G. Aloupis, J. Cardinal, S. Collette, S. Langerman, and S. Smorodinsky, Coloring geometric range spaces, *Discrete & Computational Geometry*, 41(2):348–362, 2009.

[8] K. Appel and W. Haken, Every planar map is 4-colorable – 1: Discharging, *Illinois Journal of Mathematics,* (21):421–490, 1977.

[9] K. Appel and W. Haken, Every planar map is 4-colorable – 2: Reducibility, *Illinois Journal of Mathematics,* (21):491–567, 1977.

[10] A. Bar-Noy, P. Cheilaris, S. Olonetsky, and S. Smorodinsky, Online conflict-free colouring for hypergraphs, *Combinatorics, Probability & Computing,* 19(4):493–516, 2010.

[11] A. Bar-Noy, P. Cheilaris, and S. Smorodinsky, Conflict-free coloring for intervals: from offline to online. In *Proc. ACM Symposium on Parallelism in Algorithms and Architectures,* pages 128–137, New York, NY, USA, 2006. ACM Press.

[12] A. Bar-Noy, P. Cheilaris, and S. Smorodinsky, Deterministic conflict-free coloring for intervals: From offline to online, *ACM Transactions on Algorithms,* 4(4), 2008.

[13] A. L. Buchsbaum, A. Efrat, S. Jain, S. Venkatasubramanian, and K. Yi, Restricted strip covering and the sensor cover problem. In *Proc. Annu. ACM-SIAM Symposium on Discrete Algorithms,* pages 1056–1063, 2007.

[14] T. M. Chan, Conflict-free coloring of points with respect to rectangles and approximation algorithms for discrete independent set. In *Proceedings of the Twenty-eighth Annual Symposium on Computational Geometry,* SoCG '12, Chapel Hill, North Carolina, USA. ACM (New York, NY, USA, 2012), pp. 293–302.

[15] P. Cheilaris, B. Keszegh, and D. Pálvölgyi, Unique-maximum and conflict-free colorings for hypergraphs and tree graphs, *arXiv:1002.4210v1,* 2010.

[16] P. Cheilaris, S. Smorodinsky, and M. Sulovský, The potential to improve the choice: list conflict-free coloring for geometric hypergraphs. In *Proc. 27th Annu. ACM Sympos. Comput. Geom.,* 2011.

[17] P. Cheilaris and G. Tóth, Graph unique-maximum and conflict-free colorings. In *Proc. 7th International Conference on Algorithms and Complexity (CIAC),* pages 143–154, 2010.

[18] K. Chen, A. Fiat, M. Levy, J. Matoušek, E. Mossel, J. Pach, M. Sharir, S. Smorodinsky, U. Wagner, and E. Welzl, Online conflict-free coloring for intervals, *Siam. J. Comput.,* 36:545–554, 2006.

[19] K. Chen, H. Kaplan, and M. Sharir, Online conflict-free coloring for halfplanes, congruent disks, and axis-parallel rectangles, *ACM Transactions on Algorithms,* 5(2), 2009.

[20] X. Chen, J. Pach, M. Szegedy, and G. Tardos, Delaunay graphs of point sets in the plane with respect to axis-parallel rectangles. In *Proc. Annu. ACM-SIAM Symposium on Discrete Algorithms,* pages 94–101, 2008.

[21] K. L. Clarkson and P. W. Shor, Application of random sampling in computational geometry, ii, *Discrete & Computational Geometry,* 4:387–421, 1989.

[22] T. H. Cormen, C. E. Leiserson, and R. L. Rivest, *Introduction to Algorithms.* MIT Press, Cambridge, MA, 1990.

[23] K. Elbassioni and N. Mustafa, Conflict-free colorings of rectangle ranges. In *STACS '06: Proc. 23rd International Symposium on Theoretical Aspects of Computer Science*, pages 254–263, 2006.

[24] P. Erdős, A. L. Rubin, and H. Taylor, Choosability in graphs. In *Proc. West Coast Conf. on Combinatorics, Graph Theory and Computing, Congressus Numerantium XXVI*, pages 125–157, 1979.

[25] G. Even, Z. Lotker, D. Ron, and S. Smorodinsky, Conflict-free colorings of simple geometric regions with applications to frequency assignment in cellular networks, *Siam. J. Comput.*, 33:94–136, 2003.

[26] M. Gibson and K. R. Varadarajan, Decomposing coverings and the planar sensor cover problem. In *50th Annual IEEE Symposium on Foundations of Computer Science, FOCS 2009, October 25-27, 2009, Atlanta, Georgia, USA*, pages 159–168, 2009.

[27] A. Hajnal, I. Juhász, L. Soukup, and Z. Szentmiklóssy, Conflict free colorings of (strongly) almost disjoint set-systems, *Acta Mathematica Hungarica*, 2011, to appear.

[28] S. Har-Peled and S. Smorodinsky, Conflict-free coloring of points and simple regions in the plane, *Discrete & Computational Geometry*, 34(1):47–70, 2005.

[29] D. Haussler and E. Welzl, Epsilon-nets and simplex range queries, *Discrete & Computational Geometry*, 2:127–151, 1987.

[30] E. Horev, R. Krakovski, and S. Smorodinsky, Conflict-free coloring made stronger. In *Proc. 12th Scandinavian Symposium and Workshops on Algorithm Theory*, 2010.

[31] A. V. Iyer, H. D. Ratliff, and G. Vijayan, Optimal node ranking of trees, *Information Processing Letters*, 28(5):225–229, 1988.

[32] M. Katchalski, W. McCuaig, and S. M. Seager, Ordered colourings, *Discrete Mathematics*, 142(1-3):141–154, 1995.

[33] K. Kedem, R. Livne, J. Pach, and M. Sharir, On the union of Jordan regions and collision-free translational motion amidst polygonal obstacles, *Discrete & Computational Geometry*, 1:59–71, 1986.

[34] B. Keszegh, Weak conflict-free colorings of point sets and simple regions. In *Proceedings of the 19th Annual Canadian Conference on Computational Geometry, CCCG 2007, August 20-22, 2007, Carleton University, Ottawa, Canada*, pages 97–100, 2007.

[35] B. Keszegh, *Combinatorial and computational problems about points in the plane*. PhD thesis, Central European University, Budapest, Department of Mathematics and its Applications, 2009.

[36] N. Lev-Tov and D. Peleg, Conflict-free coloring of unit disks, *Discrete Applied Mathematics*, 157(7):1521–1532, 2009.

[37] J. W. H. Liu, Computational models and task scheduling for parallel sparse cholesky factorization, *Parallel Computing*, 3(4):327–342, 1986.

[38] J. Matoušek, *Lectures on Discrete Geometry.* Springer-Verlag New York, Inc., Secaucus, NJ, USA, 2002.

[39] R. A. Moser and G. Tardos, A constructive proof of the general lovász local lemma, *J. ACM,* 57(2), 2010.

[40] J. Pach and G. Tardos, Conflict-free colourings of graphs and hypergraphs, *Comb. Probab. Comput.,* 18(5):819–834, 2009.

[41] J. Pach and G. Tardos, Coloring axis-parallel rectangles, *Journal of Combinatorial Theory, Series A,* in press, 2009.

[42] J. Pach, G. Tardos, and G. Tóth, Indecomposable coverings. In *The China-Japan Joint Conference on Discrete Geometry, Combinatorics, and Graph Theory (CJCDGCGT 2005), Lecture Notes in Computer Sceince,* pages 135–148, 2007.

[43] J. Pach and G. Tóth, Conflict free colorings, *Discrete & Computational Geometry, The Goodman-Pollack Festschrift,* pages 665–671, 2003.

[44] J. Pach and G. Tóth, Decomposition of multiple coverings into many parts, *Comput. Geom.: Theory and Applications,* 42(2):127–133, 2009.

[45] A. A. Schäffer, Optimal node ranking of trees in linear time, *Information Processing Letters,* 33(2):91–96, 1989.

[46] A. Sen, H. Deng, and S. Guha, On a graph partition problem with application to vlsi layout, *Information Processing Letters,* 43(2):87–94, 1992.

[47] S. Smorodinsky, *Combinatorial Problems in Computational Geometry.* PhD thesis, School of Computer Science, Tel-Aviv University, 2003.

[48] S. Smorodinsky, On the chromatic number of some geometric hypergraphs, *Siam. J. Discrete Mathematics,* 21:676–687, 2007.

[49] S. Smorodinsky, Improved conflict-free colorings of shallow discs, manuscript. 2009.

[50] R. P. Stanley, *Combinatorics and Commutative Algebra, Second Edition.* Birkhuser, Boston, Inc., 1996.

[51] V. Vizing, Coloring the vertices of a graph in prescribed colors (in russian), *Diskret. Analiz.,* Metody Diskret. Anal. v. Teorii Kodov i Shem 101(29):3–10, 1976.

[52] D. B. West, *Introduction to Graph Theory.* Prentice Hall, 2ed edition, 2001.

Shakhar Smorodinsky

Mathematics Department,
Ben-Gurion University,
Beer Sheva,
Israel

http://www.math.bgu.ac.il/~shakhar/
e-mail: shakhar@math.bgu.ac.il

Készült: Regiszter Kiadó és Nyomda Kft., Budapest